# MANAGEMENT BY MENU

FOURTH EDITION

LENDAL H. KOTSCHEVAR

DIANE WITHROW

John Wiley & Sons, Inc.

This book is printed on acid-free paper. ♾



Published by John Wiley & Sons, Inc., Hoboken, New Jersey.

Published simultaneously in Canada.



For general information on our other products and services, or technical support, please contact our Customer Care Department within the United States at 800-762-2974, outside the United States at 317-572-3993 or fax 317-572-4002.

Wiley also publishes its books in a variety of electronic formats. Some content that appears in print may not be available in electronic books.

For more information about Wiley products, visit our Website at http://*www.wiley.com*.

**Library of Congress Cataloging-in-Publication Data**

Kotschevar, Lendal Henry, 1908–2007
Management by menu / by Lendel Kotschevar & Diane Withrow.—4th ed.
p. cm.
Includes bibliographical references and index.
ISBN: 978–0–471–47577–4
1. Foodservice management. I. Withrow, Diane. II. Title.
TX943.K66 2007
647.95068—dc22 2006034746

Printed in the United States of America

10 9 8 7 6 5 4 3 2

*This book is dedicated to the memory of*
*Lendal H. Kotschevar, a pioneer in the field*
*of hospitality and an inspiration to generations*
*of educators, students, and professionals*

# Contents

# Preface

***Management by Menu*** has had a successful tenure since publication of its first edition. This text is widely used in schools and for self-study courses, and as a source for on-the-job training. It is also used as a reference book for foodservice managers. It gives us great pleasure to present to you the long awaited ***Fourth Edition*** of ***Management by Menu*** which continues to be unique in its presentation of the menu as a central theme that controls and influences all foodservice functions. Other books explain how to set up a menu but none of them tie the menu to overall management principles. ***Management by Menu, Fourth Edition*** begins with an introduction to the industry and then develops the theme of shaping the menu to best perform its function of controlling and directing the foodservice operation. This ***Fourth Edition*** has gone through an extensive revision to ensure it is up to date with regard to industry trends that have evolved since the third edition was published. As with previous editions, the goal of this fourth edition continues to be to illustrate how the menu serves as the guiding tool in managing foodservices.

***Management by Menu, Fourth Edition*** is written for both those who are just being introduced to foodservice management and those who already have experience in foodservice. This book is suitable for use as an introduction to the foodservice industry as well as a text on menu construction and formulations. Instructors have also used it successfully in teaching foodservice management, since the menu is at the central core around which management revolves.

## New to this Fourth Edition

Several new industry developments have been added to the discussions in this revision and we have included more real-world application questions, projects, and mini cases at the end of each chapter to provide a more hands-on experience for students. Growth in the foodservice industry has been enormous in recent years. Desktop publishing has made designing and printing menus much faster and less expensive for operators. Concern about responsible alcohol service has given new focus to liquor menu promotions. Consumer interest in nutrition, legislation governing health claims, and the evolution of the new Food Pyramid, have all played a role in adding to the responsibility of menu planners to offer nutritious items. Exercises on the new Food Pyramid's application including various diets, vegetarian, vegan, allergies etc. have been included and their impact on menu planning has

been identified. Fad diets, obesity claims, and our love/hate relationship with healthy eating have also been included along with information on standard recipes.

It goes without saying that technological advancements relating to so many foodservice management functions has put immeasurable information literally at managers' fingertips. We have emphasized the impact of technology and its contribution to the efficiency of the foodservice industry wherever applicable.

## Chapter-Specific Significant Changes

- Each chapter of the ***Fourth Edition*** attempts to aid learning by presenting the learning objectives initially, and then, providing an in depth discussion of a distinct facet of menu development. Questions, projects, and mini cases are included at the end of each chapter to help emphasize the main points and reinforce student learning.
- Nutrition is addressed throughout text, as appropriate, rather than in a single chapter as in previous editions of this text, since it is such an important topic. As mentioned previously, we have included information on the new food pyramid and exercises focusing on various diets, vegetarian, vegan, allergies to help illustrate the new food pyramid's role in planning menus for a variety of circumstances. Fad diets, obesity claims and our love/hate relationship with healthy eating should be included along with info. on standard recipes
- Chapter 1 on the evolution and history of the foodservice industry highlights the contributions of Dave Thomas and Rich Melman, famous entrepreneurs who have been extremely successful and made significant contributions to the foodservice industry.
- Chapter 2 focuses on the advancement of the foodservice industry during the past 10 years using demographic data when appropriate. This chapter sheds light on where and when the industry exploded in popularity and revenue growth. This chapter discusses Quick Service, Family Service, Fine Dining, and Institutional heavy revenue generators as well as societal and economic factors that have impacted the popularity of various segments of the industry. Information on Point of Sale data and its impact on menu development is provided as well as content on printing menus on site. Pricing formulas and labor formulas have been revised as necessary so they are realistic in relation to today's industry. Finally, we've provided industry and demographic projections to help illustrate that they play a key role in making reasonable projections about growth in particular ethnic cuisines, identifying potential strong growth markets, and predicting slower growth markets.
- Chapter 6 trends in pricing strategies, including non-cost approaches, are explored in detail. New menus reflecting today's prices are provided to help illustrate the pricing strategies. We discuss several operators that have successfully used various approaches to menu pricing such as sales potential and marginal analysis pricing
- Chapter 7 provides information on menu mechanics and has been updated to include information on menu building software and examples that illustrate how particular software works.
- Chapter 10, previously Chapter 11 in the 3rd edition text, focuses on producing the menu and emphasizes the relationship between planning the menu and the role of the purchaser. In this chapter, we identify the ways in which purchasing needs are determined and

methods of meeting those needs are found. We also discuss the purpose and benefits of purchasing controls and provide examples of procedures used. The relationship between planning the menu and the role of production is also discussed.

- Chapter 11, formerly Chapter 12 in the 3rd edition text, covers service and the menu. In this chapter, we identify the importance of service in fulfilling the objectives of the menu; we identify and explain the concept of service as it relates to the hospitality industry; and we explain the methods in which guest payments are secured such as manual checks and point of sales systems.
- Chapter 12, which was Chapter 14 in the previous edition concentrates on the relationship between the menu and the financial plan. In this chapter, we discuss the feasibility study and business plan software. Factors contributing to the success or failure of a restaurant are also identified and we discuss ways in which the positives can be cultivated and negatives avoided.
- We have written a **NEW** chapter, Chapter 13, on Ethical Leadership in Restaurant Management. This chapter includes real world cases and discusses the role of industry-related agencies such as the FTC, FDA, USDA, and local health officials as appropriate. We also discuss operators' responsibility to provide healthful options within their menus for the health conscious of today. Finally, we discuss the role of the manager as a leader at the work place and how ethics plays a role in being an effective leader.
- End of Chapter Material: Include mini cases with questions that include mathematical calculations and discussion questions that build upon the concepts discussed in the chapter on a chapter by chapter basis as best suits that chapter
- Five appendices present additional support material relating to content in specific chapters of the text.
- Additional resources also appear throughout the text as exhibits, such as an abundant collection of newly added menus from different types of foodservice operations, to further enhance information given in the discussions.
- A glossary is also included to assist in defining many of the terms mentioned throughout the text.

## Additional Resources

An ***Instructor's Manual*** (ISBN: 978-0-470-16765-6) including chapter outlines, lecture outlines, quizzes, suggested answers to in-text review questions and mini cases along with a test bank that includes multiple-choice, true/false, and short answer questions is available. Please contact your John Wiley & Sons representative for a copy.

A ***Companion Website*** *(www.wiley.com/college/kotschevar)* provides readers with additional resources as well as enabling instructors to download the electronic files for the Instructor's Manual, PowerPoint slides, and Test Bank.

A ***Study Guide*** which has been newly created for this edition (ISBN: 978-0-470-14053-6) includes chapter objectives, detailed chapter outlines, and concept review and discussion questions to help students to reinforce and test their understanding of the key concepts and features within the text.

# ACKNOWLEDGMENTS

We would like to thank the following instructors for their insightful feedback during the course of their review of our revised ***Fourth Edition*** manuscript:

Bill Barber, Orange Coast College

Steve Bergonzoni, Burlington County College

Roger Gerard, Shasta College

Steve Nigg, Southwestern Illinois College

Susan Torrey, Southern New Hampshire University

Odette Smith-Ransome, Art Institute of Pittsburgh

LENDAL H. KOTSCHEVAR, PH.D.
*(1908-2007)*
*Professor Emeritus*
*Florida International University*
*Educational Foundation 1992 Ambassador of Hospitality*

DIANE WITHROW
*Program Coordinator, Hotel/Restaurant Management*
*Cape Fear Community College*

# 1

# A LOOK BACK AT THE FOODSERVICE INDUSTRY

## Outline

Introduction
Ancient Foodservice
Foodservice in the Middle Ages
Early Renaissance—The Development of Haute Cuisine
- French Cuisine
- The Coming of the Restaurant
- Other National Cuisines

The Industrial Revolution
The Advancement of Science
The Golden Age of Cuisine
- Carême
- Escoffier and Ritz

Foodservice in the United States
- The Early Years
- Postwar Expansion

Continual Growth
Summary
Questions
Mini-Cases

## Objectives

After reading this chapter, you should be able to:

1. Describe the historical context in which the foodservice industry began and evolved.
2. Identify the major contributions that industry leaders and innovators have made to enhance the growth of this industry.

# Introduction

The story of humans and food is the story of life itself, and it would ultimately lead us back to the metabolizing of nutrients by a one-celled organism in some warm, shallow sea millions of years ago. Our account is a little less ambitious, seeking only to deal with how the foodservice industry developed to provide food for those who eat away from home—more specifically, how this vast industry is influenced by one particular document to the point that it dominates the industry's management. This document is the *menu,* a list of foods offered along with their prices. This book will explain how to manage foodservice using the menu as the controlling factor.

It is said that a science is an organized body of knowledge, and art is the application of the science. We shall try to point out the principles and theories of the science of menu planning and its implementation, as well as the art required. However, first we will trace the beginnings of the foodservice industry, show what it is today, and explain how it consists of a large number of different segments that require different menus to suit various market needs. A single foodservice operation is a member of the larger foodservice industry. The more one knows about this field, what it is and how it grew, the more one can adjust to the overall conditions that affect the foodservice industry as a whole and, in turn, the individual operation.

# Ancient Foodservice

The foodservice industry is both old and new. It is old in that humans have prepared and eaten food in groups since the earliest times. It is new in that it has changed considerably in the past 150 years. There is evidence that before 10,000 B.C., tribes in Denmark and the Orkney Islands off the coast of Scotland cooked food in large kitchens and ate together in large groups. Swiss lake dwellers left records that show that they dined in groups around 5000 B.C. Pictorial evidence in the tombs and temples of the ancient Egyptians also show that people knew how to prepare and serve food for large groups. There is also evidence in these pictures that prepared food was sold in their marketplaces, just as it is today. Vendors also sold foods in the streets, the same as a mobile unit might today, though lacking understanding of sanitary principles and the regulations that govern sales of food.

We get much knowledge also from ancient tombs. For instance, the tomb of the Egyptian Pharaoh Tutankhamen contained many of the foods this king would need in the afterworld, many of which were surprisingly like the foods we have today. (Some wheat kernels left there to make flour were later planted, and they grew.) The Ancients also had recipes. The oldest recipe for beer was left on an Assyrian tablet found near Mount Ararat. The Assyrians put their wine and beer in animal-skin containers. The poor drank beer while the wealthy drank wine.

Ancient Chinese records indicate that travelers ate and stayed in roadside inns. In large urban cities, restaurants existed in which cooked rice, wine, and other items were sold. In India, the operation of roadside inns, taverns, and foodservices was so prevalent that ancient laws were passed to control them.

In ancient Mohenjo-Daro, a recent excavation in Pakistan shows evidence that people ate in restaurant-type facilities that were equipped with stone ovens and stoves for quantity food preparation.

The Bible gives many accounts of a quantity mass-feeding industry. For instance, accounts tell of Xerxes, the Persian king, giving a banquet that lasted 180 days, and of Solomon butchering 22,000 oxen for a public feast. Sardanapalus, the Assyrian king, was a patron of the art of eating, and he loved huge feasts. He organized a cooking contest at which the top professional cooks vied for honors, much as they do today at the Culinary Olympics held in Germany every four years.

The ancient Greeks had a high level of public dining, and much of their social lives took place around banquets at home or at public feasts. Inns and foodservice operations existed. Greece was the land of Epicurus, who spread the philosophy of good eating and good living. The Grecians went all out for their feasts. The bacchanal feast in honor of the god of wine, Bacchus, was a lavish outlay of food, drink, and revelry. Professional cooks in Greece were honored people and had important parts in plays, where they declaimed their most famous recipes. It was even possible in ancient Greece to copyright a recipe.

The ancient Greeks loved to gather to discuss matters of interest and drink and eat snack foods. These foods were appetizers of different kinds. They learned about these foods from the peoples of Asia, and even today, the Near East and Greece follow the custom of having an array of snack foods with beverages.

The Romans also loved feasting. In fact, several of the emperors were so fond of banquets they bankrupted the nation as a result of giving them. Emperor Lucullus, a Roman general, loved lavish banquets, and today whenever the word *Lucullan* is used, it means lavish and luxurious dining. A special rich sauce used to grace meat is called *Lucullus sauce*. It is perhaps one of the richest sauces used, and has a garnish of cocks' combs. Marc Antony was so pleased with the efforts of Cleopatra's cook that he presented her with a whole city.

*Tabernas,* from which we get the word *tavern,* were small restaurants in ancient Rome where one could get wine and food. We can see one such restaurant almost intact in the ruins of ancient Pompeii. It had a large service counter where huge urns of wine were kept. In the back area, a huge brick oven and other cooking equipment still stand. These small tabernas were the forerunners of the *trattorias,* or small community restaurants of modern Italy. The Romans had a number of laws regulating the sale of foods and the operation of foodservices.

The first known cookbook, titled *Cookery and Dining in Imperial Rome,* was once thought to be written by a Roman epicure named Apicius. However, it is now thought that the manuscript was written several centuries after Apicius. It is still an interesting revelation of how the upper class dined in those times. From the book we learn that the Roman feast consisted of three courses, during each of which a number of foods were served: (1) the *gustatio,* a group of appetizers that turned into the Italian antipasto of today; (2) meats and vegetables of different kinds, many of which were rare items imported from foreign lands; and (3) fruits and sweets. Plenty of wine was served throughout. The Romans did not sit at their meals but, rather, reclined on couches. A lot of the Roman culinary art was preserved during the Dark Ages and emerged as the basis of Italian and French cuisine that later influenced all of European cooking. The Romans had no stoves, but cooked over open fires or in fireplaces, and also baked in brick ovens. Apicius' life ended tragically; he committed suicide in remorse when he found himself bankrupt after giving a lavish banquet.

## Foodservice in the Middle Ages

Following the disintegration of the Roman Empire, group eating became somewhat less lavish. Public eating virtually went underground in the Dark Ages. Some inns functioned along the most protected and traveled highways. We can read about these in the tales of the Crusaders. Chaucer's *Canterbury Tales* revolves around stories told by a group of travelers as they stayed in inns. Quantity foodservice in its highest form was practiced in the monasteries and abbeys by monks or friars. They considerably advanced the knowledge of baking, wine and beer making, and cooking. Many of the master craftsmen who later formed the various foodservice guilds gained much of their knowledge in these religious communities. Some recipes originated by these friars are still used today, such as pound cake and many meat dishes. It was during that time that Benedictine, Cointreau, Grand Marnier, Chartreuse, and other famous liqueurs were developed. These are still made by formulas held secret by the makers.

Eating was quite crude by our standards. The *trencher,* a large, shallow, oval-shaped wooden bowl, was the main food container from which people ate. It was often filled with a soup, stew, or ragout consisting of meat and vegetables. Bread was used to sop up the liquid, and one's own dagger was used to cut up and spear meats and vegetables.

Toward the end of the Middle Ages, eating became somewhat refined, and a distinct pattern of courses began to emerge. The French are credited with bringing about this change. Their menus had more appetizers, and soups and salads began to appear as first courses. Lighter foods were served at the beginning of the meal, while the heavier ones were served later, followed by desserts.

The English also improved their dining during this period. We know from records left by the court of Henry VIII that they served many elaborate foods and, while the course structure of the meals was not as advanced as the French, they had a pattern somewhat similar to it. The menus emphasized game, and most meat, fish, or poultry was spit-roasted. They ate a variety of soups, pastries, and puddings, and often had a sweet meat dish or two at every meal.

During the Middle Ages, various *guilds* arose to organize foodservice professionals. The *Chaine de Rotissieres* (Guild of Roasters) was chartered in Paris in the twelfth century. This charter is owned today by the gourmet society of that name. A guild had a monopoly on the production of its specialties and could keep others from manufacturing them.

The guilds developed into the classic kitchen organization of chef and entourage, and codified many of the professional standards and traditions that are still in existence. It was at that time that the chef's tall hat, the *toque,* became a symbol of the apprentice. Later the black hat became the mark of a master chef nominated by his peers as having the right to wear it. The hat was a small, round one made of black silk and could be worn only by chefs elected to wear it. (Black in medieval times was the color indicating nobility.) The modern *Society of the Golden Toque,* an organization of master chefs elected by their peers (as in the French group), was designed to duplicate this honor for great chefs in the United States.

# Early Renaissance—The Development of Haute Cuisine

The Renaissance saw a rise in elegant dining as well as the arts. While the renaissance in dining really began in Italy, its move to France brought fine dining to its highest form. As usual, royalty led the way.

## French Cuisine

France has not always been known for its fine food. In medieval times, its food was coarse and plain. However, with the marriage in 1533 of Henry II of France to Catherine de Medici of Florence, Italy, France started its ascendancy as the country of *haute cuisine,* or high food preparation. In Italy, the Medicis not only were great patrons of such artists as Michelangelo, but also served the finest food and drink in their households. When Catherine came to France, she brought the master Medici cooks with her, and established herself as dictator of Henry's table and court. Foods never before known in France were soon being served, much to the delight of Henry and his court. Catherine introduced ice cream and many other great dishes that became part of French culinary accomplishments.[1]

All this started the French on their way toward a great improvement in the quality of foods appearing on the tables of the court and elsewhere. Besides fine food, Catherine taught the French to eat with knives, forks, and spoons instead of using their fingers and daggers. She brought these utensils from Florence and introduced them to the nobility. Soon it became a custom for guests to carry their own eating utensils when they went to dine outside their homes.

Henry's nephew, Henry of Navarre, who became Henry IV after his uncle's death, visited the court frequently and became quite fond of a good table. During his reign from 1589 to 1610, he continued to promote the service of fine foods at his own table, and encouraged the more influential households in France to do the same. Henry IV became known in history as a great gourmet, and today we have a famous soup named after him, *Potage Henri IV,* which is dished into a large tureen and has big pieces of chicken and beef in it.

After Henry IV, the court and kings of France continued their interest in food and dining. It was considered the mark of gentility to set a good table and to encourage the development of top chefs and culinary personnel, as well as the development of fine recipes. In the 1600s, the courts of the Bourbons, Louis XIII to Louis XV, continued to develop a knowledge of cuisine and to encourage the training of top chefs. Louis XIV, who reigned from 1643 to 1715, was known for his ostentatious and luxurious manner of living, and was very active in the development of good schools where chefs and cooks could be trained. A number of the nobles of the court also became famous for their tables and had fine dishes and sauces named after them. The fine white sauce, *Béchamel,* was named for the count of that name, and *Sauce Mornay* was named after Count Mornay. A fine sauce highly seasoned with onions was named after Count Soubisse.

Louis XV, who reigned from 1715 to 1774, continued his predecessor's advancement of the science and art of cooking. Maria Leszczynska, his wife and the daughter of Polish King Stanislaus I, who reigned from 1704 to 1735 and was himself a gourmet and cook, duplicated Catherine de Medici's supervision of the kitchen and set standards for great quality and elaborateness in foodservice. In addition, Louis XV's mistresses, Madame Pompadour and Madame du Barry, were not only lovers of fine food but were proficient cooks. Today many fine dishes are named after them. The king considered Madame du Barry such an excellent cook that he had her awarded the *Cordon Bleu,* an award given only to the best chefs.

The menus of this time were very elaborate. The French had three courses, but as many as 20 or more dishes would be served in one course. The methods of preparation were also often very elaborate. However, toward the end of this period, the menus became more simplified, with fewer dishes in each course.

Although the French Revolution ended the reign of the Bourbons in 1792, it did not stop the French love of food and drink and French dominance in the art of fine dining. Some of the servants of the nobility and wealthy began to use their cooking skills in restaurants. Some of the nobility that had not lost their heads or their servants, but had lost their fortunes, opened up their homes and made their living by serving meals. These operations continued the high standards that had previously existed in the homes of the famous. A group of prominent gourmets now came to the fore to support these operations. Some were writers who began to build literature in the art of cooking and eating. Such writers as Jean-Anthelme Brillat-Savarin, who wrote *The Physiology of Taste;* Grimrod de la Reyniere, editor of the world's first gourmet magazine; Alexandre Dumas père, compiler of *The Grand Dictionaire de Cuisine;* and Vicomte de Chateaubriand, after whom the famous steak dish was named, left writings about the foods of the time that mark their period as one of the greatest in fine dining. With the crowning of Napoleon, the tables in the homes of the wealthy, nobility, and others assumed their former high level. It was almost as if the time of the Louis kings had returned.

## The Coming of the Restaurant

During the development of haute cuisine in Europe, the eating habits of common people away from home continued to be spotty and casual. Crude inns and taverns existed along the main roads where coaches traveled. When people of wealth or high rank traveled and stayed in these inns, they often had their own servants prepare the food. Religious orders continued to care for travelers, but places for common people to go out and dine did not exist. People ate largely in private homes. The common people had neither the resources, equipment, facilities, nor know-how to provide more than simple food. Those in prisons or hospitals were served fare barely above minimal needs.

About 1600, an important development occurred that would influence the growth of the modern foodservice industry. The first coffeehouses (cafés) appeared in France, and their spread was rapid in the great cities of Europe. Not much food was served; the fare consisted mostly of coffee and cocoa or mild alcoholic beverages, such as wine. These coffeehouses were largely places where the local gentry and others could go to get the latest news and gossip, discuss matters of interest, and have a good beverage. The coffeehouse was a forerunner of the modern restaurant.

Another important event occurred later in France. In 1760, during the reign of Louis XV, a man named Boulanger opened an eating place that served soups that were believed to be health restorers. These soups were claimed to be highly nutritious and filled with foods that brought about the cure of many ailments. One soup was distinguished by a calf's foot floating in it. Boulanger called his health restorers *restaurers,* and he called his enterprise a *restorante.* We can readily recognize that this became the word *restaurant.*

The powerful guilds, *Chaine de Rotissieres* and *Chaine de Traiteurs* ("caterers," from the French verb *traiter,* "to treat"), opposed this infringement of Boulanger on their rights as cooks and developers of new dishes. (The guild of bakers also disapproved because it was a threat to their sovereignty.) The guild cooks claimed they had sole right to serve food of this kind, and Boulanger was not a member of their guild. The case gained wide notoriety. Boulanger, however, was adept at public relations, and got leading gourmets, the French legislature, King Louis XV, and other influential individuals on his side. Boulanger won his right to operate as a restaurateur, and this decision lessened the powers of the guilds. Boulanger enlarged his menu and included a much wider list of foods that met with great success. Many other coffeehouses followed Boulanger's example, becoming restaurants. Within a 30-year period, Paris had over 500 of these, and the beginnings of the modern foodservice industry.

## Other National Cuisines

Although France was the leader in the development of serving fine foods, it was not alone. Other nations also developed high levels of food preparation and service.

England was one of these. English cuisine never approached that of the French in reputation, but its tables were worthy of note. Much of the food was of local origin, but because England was a great seapower, some fine additions came from foreign lands. The English blended their food standards with those of Spain and Portugal when Prince Philip, heir to the Spanish throne, married Mary, daughter of Henry VIII and Catherine of Aragon. Mary later became Queen of England, known in history as Bloody Mary. Philip brought a large retinue of servants with him when he moved to England, many of whom were fine cooks. They introduced items into the English cuisine such as sponge cake, famous Spanish hams and bacons, and sherry and port. Even today the British consume these foods to a point that some of them are thought of as original English foods. Philip stayed in England long enough to leave a Spanish flavor to English food. Troubles in the Netherlands, which Spain ruled, took him there, and he never returned.

Russia also developed an original and fine cuisine, Catherine the Great thought highly of the French; in fact, French became a language spoken by many in the Russian court. French chefs were imported—later Czar Alexander hired the great French chef, Antoine Carême, to rule over the royal kitchens. This Russian cuisine featured many game animals, fish, and vegetables of Russian origin. The Russians were more robust eaters and drinkers than the French, and their style of eating and drinking reflected this, but often done in an original way. For instance, instead of having a seated appetizer at the table, they liked a large gathering where plenty of vodka was served (it was a much more potent product in those times), along with what they called *flying dishes* (appetizers), so named because servants carried them around to serve to the guests on platters held high over their heads to allow them to pass through the number of people gathered. The Russians left a rich heritage of foods, some of which were dishes like *stroganoff* (borrowed from Poland), caviar, borscht, and vodka.

The Italians had their own distinctive cuisine. The fine food tradition of the Medicis continued in northern Italy, but because Italy was a land divided into many small, independent *duchies* (headed by dukes) and other political units, a true national cuisine did not develop until later. When it did arrive, it was dominated by pasta, and today Italian pastas are world renowned. At one time Henry Sell, renowned editor of *Town and Country,* at a dinner in Le Pavilion in New York City, was asked where the greatest food was served in our day. He answered, "In Florence; their great traditions still hold."

Every cuisine is distinguished by individual characteristics from the region where it first took root, and Italy's cuisine is no different. Regional foods greatly influence its cooking. Italy is a long, narrow peninsula jutting out into the Mediterranean, and with its extensive seacoast, has plenty of fish and other seafood. Thus, Italian menus will often offer more fish than meat dishes. Olive oil is the main cooking fat, and wine is often served at meals in place of water.

Southern Italian cooking is distinguished by its heavy emphasis on tomato sauces and tomatoes in dishes, especially in those using pasta. More fruits are used, especially citrus. The North places less emphasis on the tomato and has more subtle and delicate seasoning. Garlic is a common ingredient in all Italian cooking, but the North uses it a bit more gently. Veal is more popular in the North, while goat's meat is often seen in the South. Thus, we can see that cuisines are largely defined by resources of the region, and Italy's cuisine is a good example of this regional influence and of a unique cooking heritage.

Largely through the royal courts of all these countries, the production of great food, the art of eating, and an interest in gastronomy grew. It was an important development, and preceded the modern foodservice industry.

## The Industrial Revolution

During the final years of the development of *haute cuisine* in France and in other countries, and the emergence of the *restorante,* another very important event was occurring. The Industrial Revolution, which started at the end of the eighteenth century, brought about great societal and economic changes. Vast industrial complexes emerged with the consequent loss of the guild system. Commercial trade became an important factor.

For a long time, the gentry, with their wealth in agricultural lands, dominated the political and economic scene in Europe. As the Industrial Revolution grew, this changed. The political upheaval caused by the fall of the Bourbons in France and the coming of Napoleon also sped up changes in European society. A new social class emerged as a result of the Industrial Revolution, a middle-class composed of entrepreneurs, shopkeepers, industrialists, and financiers. (For instance, Baron Rothschild, a British financier, became the wealthiest man in the world, with much more money and more influence on society than any nobility had.) This new class began to dominate and affect the social and economic spheres in European society. The newly wealthy demanded a food standard as high as that in the homes of the nobility. Great chefs and retainers were hired. Food was served in exclusive clubs for the wealthy entrepreneurs. Even the lower-income middle class began to ask for food prepared by competent people. Dining out became more popular because these middle-class people could afford it. A true foodservice industry was emerging.

# The Advancement of Science

Another factor in the development of the foodservice industry, and probably one of the stimulants of the Industrial Revolution, was the development of modern science. Until the seventeenth century, science was influenced by the Greek philosophers, who based scientific theories on philosophical ones. Modern scientists began to use the *inductive method,* in which scientific theories are based on observable phenomena. It was the development of this scientific method that led to the great enlargement of knowledge and to the advancement of society in general.

Such great men as Galileo, Bacon, Descartes, Pasteur, and others appeared on the scientific scene and developed the kind of knowledge that advanced technology and people's standard of living. This technology affected food processing. Nicolas Appert discovered canning and earned a 2,000-franc reward from Napoleon I because it helped him feed his vast armies on his march to conquer Europe. Appert and other scientists made technological discoveries that advanced our ability to produce, preserve, and manufacture food. The ability to preserve foods much in their original state, in addition to methods for holding fresh foods longer, considerably enlarged the available year-round food supply. Never before had humans had such a surplus of food. Mass starvation, common in the Middle Ages, appeared to end. The resources needed for the development of a foodservice industry and feeding people away from home were now available.

# The Golden Age of Cuisine

Several developments that occurred in France ushered in a new era that perfected dining standards and helped make dining out a central social activity. This era was called the *Golden Age of Cuisine.*

## Carême

The Golden Age of Cuisine began around 1800 with the rise of Marie-Antoine Carême, who was one of the world's most famous chefs. The century ended with Georges August Escoffier, another chef of equal eminence, who died in 1935. Carême worked as a chef for the French statesman Talleyrand, Czar Alexander I of Russia, and the banking giant, Baron Rothschild,[2] Carême wanted to become an architect, but was never able to do so. Instead, his father apprenticed him as a small boy to Carême's uncle, who operated a small restaurant. Here, Carême learned the basic rudiments of cooking. Originally, Carême was trained to be a pastry chef, and he developed a number of famous dishes in this area, but he branched out and became highly proficient in the other areas of cuisine. He trained many renowned chefs, some of whom became chefs at the famous Reform Club of London, considered the apex of jobs for chefs—Carême himself and the famous Escoffier were also rulers of its kitchens.

In his teens, Carême traveled to Paris, where he quickly progressed through the various food production sections to become a chef. He soon attained a position of prominence and was sought by the leading gourmets of the time to prepare foods for them. With them Carême developed many of the basic concepts for the progression of courses in a dinner and the sequence of the proper wines to accompany them. Carême perfected the very delicate soup, *consommé,* which took its name from the word *consummate,* which means "to bring to completion, perfection, or fulfillment." After he introduced it at a dinner as the first course, Grimod de la Reyniere exclaimed in his approval, "A soup served as the first course of a meal is like the overture to the opera or the porch to the house; it should be a proper introduction to that which is to follow." Carême developed many fine French sauces and dishes. He also originated *pièces mountées* such as ice carvings, tallow pieces, and highly decorated foods, that were used as displays, evidently working out his love for architecture. However, Carême's greatest claim to fame, and perhaps his greatest contribution to food preparation, was that he trained a large number of famous chefs who became his disciples and followed him in holding some of the most prominent cooking positions in clubs, restaurants, and hotels. Carême also gave a considerable amount of his time to writing, and there is still a rich legacy of his ideas on foods. Undoubtedly, Brillat-Savarin and other great gourmets of the time were considerably influenced by him.

## Escoffier and Ritz

Georges Auguste Escoffier, like Carême, was an innovator of fine foods. He was sought by royalty and other society leaders, and at one time was the executive chef of London's famous Reform Club. Later in his life, to spread his talents, he became the supervising chef at a number of leading hotels and clubs in London and continental Europe.

It was Escoffier who perfected the classical, or continental, organization of workers in the kitchen and precisely defined the responsibility of each one. Escoffier was the first to use a food checker and to establish the close coordination that an executive chef must have with the chief steward. Escoffier insisted that his men (kitchen workers at the time were exclusively male) dress neatly, never use profanity, and work quietly, with decorum and gentlemanliness. To reduce the amount of noise and talking in his kitchen, he introduced the *aboyeur* (announcer), who took orders from service personnel and in a clear, loud voice called out the foods ordered to the various production centers. Escoffier wrote many articles and several books, one of which is a cookbook still used all over the world, both in homes and foodservices. He liked to name his food inventions based on the names of famous people, like Peach Melba and Melba Toast named for the famous singer Nellie Melba.

Escoffier was a true scientist, giving careful observations to the reactions of foods to preparation procedures. From these he developed sound rules for the preparation of foods in quantity. He teamed up with the famous hotelier, Cesar Ritz, to operate many of Europe's finest hotels; Escoffier ran the back of the house, while Ritz tended to the front. Perhaps at no other time has the level of dining and staying away from home been raised so high as under these two men. Even today the word *ritzy* means "elegant, ostentatious, fancy, or fashionable."

It was the highly social Edward VII (1841–1910), Prince of Wales and the playboy of Victorian England, who remarked, "Where Mr. Ritz goes, there I go." The informal leaders of culture, society, politics, the arts, and sciences became patrons of these two men. When Ritz and Escoffier died, an era died with them. However, another era was about to be built on their accomplishments. The new era belonged to working people in general, who were beginning to emerge as important

**Exhibit 1.1**
White House Dinner Menu

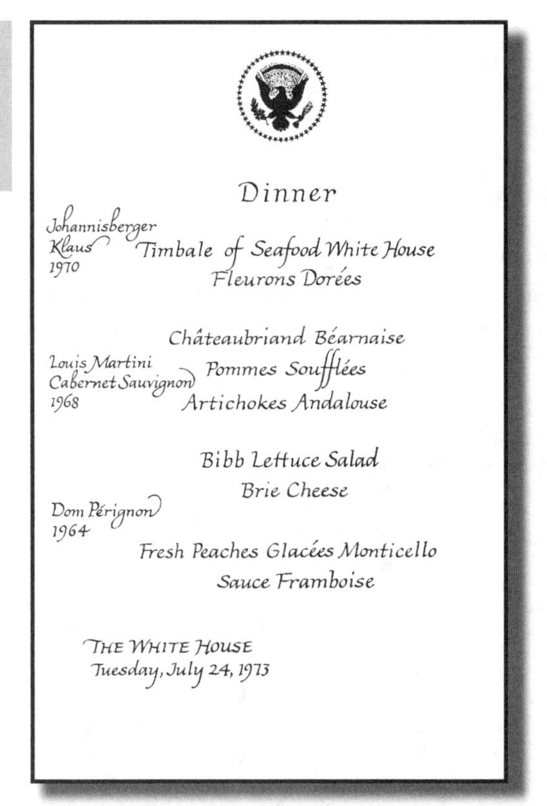

Dinner

Johannisberger Klaus 1970 — Timbale of Seafood White House
Fleurons Dorées

Châteaubriand Béarnaise
Louis Martini Cabernet Sauvignon 1968 — Pommes Soufflées
Artichokes Andalouse

Bibb Lettuce Salad
Brie Cheese

Dom Pérignon 1964 — Fresh Peaches Glacées Monticello
Sauce Framboise

The White House
Tuesday, July 24, 1973

*Courtesy of Chefs Haller and Bender of the White House*

figures in society since their wages could help a new foodservice industry develop and survive. A substantial middle class was also coming into its own, and the new market would be shaped largely for them.

Another important contribution of Escoffier was simplifying the menu. He felt that too many foods were served at each course, so he often reduced this to just one food item with an accompaniment. It was he, following Carême and others, who developed the progression of courses in a formal meal to: a light soup, fish, poultry, entree, salad, cheese, and dessert. Each course might have its accompaniment and garnish; the entree would be the heaviest course, with a roast beef meat accompanied by a vegetable and starch food, often a potato. After this course, the meal was to drop in intensity. The salad gave a refreshing respite, with the cheese following to keep up the flagging appetite so the dessert could be appreciated. A formal progression of wines accompanied the courses. We will see this progression of courses later in our discussions; its purpose here is to point out that it was Escoffier who stamped it indelibly into our eating culture. However, Escoffier also did not hesitate to serve very simple luncheons and dinners. He felt that foods above all had to be supremely prepared and not just be exciting to the eye, and food that good should be served in an adequate quantity, and satisfy the appetite.

**Exhibit 1.1** shows a fine dinner menu used at the White House when the famous Chef Haller supervised the kitchens. Its simple elegance reflects Escoffier's influence.

# Foodservice in the United States

## The Early Years

During colonial times in the United States, foods served to those away from home were served much the same way as in Europe. For a considerable period of time, travelers were cared for at roadside inns, some of which still exist. Coffeehouses operated in New York, Philadelphia,

Boston, and other large cities. Taverns and *eateries (beaneries* in Boston) also served food (see **Exhibit 1.2**). Some clubs existed that also served a fairly high standard of food, usually of a local character.

### *Institutional Feeding*

Orphanages, hospitals, prisons, and other institutions cooked food in quantity over fireplaces or over beds of coals. A form of cooking sometimes called *ground cooking* was also used to cook foods. A fire was built in a hole in the ground containing some stones or in a clay or brick oven. When the stones were very hot, the fire was put out, the food was put in, and the oven was closed or the hole covered, and the food then cooked for a long time in the stored heat.

**Exhibit 1.2** Delmonico's Menu

Consumme de Volaille
Huitres à la Poulette

Saumon Truites

Au Beurre de Montpellier
Filets de Boeuf à la Bellevue Galantines de Dinde à la Royale
Pâtes de Gibiers à la Moderne Cochons de Lait à la Parisienne
Pains de Lièvres Anglais Histories Terrines de Nerac aux Truffes
Jambons de Westphalie à la Gendarme
Langues de Boeuf à l'Escarlate

Mayonnaises de Volailles Salades de Homards à la Russe
Grouses
Bécassines Bécasses
Faisans
Gelées au Madère Macédoines de Fruits
Crèmes Francaises Glaces à la Vanille et Citron
Petits Fours Charlotte Russe
Pêches, Poires, Raisins de Serre, etc.

PIECES MONTÉES
La Reine Victoria et le Prince Albert
Le Great Eastern La Vase de Flora
Silver Fountain

This is the Menu from the 1860 ball given in honor of the Prince of Wales.
Delmonico's catered this event for four thousand guests, held at the Academy of Music.

*Source: Root W, de Rouchemont R,* Eating in America, A History. *NY: Ecco Press 1981 2nd edition p329.*

Some universities had a dining service where the students came to eat family style. Food was brought to the table in serving dishes and the students passed the food. This service style is still used today in some schools—the U.S. Naval Academy is one such place. Other schools offered table service. These dining rooms were often called *commons.* Still other schools had small apartments for the students and each student brought a cook from home, usually a slave, who would cook and care for the student. These quarters still stand on the University of Virginia campus, designed by Thomas Jefferson.

### *Hotels Appear*

With the continued growth of the United States, a need arose for hotels, which began to appear in the larger cities. In 1818, New York City had eight hotels. By 1846, there were more than 100. In 1850, Chicago boasted 150 hotels. Some of these hotels were built to provide great luxury. Famous chefs were brought from Europe, along with their staffs, to cook for guests. The Astor House in New York City, the Palace in San Francisco, the Brown Palace in Denver, the Butler in Seattle, and the Palmer House in Chicago were as elegant and fashionable as any hotel operated by Ritz and Escoffier. Many of these hotels became the cultural and social centers of the cities where they were located.

The Gold Rush to California and later gold discoveries in Montana, Alaska, and elsewhere, along with the discovery of silver lodes in Colorado and Nevada, drew many people to the West. Many became wealthy and could then afford fine dining.

After the Civil War, railroads were rapidly built all over the country. In communities served by railroads, small hotels were built near railroad stations to feed and shelter travelers. Restaurants also opened, but these were patronized largely by affluent people or those away from home for a short period of time.

### *Dining Trends*

The trend toward elegance and luxury found in the best hotels also began to appear in restaurants. Lorenzo Delmonico (1813–1881) started his famous restaurant, Delmonico's, near Wall Street in New York City for the financial tycoons there. It developed an international reputation. He and his brothers started other restaurants in New York City, beginning what was probably one of the first restaurant chains. However, the era of the working class was approaching in the United States, and with it a need for a type of foodservice that did not have the elegance, fashion, or high cost of operations, such as Delmonico's. Delmonico's was the "first" respectable restaurant that allowed women of "good standing to dine out."[3]

At the turn of the century, people began to leave their homes to work in factories, office buildings, stores, hospitals, schools, and commercial centers. They needed meals, particularly lunch, and many coffee shops and restaurants sprang up, including Child's, Schrafft's, and Savarin. Popular foods were served at nominal cost. Horn and Hardart, a cafeteria system, introduced the *nickelodeon* automatic foodservice. Food portions were loaded into small compartments, some of which were heated by steam. Patrons would insert the required coins and a door would open to allow removal. When the door closed, it locked again, and empty compartments would be replenished. In some factories and office buildings, employee feeding took place on a restricted basis. Many facilities built kitchens and dining areas to serve their workers. The food was usually very good and economical. It also became common to see mobile kitchens parked outside factories and other buildings where personnel worked waiting to serve them light meals or snacks during the day.

Other changes were occurring. The average American began to have more disposable income. The automobile arrived, giving the population much more mobility. Working hours were shortened, providing more leisure time. Electricity became available on a wider scale. Foodservice operations now could have refrigerators and freezers rather than iceboxes. Mixers, dishwashers, and other electrical equipment became available, reducing many laborious tasks in the kitchen.[4]

## POSTWAR EXPANSION

After World War II, the foodservice industry began to grow rapidly. While the normal rate of growth in retailing during that time was about 6 percent per year, the foodservice industry saw an annual 10 to 11 percent growth. Some of this growth was prompted by the expansion of institutional feeding. Factories and office buildings put in their own foodservice units for workers. In 1946 the federal government passed the National School Lunch Act, which started a vast school feeding program. Colleges and universities put in extensive dining services. Dietitians, trained to operate foodservices, began to be sought out, not only by health services, but by commercial operations, such as Stouffer's. Margaret Mitchell, a vice-president and one of the dominant forces behind Stouffer's, was herself a dietitian. Cornell University, under Professor Howard Meek, introduced the first hotel school, which branched out into restaurant and institutional curricula, and later added tourism.

Vast hotel chains were being formed, with Ellsworth Statler leading the way. People began to eat out, not only as a necessity, but because they wanted recreation, entertainment, and a change from eating at home. The desire was not just for food to satisfy nutritional and hunger needs, but for an environment that would help meet social and psychological desires.

### *Quick Service and Corporate Concepts*

Two very important changes in the foodservice industry occurred simultaneously and were related. These were the development of the *fast-food*, or *quick-service*, concept, and the birth of multiple-operation foodservice groups, or *chains*. The first of the units serving food that could be prepared and eaten quickly were the White Castle hamburger units, which appeared in the 1930s. The idea grew slowly until after World War II. Then in the late 1940s and 1950s, many hamburger chains were started. They were immensely popular, serving an item that had wide acceptance at a nominal price with rapid service. Another item that found acceptance as a quick-service entree was chicken. To Americans on the go, fast food was an affordable luxury. Vast chains, such as Kentucky Fried Chicken, built by Colonel Harlan Sanders, and McDonald's, led by Ray Kroc, soon appeared all over the country, and were faced with competitors seeking to break into this lucrative market. The low price and lower margin of profit per sale was compensated for by big volume. A limited menu made it possible to simplify operations and use personnel with less experience and skill in food preparation and service. This chain segment of the industry has grown tremendously. The National Restaurant Association says it now makes up nearly 30 percent of all commercial foodservice units. The quick-service segment's percentage of entire industry sales is much higher than that.

### *Take-out and Curbside Service*

Take-out has become a significant part of foodservice distribution, with continued growth expected in this industry segment. Households with both spouses working, fully scheduled family activities; a growing population of older individuals with higher disposable income; busy people that need food prepared quickly; and the convenience of not having to cook and clean up are the motivating factors behind the usage of take-out and curbside foodservices. Take-out has been the fastest-growing segment of the foodservice industry. It is estimated that over 40 percent of the food consumed today is take out. This includes quick-service restaurant fare from traditional burger drive-throughs, ethnic food, pizzas, and many others.

Grocery stores have also entered this market with extensive deli offerings. Some of these items are immediately consumed at tables located in the store; others are taken away. Food bars or packaged offerings are available in many locations.

Morrison's Cafeteria and Golden Corral Buffet are two companies that have developed substantial take-out business. Golden Corral weighs the container and the customer pays a per-pound charge. This pricing method allows the introduction of *to-go* business and the *doggie bag* to the traditional *all-you-can-eat* buffet.

Fine-dining establishments have gotten into the game with curbside service whereby preordered food, sometimes in family-size portions, is delivered to the cars of waiting customers, parked in designated spaces just outside the restaurant. The popular Manhattan restaurant Woods has two take-outs. They're called "Out of the Woods." Consumers are hungry for this service.

Take-out is growing rapidly in other segments as well. A sizable number of operations now exist to prepare complete dinners, lunches, and other prepared foods. Some sell almost all of their food as take-out. "Take-out taxi" companies pick up food from an established list of restaurants that typically do not offer delivery service. Roma's has stated that it is trying to develop 5 to 7 percent of its business in take-out. The growth in take-out and curbside business is expected to continue and become an even more significant part of the industry in response to guest desires. About 44 percent of the food sold by foodservices is considered take-out.

**Photo 1.1** Dave Thomas, founder of Wendy's International, 1932–2002
*Courtesy of Wendy's International, Inc.*

What type of individual does it take to build a successful foodservice empire? Men like Dave Thomas, (1932–2002), founder of Wendy's International, and Richard Melman of Lettuce Entertain You Restaurant Corporation are good examples. First of all, these entrepreneurs had to have a differentiated product, a food that had something distinct about it that made customers want it. Second, they had to be individuals with a lot of aggressive drive to develop a market for their product.

Dave Thomas got his first job at the age of twelve in a restaurant. At fifteen he dropped out of school to work full time at the Hobby House restaurant and live at the YMCA when his adopted family moved out of Fort Wayne, Indiana. Through this work he got to know Colonel Sanders, the founder of Kentucky Fried Chicken, and got the opportunity to turn around four failing Ken-

**Photo 1.2** Rich Melman, founder of Lettuce Entertain You Enterprises Inc.
*Courtesy of Lettuce Entertain You Enterprises Inc.*

tucky Fried Chicken enterprises, owned by his Hobby House boss. In four years he sold the restaurants back to KFC and was a millionaire at 35.

In 1969 he made his dream of owning a hamburger restaurant come true. He opened the first Wendy's Old Fashioned Hamburgers in Columbus, Ohio. His hamburger was fresh, not frozen, meat. Each hamburger was freshly made to order, not stock-piled. His salad bar and baked potato were the first in a quick-service setting, and they played a big role in the success of the operation. He also created the first modern-day drive-through window.

Dave felt he was he was an American success story and advised all to seize their opportunities. His "rags to riches" story earned him the Horatio Alger award. Despite his success, he referred to himself as "just a hamburger cook." A big factor in Dave's success was his concept that it all came down to customers. Taking care of customers by serving the best food at a good value in clean comfort was his underlying philosophy. (Reference Wendy's Website).

A more modern illustration of success in the restaurant business is *Lettuce Entertain You Enterprises* (LEYE), a Chicago-based corporation, owning at this time more than 50 successful units operating nationwide. The creative genius behind the building of this company is Richard Melman, founder and chairman. Through Melman, Lettuce has earned a reputation for the ability to be prolific in original, successful menu creation with over 30 concepts operating at this writing.

Like many entrepreneurs before him, Melman began his working career as a teenager in a restaurant. The business got in his blood, and he and his partner Jerry A. Orzoff opened R. J. Grunts in 1971 in Chicago's Lincoln Park. Food was presented with a sense of humor and fun, leading the trend toward dining out as entertainment (see **Exhibit 1.3**). Orzoff died in 1981, leaving Rich Melman with a philosophy based on the importance of partners, which has remained a cornerstone of the Lettuce philosophy. People are the core of Lettuce with former employees referred to as *alumni*. Melman is credited with being a prodigy at concept development. He sees his success as being able to listen to people, giving them what they want almost before they knew they want it. (Reference LEYE Website) At this writing, LEYE is delving into retail foodservice with a concept called Shanghai Circus Stir Fry and Steam Show, offering Asian cuisine and Osteria Via Stato, which embraces the lifestyle of Italian dining.[5]

## Continual Growth

In a period of about 40 years, the foodservice industry has changed from a rather small one in our total industrial economy to one of our major ones. Growth and changes in feeding habits will not stop. Eating away from home should become an even more important factor in the lives of people than it is now. Some say that in the not too distant future, we will be eating more than 50 percent of our meals outside the home. If past growth reflects the future, this seems quite possible.

**Exhibit 1.3**
R.J. Grunts
Menu

# R.J. GRUNTS®

LUNCH 11:30 to DINNER

## THE ORIGINAL 'LETTUCE ENTERTAIN YOU' CREATION

GOOD AFTERNOON...
If this were Sunday, you'd be reading the brunch menu... But it isn't.. So, you're not. Now if I can just remember where I left my sneakers.

### SANDWITCHES

senior, R.J., and Jr. burger it up in the kitchen

**1. HAMBURGER**
Juicy chopped sirloin charbroiled to your taste and served on a lightly toasted sesame bum. (And to add a touch).. it's graced by a heaping portion of cottage fries, our own home made cole slaw and (of course) a dull pickle. $3.25

**1a.* CHEESEBURGER**
Absolutely the same as above with one gooey exception. Cheese.
* Your choice at no extra charge $3.25

**2. GRUNTBURGER** ★★★★★
The cornerstone of Burgerdom. A charbroiled chopped sirloin topped w/fried onion and blue cheese dressing. Plus.. cottage fries, cole slaw & pickle $3.25

**3. GOURMET BURGER**
This burger is topped with mush rooms, onions, Swisss cheeeze, a rich brown sauce, and served on a toasted sesame bun with cole slaw, cottage fries and a pickle. $3.35

**4. SORRY HUN-NO BUN**
Sauteed vegetables nestled upon a burger with Italian seasonings, sprinkled w/ parmesan and covered with mozarella cheeze. $3.35

**5. MARK'S RYE BURGER**
When Mark got tired of the usual employee lunch he took our hamburger, added cooked onions, swiss cheese, worchester sauce and grilled it all on rye served with cottage fries $3.35

**6. ROAST BEEF AW JUICE** ★★
Thinly sliced roast beef brisket steeped in its own natural gravy and piled high on a toasted sesame bun. ..complimented by tasty fried onion slices, cottage fries and a very dill pickle $3.75 Try this on Italian bread.

**7. ROAST BEEF BE QUE** ★★★
Very much (exactly) as #6 with R.J.s own B.B.Q sauce 10¢ EXTRA

**8. STEAK TERIYAKIWICH**
Charcoal broiled steak that's served on garlic bread with a side of teriyaki sauce and fries $3.95

**9. THE TACO THAT ATE CHICAGO**
Don't just sit there. Bite it back. With chopped beef, lettuce, tomatos, cheese and a little guacamole $3.25
* Ask your bus boy.

**10. THE UN-BURGER**
A yummy marinated & char-broiled, boneless Breast of Chicken sandwich with cottage fries, cole slaw & pickle. (It's UN in a million.) $3.25

### EGGS, EGG-CETERA

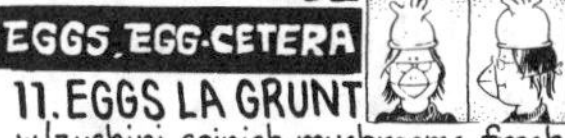

**11. EGGS LA GRUNT**
w/zuchini spinich, mushrooms, fresh tomatoes, melted cheese. & served w/a Grunt Treat. $2.95
CHEESE OMELETE AVAILABLE AT SAME PRICE

**12. THE "COULD-BE QUICHE"**
R.J.'S Chef, Peter, changes his mind a lot. That's why our light & cheesy Homemade Quiche of the day, served with a veggie, is always different. $3.50

Constant testing and endless research personally conducted by Himself."

**LUNCHEON SALAD BAR** ★★★★★
From 11:30 till DINNER, Monday through Saturday you may help yourself to the Salad table w/your sandwitch.. $1.35
Now, if all you want is salad........ $3.25

w/onion soup, cuppa Fish soup, or chili, salad is $1.95 extra

### ★★ OTHER ★★

**13. FISH SOUP ALA GRUNT**
Scalllopps, halibut, clams & shrimps enjoy communal bath in warm spicy tomato base. $3.50

**14. MEAL IN A BOWL**
Large bowl of our homemade vegetable beef soup (w/ beef), with crouton and served with cobblestone bread and whipped butter $3.25

**15. ONION SOUP EPISODE** ★★★
All we can say at this time is.. Home made onion soup covered w/croutons and some melted cheeses and it's served in a crock! $1.95

**16.* TUNA TERRIFIC**
Tuna w/tomato, mushrooms, melted cheese & a side of vegetables open faced n'cold, or hot $3.10
*available in casserole.

**17. "CHICAGO'S BEST" CHILI** ★★
Winner of "The Great Chicago Chili Cook-Off" served with grated cheddar cheese and a dallop of sourcream on the side $1.95

### LIQUID ★

You never know who you'll run into at Grunts.

**SPIRITS** ★★★★★★★★★★★★
For those of you who prefer your beer fresh squirted from a cold keg let us suggest Michelob. If (on the other hand) bottles do it for you try Heinekens, light or dark or R.J's favorite lite.

♥♥♥ **THE GRUNTS GIRLS** ♥♥♥♥
A special thank you to Amy, Elaine, Paulette, Nancy, Susan, Ilene, Murph, House, Elinor, Arlene, Ingrid, Andrea, Liza, Karen, Vickie, Lynda, Kitty, France, Debbie, Deborah, Holly, Elizabeth, Jennifer, Moo, Paula, Sue x 3, and Pat x 2. ♥♥♥ I Love You, R.J.

☆☆☆☆☆☆☆☆☆☆☆

**GIANT MALT & SHAKES**
Peanut butter, Banana, Chocolate, Raspberry, Strawberry, CoCo Nut, pineapple, Mocha Malt.................. $1.50
Raw egg or wheat germ added 25¢

SUNDAY BRUNCH

**SOFT DRINKS** ★★★★★★★*
Lets see now... theres coke, seven up, uh.. gingerale, milk, ..iced tea, tab, Diet 7↑.. root beer or lemonade 60¢
* 7-up available at diet level. See: Diet 7↑ in Soft Drink Sec.

**FRUIT JUICES**
Fresh squeezed orange juice, apple*, coconut*, banana*, Rrraszpberrie*, strawberry*, pineapple* or grapefruit* blended into any rediculous combination. $1.10
* NOT SO FRESH SQUEEZED.

**BLACK COW**
Root Beer and Ice cream........ $1.50

**2056 LINCOLN PARK WEST • CHICAGO, ILLINOIS • TEL 929-5363**

Rick Rogers

WE NOW ACCEPT THE FOLLOWING PLASTIC CURRENCY.
AMERICAN EXPRESS, VISA and MASTERCHARGE.

*Courtesy of Lettuce Entertain You Enterprises Inc.*

## Summary

The seeds of our foodservice industry were sown many centuries ago when a cluster of cave dwellers gathered together and cooked food over an open fire. Many ancient cultures had foodservice operations that served food to travelers, prepared foods for feasts, and served meals to ordinary people at public gatherings. During the Middle Ages, monasteries maintained quantity cooking and service and even acted as hostelries for travelers. Inns and taverns also opened in cities and along travel routes. People of wealth and royal blood had kitchens and dining areas designed to serve food in quantity, and many of their best cooks advanced our knowledge of food preparation and service.

The restaurant industry really began in Paris just after the French Revolution and rapidly spread throughout Europe. An aristocracy of wealth appeared and provided patronage for fine, elaborate restaurants, clubs, and hotel dining services.

The development of the foodservice industry in the United States did not occur to any degree until after the Civil War. Until that time, it functioned on the limited scale that was typical in Europe. As the United States started to industrialize, people had more discretionary income, and they began to want food away from home. Along with the coming of a vast network of railroads across the nation came a demand for hotels offering meals, as well as lodging. After World War I, factories and office buildings began to provide meals for workers. A large number of cafés, restaurants, and cafeterias began to appear in highly populated areas. School lunches started in 1946 as a federally sponsored project.

After World War II, the foodservice industry leapt forward. One of the reasons for this was the introduction of the quick-service operation, in which certain foods could be offered at moderate prices. These establishments became a significant segment of the foodservice industry. With continued development and the increase of chains and franchises, the foodservice industry has developed into one of the largest industries in the nation. The growth of this huge industry is expected to continue. Chapter 2 will discuss both the present and future of the foodservice industry.

## Questions

1. What was the significant contribution made to foodservice by the ancient Greeks? By the Romans? By Europeans in the Middle Ages?
2. What characterized French cuisine during the reigns of Henry II and Henry IV? Of Louis XIII and XIV?
3. What were the significant contributions made to foodservice by Carême? By Escoffier?
4. What historical events and forces led to the rise of hotels and restaurants in the United States?

## Mini-Cases

1.1 You are a member of a crew doing an excavation of an ancient city and come across what appears to be a communal kitchen and dining area. What would you hope to find and learn in your forthcoming excavation work?

1.2 What factors shape the destiny of the foodservice industry? Which factors do you think you can use to influence the operation of a restaurant that you manage? How can you use that factor or factors?

1.3 You are a consultant in mass feeding to the United Nations and have been sent to visit a group of the Near East undeveloped nations to assist them in improving their foodservice industry. These countries have asked the UN for this service.

The UN will give technical assistance, food, supplies, and even some money to support projects you develop with them. Technicians and others sent will run programs for a period of one to two years, while at the same time will train a native person to take over the technician's job before leaving. After that, from time to time experts will be sent to observe and render any assistance needed. You will have diplomatic credentials that introduce you to the highest officials of the various agencies of the country. These individuals have been informed by the UN of your mission and the program you are there to set up for them. How would you start and then follow through on your job to try to help these countries improve their foodservices to a point where they improve the food of the people and give support to the economy?

1.4 You are planning a menu and know that about 40 percent of your clientele is nutrition and/or value minded. The other 60 percent pay little attention to these factors. Shape a menu that will satisfy both groups.

# 2

# PROFILE OF THE MODERN FOODSERVICE INDUSTRY

## Outline



## Objectives

After reading this chapter, you should be able to:

1. Characterize the distinguishing features of various types of foodservice operations and describe situations where categories may blur.
2. Differentiate between major segments of the foodservice industry: commercial operations, institutional operations, transportation foodservices, health services, clubs, military feeding, and central commissaries.
3. Identify economic, social, labor, health, and technological trends that are likely to impact the future of the foodservice industry.
4. Speculate on industry segments that may grow or recede in popularity and provide justification for your predictions.

# Introduction

The foodservice industry consists of many diverse operations that provide food and drink for people away from home. This somewhat simplistic definition covers a wide variety of operations and markets. The industry is huge; in 2001 it was made up of 877,204 operations, and in 2007 it was projected that the sales value of the food and drink served would total nearly $537 billion. (See **Exhibit 2.1.**) It hires more people than any other industry in the world. It is estimated that this industry takes about 47.5 percent of the food dollar spent in this country. Today, foodservice is the nation's largest private sector employer, employing 12.8 million people.

The foodservice industry has been one of our country's most consistent growth industries. After World War II, the industry, then relatively unimportant in our economy, started to grow rapidly. By 1970, it was doing more than $40 billion a year in sales. This growth continues to the present. **Exhibit 2.2** shows the percent change in restaurant-industry food-and–drink sales per year in both current-dollar and real (inflation-adjusted) growth from 1971 to 2007. The projected total industry sales in 2007 show continued growth and a 5 percent increase from 2006 to 2007.

# A Diverse Industry

The types of units differ greatly from each other in the food and drink offered, and in their preparation and service. They serve different clientele, merchandise their items differently, and run their businesses differently. Yet they all exist to serve market needs and patron desires in food and drink.

Eating and drinking establishments make up most of the units in the industry and do most of the sales. In 2007 this segment is projected to do about $379 billion (71 percent) of the sales in the restaurant industry with 5.2 percent growth. At one time, managed services lagged behind other segements in total sales, but in the last several decades have made a larger contribution to sales, projected in 2007 to be $36 billion, or nearly 7 percent of all sales. Lodging foodservices will probably do about $26 billion in 2007. The group, called *retail hosts,* is just below that figure in projected sales, at $25 billion registering faster growth than full- or quick-service restaurants at 5.3 percent. Noncommercial restaurant services, or Group II, are projected to do nearly $44 billion, or 8.15 percent of total sales in 2007. Military restaurant services, or Group III, were projected in 2007 as doing 0.35 percent of the business, or $1.9 billion.

# Income and Costs

Few foodservice operations operate successfully if their combined food, beverage, and labor costs are above 65 percent—the most common division being 35 percent food and beverage and

**Exhibit 2.1** Restaurant Industry Food-and-Drink Sales Projected through 2007

| | 2004 Estimated F&D Sales ($000) | 2006 Projected F&D Sales ($000) | 2007 Projected F&D Sales ($000) | '06–'07 Percent Change | '06–'07 Percent Real Growth Change | '04–'07 Compound Annual Growth Rate |
|---|---|---|---|---|---|---|
| **Group I—Commercial Restaurant Services[1]** | | | | | | |
| **EATING PLACES** | | | | | | |
| Fullservice restaurants[2] | $156,856,918 | $172,769,112 | $181,580,337 | 5.1% | 2.1% | 5.0% |
| Limited-service (quickservice) restaurants[3] | 129,520,592 | 142,932,450 | 150,079,072 | 5.0 | 2.1 | 5.0 |
| Cafeterias, grill-buffets and buffets[4] | 4,968,152 | 5,223,714 | 5,448,334 | 4.3 | 1.4 | 3.1 |
| Social caterers | 4,936,007 | 5,667,078 | 6,069,440 | 7.1 | 4.1 | 7.1 |
| Snack and nonalcoholic beverage bars | 15,741,864 | 18,531,699 | 20,162,488 | 8.8 | 5.9 | 8.6 |
| TOTAL EATING PLACES | $312,023,533 | $345,124,053 | $363,339,871 | 5.3% | 2.3% | 5.2% |
| Bars and taverns | 14,438,492 | 15,302,924 | 15,792,618 | 3.2 | −0.2 | 3.0 |
| TOTAL EATING-AND-DRINKING PLACES | $326,462,025 | $360,426,977 | $379,132,289[5] | 5.2% | 2.2% | 5.1% |
| **MANAGED SERVICES[6]** | | | | | | |
| Manufacturing and industrial plants | $6,281,258 | $6,905,276 | $7,099,256 | 2.8% | −0.1% | 4.2% |
| Commercial and office buildings | 2,129,879 | 2,334,774 | 2,430,500 | 4.1 | 1.2 | 4.5 |
| Hospitals and nursing homes | 3,361,643 | 3,833,756 | 4,065,177 | 6.0 | 4.2 | 6.5 |
| Colleges and universities | 8,594,981 | 10,034,080 | 10,897,011 | 8.6 | 4.8 | 8.2 |
| Primary and secondary schools | 3,766,072 | 4,602,517 | 5,012,141 | 8.9 | 6.2 | 10.0 |
| In-transit restaurant services (airlines) | 2,061,488 | 1,921,471 | 1,948,372 | 1.4 | −1.6 | −1.9 |
| Recreation and sports centers | 4,019,672 | 4,368,555 | 4,560,771 | 4.4 | 1.4 | 4.3 |
| TOTAL MANAGED SERVICES | $30,214,993 | $34,000,429 | $36,013,228 | 5.9% | 2.9% | 6.0% |
| **LODGING PLACES** | | | | | | |
| Hotel restaurants | $22,166,551 | $25,047,493 | $26,425,105 | 5.5% | 2.5% | 6.0% |
| Other accommodation restaurants | 359,827 | 396,774 | 412,248 | 3.9 | 1.0 | 4.6 |
| TOTAL LODGING PLACES | $22,526,378 | $25,444,267 | $26,837,353 | 5.5% | 2.5% | 6.0% |
| Retail-host restaurants[7] | 21,372,399 | 23,788,291 | 25,053,348 | 5.3 | 2.4 | 5.4 |
| Recreation and sports[8] | 10,907,425 | 11,918,364 | 12,626,743 | 6.0 | 3.0 | 5.0 |
| Mobile caterers | 866,738 | 951,543 | 980,489 | 3.0 | 0.1 | 4.2 |
| Vending and nonstore retailers[9] | 9,402,932 | 10,219,106 | 10,556,336 | 3.3 | 1.3 | 3.9 |
| TOTAL—GROUP I | $421,752,890 | $466,746,977 | $491,199,786 | 5.2% | 2.3% | 5.2% |
| **Group II—Noncommercial Restaurant Services[10]** | | | | | | |
| Employee restaurant services[11] | $623,425 | $470,160 | $478,761 | 1.8% | −0.1% | −8.4% |
| Public and parochial elementary, secondary schools | 5,206,321 | 5,382,626 | 5,362,423 | −0.4 | −3.1 | 1.0 |
| Colleges and universities | 5,624,963 | 5,700,734 | 5,718,953 | 0.3 | −3.5 | 0.6 |
| Transportation | 1,459,767 | 1,724,465 | 1,856,497 | 7.7 | 5.5 | 8.3 |
| Hospitals[12] | 11,885,883 | 12,634,182 | 12,962,671 | 2.6 | 0.8 | 2.9 |

*National Restaurant Association 2007 Forecast*

## Exhibit 2.1 *(Continued)*

| | 2004 Estimated F&D Sales ($000) | 2006 Projected F&D Sales ($000) | 2007 Projected F&D Sales ($000) | '06–'07 Percent Change | '06–'07 Percent Real Growth Change | '04–'07 Compound Annual Growth Rate |
|---|---|---|---|---|---|---|
| Nursing homes, homes for the aged, blind, orphans and the mentally and physically disabled[13] | 6,087,948 | 6,433,948 | 6,581,929 | 2.3 | 0.6 | 2.6 |
| Clubs, sporting and recreational camps | 7,594,885 | 8,553,332 | 9,061,235 | 5.9 | 2.8 | 6.1 |
| Community centers | 1,579,202 | 1,703,121 | 1,754,215 | 3.0 | 1.3 | 3.6 |
| TOTAL—GROUP II | $40,062,394 | $42,602,568 | $43,776,684 | 2.8% | 0.3% | 3.0% |
| TOTAL—GROUPS I AND II | $461,815,284 | $509,349,545 | $534,976,470 | 5.0% | 2.1% | 5.0% |
| **Group III—Military Restaurant Services[14]** | | | | | | |
| Officers' and NCO clubs (Open mess) | $1,125,810 | $1,247,121 | $1,313,218 | 5.3% | 2.3% | 5.3% |
| Military exchanges | 521,465 | 562,764 | 601,032 | 6.8 | 3.9 | 4.8 |
| TOTAL—GROUP III | $1,647,275 | $1,809,885 | $1,914,250 | 5.8% | 2.8% | 5.1% |
| GRAND TOTAL | $483,462,559 | $511,159,430 | $536,890,720 | 5.0% | 2.1% | 5.0% |

**Restaurant-Industry Percent Change in Dollar Sales**

Industry real growth in 2007 is projected to be 2.1 percent

Percent change in current dollars

Percent change in real dollars

| **Total Industry** | **Fullservice Restaurants** | **Quickservice Restaurants** |
|---|---|---|
| 2006/2007 5.0% | 2006/2007 5.1% | 2006/2007 5.0% |
| 2005/2006 5.0% | 2005/2006 4.7% | 2005/2006 5.1% |
| 2004/2005 5.0% | 2004/2005 5.2% | 2004/2006 5.0% |
| 2006/2007 2.1% | 2006/2007 2.1% | 2006/2007 2.1% |
| 2005/2006 1.8% | 2005/2006 1.4% | 2005/2006 2.0% |
| 2004/2005 1.9% | 2004/2005 2.0% | 2004/2005 1.9% |

1. Data are given only for establishments with payroll.
2. Waiter/waitress service is provided, and the order is taken while the patron is seated. Patrons pay after they eat.
3. Patrons generally order at a cash register or select items from a food bar and pay before they eat.
4. Formerly commercial cafeterias.
5. Food-and-drink sales for non-payroll establishments should total $13,770,774,000 in 2007.
6. Also referred to as onsite food-service and food contractors.
7. Includes drug and proprietary-store restaurants, general-merchandise-store restaurants, variety-store restaurants, foodstore restaurants and grocery-store restaurants (including a portion of delis and all salad bars), gasoline-service-station restaurants and miscellaneous retailers.
8. Includes movies, bowling lanes, recreation and sport centers.
9. Includes sales of hot food, sandwiches, pastries, coffee and other hot beverages.
10. Business, educational, governmental or institutional organizations that operate their own restaurant services.
11. Includes industrial and commercial organizations, seagoing and inland-waterway vessels.
12. Includes valuatory and proprietary hospitals, long-term general, TB, nervous and mental hospitals; and sales or commercial equivalent to employees in state and local short-term hospitals and federal hospitals.
13. Sales (commercial equivalent) calculated for nursing homes and homes for the aged only. All others in this grouping make no charge for food served either in cash or in kind.
14. Continental United States only.

*Source: National Restaurant Association*

## Exhibit 2.2 Restaurant-Industry Food-and-Drink Sales, 1971–2007 (Percent Change): Historical Perspective

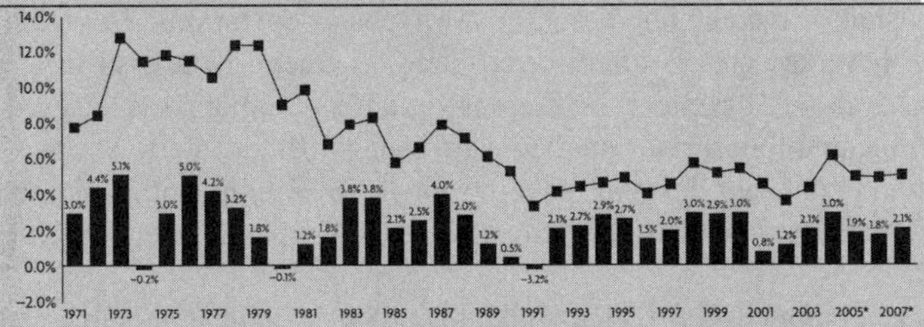


*Growth rates are estimated for 2005 and projected for 2006–2007. Providing final estimates for restaurant-industry sales from previous years is an ongoing process. The National Restaurant Association's Restaurant TrendMapper offers updated sales estimates as they become available. Visit www.restaurant.org/trendmapper to learn more.
*Source:* National Restaurant Association

### At a Glance

**Key Restaurant-Industry Segments, 2003–2007**

National Restaurant Association economists project restaurant-industry sales each year and also take a close look back, in an ongoing process of providing detailed final estimates for restaurant-industry sales in previous years. This chart shows the NRA's updated estimates of sales for key restaurant-industry segments from 2003 through 2005, along with projections for 2006 and 2007.

**Restaurant-Industry Food-and-Drink Sales, 2003–2007 ($000)**

| | 2003 | 2004 | Percent Change | 2005* | Percent Change | 2006** | Percent Change | 2007** | Percent Change |
|---|---|---|---|---|---|---|---|---|---|
| COMMERCIAL RESTAURANT SERVICES | $396,831,067 | $421,752,890 | 6.3% | $443,657,105 | 5.2% | $466,746,977 | 5.2% | $491,199,786 | 5.2% |
| Eating Places | 292,768,946 | 312,023,533 | 6.6 | 328,378,609 | 5.2 | 345,124,053 | 5.1 | 363,339,671 | 5.3 |
| Fullservice restaurants | 148,257,957 | 156,856,918 | 5.8 | 165,013,478 | 5.2 | 172,769,112 | 4.7 | 181,580,337 | 5.1 |
| Limited-service restaurants | 120,484,272 | 129,520,592 | 7.5 | 135,996,622 | 5.0 | 142,932,450 | 5.1 | 150,079,072 | 5.0 |
| Cafeterias, grill-buffets & buffets | 5,059,218 | 4,968,152 | –1.8 | 5,022,802 | 1.1 | 5,223,714 | 4.0 | 5,448,334 | 4.3 |
| Social caterers | 4,604,484 | 4,936,007 | 7.2 | 5,281,527 | 7.0 | 5,667,078 | 7.3 | 6,069,440 | 7.1 |
| Snack & nonalcoholic beverage bars | 14,363,015 | 15,741,864 | 9.6 | 17,064,180 | 8.4 | 18,531,699 | 8.6 | 20,162,488 | 8.8 |
| Bars and taverns | 13,950,234 | 14,438,492 | 3.5 | 14,857,208 | 2.9 | 15,302,924 | 3.0 | 15,792,618 | 3.2 |
| TOTAL EATING-AND-DRINKING PLACES | 306,719,180 | 326,462,025 | 6.4 | 343,235,817 | 5.1 | 360,426,977 | 5.0 | 379,132,289 | 5.2 |
| NONCOMMERCIAL RESTAURANT SERVICES | 38,199,462 | 40,062,394 | 4.9% | 41,345,484 | 3.1% | 42,602,568 | 3.0% | 43,776,684 | 2.8% |
| MILITARY RESTAURANT SERVICES | 1,576,339 | 1,647,275 | 4.5% | 1,719,020 | 4.4% | 1,809,885 | 5.3% | 1,914,250 | 6.8% |
| TOTAL INDUSTRY SALES | $436,606,868 | $463,462,559 | 6.2% | $486,731,609 | 5.0% | $511,159,430 | 5.0% | $536,890,720 | 6.0% |

* estimate ** projection
*National Restaurant Association 2007 Forecast*

30 percent labor. Operations with higher costs that operate well are usually subsidized in some way. This 65 percent figure was not always the rule. At one time, the combined total could be 75 percent and still leave a profit, but taxes, employee benefits, and other costs have risen considerably.

**Exhibit 2.3** shows the division of income and costs for two types of operations. To remain profitable, the combined food-beverage cost is often lower than 35 percent. Note also in two operations that the labor costs are above 30 percent. This represents a trend. Labor costs are gradually encroaching on profitability, and foodservice operators are trying to find ways to use technology to improve efficiency and cut other costs to make up for it. Operations with a limited menu and no table service do best in profitability, indicating that when labor can be reduced, it is easier to make a profit.

Combined food, beverage, and labor costs are often higher than 65 percent in noncommercial restaurants that do not need to make a profit. Also, other costs are often not as great. Costs such as music, entertainment, and marketing usually are nonexistent, and rent, if charged, is lower and repairs and maintenance are less. Commercial foodservices have the expense of maintaining attractive, well-decorated surroundings, with suitable furnishings, linens, and tableware. Some need a wait staff to serve patrons. All these factors are usually added costs in noncommercial restaurants.

**Exhibit 2.3** Highlights from the Restaurant Industry Dollar

- Full Service Restaurants reported income before income taxes of approximately 4% of total sales.
- Limited Service Restaurants reported income before income taxes of approximately 7% of total sales.
- At Full Service Restaurants and Limited Service Restaurants salaries and wages were approximately 33% and 30% of total sales, respectively.

The Restaurant Industry Dollar*

| | **Full Service Restaurants** | **Limited Service Restaurants** |
|---|---|---|
| **Where it came from:** | | |
| Food and Beverage Sales | 100% | 100% |
| **Where it went:** | | |
| Cost of Food and Beverage Sales | 33 | 31 |
| Salaries and Wages* | 33 | 30 |
| Restaurant Occupancy Costs | 6 | 7 |
| Corporate Overhead | 3 | 4 |
| General & Administrative Expenses | 3 | 2 |
| Other | 18 | 19 |
| Income Before Income Taxes | 4% | 7% |

*Includes Employee Benefits

*Source: National Restaurant Association and Deloitte Operations Report*

Nevertheless, the cost squeeze has also hit institutions, and many find they are challenged to operate successfully under the monetary restrictions imposed. In fact, some, like hospitals, are looking to their foodservices to become revenue producers to help pay expenses elsewhere.

Beverages give significant dollars for full-menu and limited-menu table service and only a small amount for other types. Predictions are that alcoholic beverage sales will drop or remain steady. Since beverage sales are usually more profitable than food sales, this means another squeeze on profits. Some operators hope that nonalcoholic beverage sales such as specialty coffee and tea drinks will supplement the bottom line.

Undoubtedly in the future, foodservice management will be increasingly challenged to make a profit. It is expected that the foodservice industry will remain fiercely competitive, and this, together with rising costs and customer resistance to paying more, will require highly competent management to ensure success.

# Segments of the Foodservice Industry

Before a menu for a particular operation is written, the menu planner should know the characteristics of the operation, so it can be written to please targeted guests and their changing desires while meeting the needs of the operation. In writing these menus, the planners must use all the technical and other information discussed in this text, as well as their own experience. Proper control of costs must be observed; the physical presentation of menu items and menu makeup must invite the diner to the table; the planner must remember to include items that are looked upon as healthful foods and in keeping with current dietary trends; and the goods needed for menu items must be available at a proper cost. These factors are not discussed in detail in the presentation of the various operating units that follow, but they must be kept in mind. Unless all of this is done, the menu will fail. A menu cannot do some things well and do other things poorly. All links in the chain must be strong. More than fifty percent of restaurants go out of business in the first year of operation due to underfinancing and the failure to pay attention to detail. Nothing is more fatal than to forget the guest.

## Commercial Operations

The commercial segment is the largest division of the foodservice industry and is responsible for most of the sales. The commercial category includes any type of operation that sells food and beverages for profit. The operation could be a stand-alone foodservice unit or part of a larger business, which is also designed for profit.

Commercial feeding operations include the following types of units: eating places, bars and taverns, food contractors, lodging places, retail hosts, recreational and sports centers, vending retailers, and mobile caterers.

### *Restaurants and Lunchrooms*

By far the largest group of units in the foodservice industry is full-service restaurants, at approximately 28 percent of the total number of commercial foodservice units in the industry. Some of these are fine-dining restaurants with seated service, where the average check is $25 or more. Sometimes these restaurants are referred to as *white tablecloth* operations. They attract special-occasion diners, tourists, fine-food aficionados, and business executives entertaining clients and associates.

Some of these units offer entertainment. They generally have a low customer turnover rate, with two seatings per dining period usually being the maximum. They generally serve lunch and dinner. Wine and other alcoholic beverages are usually offered in a menu separate from the food menu. While food cost in a fine-dining operation may be average, labor cost is usually higher, and the restaurant must compete for a limited number of customers.

Alcohol consumption in this country is dropping. In 1981, it peaked at 2.76 gallons of alcohol per person per year over the age of 14. In 2004, this had dropped to 2.23 gallons. Spirits accounted for .68 gallons of the total consumption, wine for .35 gallons and beer for 1.21 gallons. The cost of insurance has also dipped into the profitability of alcohol sales. Court decisions on third-party liability in serving alcoholic beverages, as well as tremendous fines, have made insurance companies raise their rates to a point where some operations cannot afford it. The foodservice industry has had to cope with this drop and try to better merchandise lighter alcoholic drinks and nonalcoholic drinks to make up for the loss. It also has set up training programs on how to safeguard against serving alcohol to minors, known alcoholics, and those who might be intoxicated. The beverage industry has also addressed the problem. Advertising programs, such as Anheuser-Busch's "We all make a difference. Thanks for Drinking Responsibly," Miller's "The Cool Spot" alcohol awareness Website for teens, and Coors's "Drive Sober, Drink Responsibly," are examples of how the beverage industry is trying to encourage responsible drinking. Federal and state governments have also stepped in and, through regulations, have tried to warn individuals of the health hazards involved in consuming excess alcohol. This, plus an increasing awareness on the part of the public of the harm that alcohol can do, has resulted in this drop in alcohol consumption, and brought about a drastic change in how operations serving alcohol must meet the problem. Restaurants have come to emphasize quality products over quantity.

Including *lunchroom* in this category might be somewhat misleading. So-called lunchrooms may also serve breakfast and dinner, but usually their main income comes from the midday period. These operations may depend largely on office, shopper, and other drop-in trade in heavy foot-traffic areas. They are often single units, although chains such as Marriott's may operate lunchrooms. The menu is simple, and may include light meals along with sandwiches, soups, salads, and beverages. Some lunchrooms may offer alcoholic beverages, but most do not. Some serve wine or beer only. Because of their patrons, they may also stress quick service. People shopping or on their lunch hour often have little time and are unable to linger. **Exhibit 2.4** shows a partial food court menu. Handheld food such as wraps, fish tacos, and grilled sandwiches are popular. **Exhibit 2.5** shows a typical lunchroom menu.

### *Family Restaurants*

*Family restaurants* are those that cater to a more casual trade, emphasizing food for the family or other groups. Many do not serve alcohol, since some families with young children will not patronize operations that do. If alcohol is offered, the menu may include only beer and wine or a limited selection of mixed drinks. The food menu will offer more popular foods, such as hamburgers, sandwiches, and "specials" (plates or meals for a moderate price). Some family operations compete for the family lunch trade that wants food such as that served in quick-service operations. Family operations expect a better turnover rate than that in fine-dining establishments; 3.5 turnovers per meal may be ideal. These operations are apt to be located in or near residential areas where the average income per household is either above average or moderate. Prices will be moderate. These facilities generally offer seated service, but may serve cafeteria or

**Exhibit 2.4** Foodlife®: Food Court Partial Menu

# foodlife

835 N. MICHIGAN, CHICAGO • 312-335-3663

## STIR FRY HEAVEN

1. Fill your bowl with your favorite vegetables.
2. Select your cooking style.
3. Add tenderloin, chicken or tofu if you wish.
4. Garnish to taste after we cook it

$7.95

**THAI BBQ SAUCE**
Complex sweet and tangy BBQ sauce with a little heat. Typical Thai flavors of lemon grass, ginger and basil.

**VEGAN MUSHROOM BROTH**
Bold flavors of shiitake mushroom and black bean sauce with mildly sweet flavors of hoisin and vegetarian oyster sauce.

**KUNG PAO SAUCE**
Mildly sweet soy based sauce with pickled ginger, garlic, rice vinegar and hoisin sauce

## COMFORT FOOD

**Specials & Favorites $4.95 and up**
Meat Loaf
Pot Roast
Gumbo
Jambalaya
Stuffed Peppers
Chicken & Biscuits
Smoked Sausage with Sour Kraut
Hot Wings
Swedish Meatballs
Beef Stroganoff
Stuffed Pork Chop
Shepards Pie
Fried Chicken
Chicken Pot Pie
Turkey Dinner
Fish and Chips
Stuffed Cabbage
Turkey a la King
Prime Rib
Country Fried Steak

**Sides $1.95**
Mashed Potatoes
Oven Roasted Potatoes
Mashed Cauliflower
Peas, Carrots & Mushrooms
Macaroni and Cheese
Sweet Potatoes
Corn on the Cob
Creamy Corn
Potato Pancakes
Country Style Green Beans

## SOUPLIFE

**Soups:**
New England Clam Chowder
Potato Florentine
Slow-Simmered Turkey Chili
Vegetarian Chili
Chicken and Crazy Pasta
Turkey Roasted Apple
Creamless Tomato
Fully Loaded Baked Potato
Tomato Basil Bisque
Cream of Potato
Italian Wedding
Tex-mex Chicken Corn Chowder
Manhattan Clam Chowder
Broccoli Cheddar
Chicken Cacciatore
With Italian Sausage, Beef & Pot Roasted Vegetable

**Dine In** (Bottomless Bowl) $5.95
**Take Away Pint** $3.95

## LA VIDA MEXICO

**BURRITOS**
With meat or veggies, black beans, pico de gallo, lettuce, guacamole, and sour cream

**QUESADILLAS**
Filled with chihuahua cheese, meat or veggies and pico de gallo

**TACOS**
Two double-thick tacos, filled with meat or veggies, lettuce, pico de gallo and cheese

**Choose from . . .**

| | |
|---|---|
| Barbacoa | 6.95 |
| Chicken al Pastor | 7.45 |
| Grilled Veggies | 6.25 |
| Ground Beef | 6.95 |
| Chips with Salsa | 1.95 |
| Chips with Guacamole | 3.95 |
| Fully Loaded Nachos | 6.95 |

## DO HOTS

**HOT HERO SANDWICHES**

**CHICKEN BRUSCHETTA**
Parmesan Chicken Breast, Marinated Plum Tomatoes, Pesto, Mozzarella & Provolone Cheeses

**ITALIAN SUPER SUB**
Salami, Mortodella, Capicol'a, Provolone Cheese, Basil, Lettuce, Tomato, Onion, Italian Dressing

**ROAST BEEF**
Sliced Roast Beef, Swiss Cheese, Lettuce, Tomato, Red Onion & Our Homemade Horseradish Sauce

**TURKEY & CHEDDAR**
Sliced Turkey, Cheddar Cheese, Lettuce, Tomato, Italian Dressing & Mayo

**SOUTHWEST CHICKEN**
Chicken Breast, Pepperjack Cheese, Lettuce, Pico De Gallo, Guacamole, Chipotle Mayo

## MIRACLE JUICE BAR

**SMOOTHIES AND SHAKES** $3.95
With Fresh Fruit Juices With Non-Fat Yogurt
**WILD BERRY**
Blueberries, Strawberries & Raspberries
**STRAW-NANA** Strawberry & Banana
**SUNSHINE** Orange, Banana & a Touch of Coconut
**STRAWBERRY** Strawberries & More Strawberries
**PEACHBERRY** Peach & Raspberries
**VEGGIE DRINKS**
Small $3.25 Large $3.75
**ORGANIC CARROT** Nature's Nectar
**SWEET JANE** Carrot & Apple
**GINGER SNAP** Carrot, Apple, Ginger
**RED VELVET** Carrot, Apple, Beet
**THE TRIP** Carrot, Apple, Beet, Celery
**WHEAT GRASS** 10 oz. $1.95
**BOOSTERS** $0.75
Ginko, Vitamin C, Spirulina, Lecithin, Ginseng, Echinacea, Flax Seed Meal, Protein Powder
**POWER POTIONS** $3.95
BREAKFAST DRINK: Banana, Orange, Apple, Protein, Granola
SHOPPERS ENERGY DRINK: Straw-Nana, Spriulina, Protein, Non-Fat Yogurt
**FRESH SQUEEZED ORANGE JUICE!**
Small $2.50 Large $3.50

*Courtesy of Lettuce Entertain You Enterprises Inc.*

**Exhibit 2.5** Chicago Flat Sammies: Menu

CHICAGO

# FLAT SAMMIES

**SOUP • SAMMICHES • PIZZA**

**SAMMICHES**

**Italian Beef**
mozzarella, sweet peppers, giardiniera, roast beef and sammie splash **5.75**

**Chicken al Pastor**
ancho mayo, pepperjack, mozzarella, chicken al pastor and sammie splash **5.45**

**Vegetable Melt**
a medley of the season's vegetables, lightly dressed and roasted in our oven **5.45**

**Grilled Pesto Chicken**
mozzarella, roma tomatoes, fresh basil, pesto chicken breast and sammie splash **5.95**

**Turkey Club**
mayo, swiss, turkey, tomato, bacon and sammie splash **5.95**

**Tuna Melt**
cheddar, mozzarella, tuna salad, tomato and sammie splash **5.45**

**"Deluxe" any sammie with soup or chips, or caesar salad! Only 95¢**

**REAL CHILI**

**Chili**
rich, thick and meaty. A meal in itself, served **2.95**

**with corn chips**

Chili Mac N'Cheese
macaroni covered with our own chili and **3.95** cheddar cheese. better than the big house

**FLAT BREAD PIZZA**

The old world recipe is rolled thin and roasted crisp with a variety of fresh toppings. all pizzas come with a small caesar.

**Margherita**
mozzarella, tomato, basil and tomato sauce **6.50**

**Southwestern**
chile marinated chicken, jack cheese, roasted tomato and caramelized onions **6.95**

**Chicken & Roasted Tomato**
roasted tomato, roasted chicken, mozzarella and basil pesto **6.95**

**South Side Special**
sausage, pepperoni, tomato sauce, mozzarella, and spicy capicola **6.95**

**Oven Roasted Vegetable**
asparagus, portbello mushrooms, roasted peppers and blue cheese **6.95**

**SALADS**

**Caesar Salad**
chopped romaine, croutons, parmesan, caesar dressing **4.95**

**Chopped Spinach Salad**
spinach, chicken, bacon, egg, tomato, mushrooms, green onion, blue cheese **5.95**

**Chopped Southwestern Salad**
al pastor chicken, romaine, pico de gallo, tortilla strips, cheddar and jack cheese **5.95**

**Cobb Salad**
chopped lettuce, roasted chicken, sammie's special dressing, crisp bacon, blue cheese, scallions, tomato, avocado and croutons **5.95**

**SNICKY SNACKS**

**Kettle Chips**
a healthy portion of thin, crispy chips **1.25**

**Maytag Chips**
chips, creamy cheese sauce, topped with Maytag blue cheese **3.50**

**Chili Cheese Chips**
spicy chili and cheddar cheese piled high on our fresh cut potato chips **3.50**

**Atomic Chips**
our fresh cut potato chips served with our top secret atomic sauce **3.50**

**SWEETS**

**Toll House Cookie**
the classic **1.75**

**Fudge Brownie**
Martha's recipe—no kidding—no nuts either **1.75**

**Ice Cream Cone**
vanilla, vanilla, or vanilla **1.50**

**Fudge Brownie Delight**
our brownie topped with ice cream and whipped cream **3.95**

**Sundaes**
chocolate, strawberry, or vanilla **2.95**

**DRINKS**

**Soda**
Coke, Diet Coke, Sprite, Diet Sprite, Orange **1.50**

**Homemade Lemonade** **195**

**Iced Tea**
china mist:
a blend of black & orange pekoe **1.25**

**Nantucket Nectars**
assorted flavors **1.75**

**House Blend Coffee**
Intelligentsia blend or decaf **1.50**

**Bottled Water**
½ liter **1.50**
1 liter **2.25**

**Milk Shakes**
chocolate, strawberry and vanilla **2.50**

**BEER & WINE**

**Domestic Beer** **3.50**

**Imported Beer** **4.00**

**Wine**
red or white **4.00**

**Margaritas** **4.95**

**Mai Tai**
Bob Chinn's recipe **4.50**

**Spiked Lemonade**
our own recipe with vodka **4.50**

**FIELD RATIONS**

Sammies version of office catering. You have the need we have the goods, and we'll deliver too. Got a crowd to feed? Just tell us how many people you are feeding and how many of each style of sammich you need.

**Italian Beef**
mozzarella, sweet peppers, giardiniera, roast beef, and sammie splash

**Chicken al Pastor**
ancho mayo, pepperjack, mozzarella, chicken al pastor, and sammie splash

**Tuna Melt**
cheddar, mozzarella, tuna salad, tomato, and sammie splash

**Vegetable Melt**
mozzarella, a medley of the season's vegetables, lightly dressed and roasted in our oven

**Grilled Pesto Chicken**
mozzarella, roma tomatoes, fresh basil, pesto, chicken breast and sammie splash

**Turkey Club**
turkey, swiss, mayo, tomato, bacon and sammie splash

Every order comes with a soda, caesar salad (a big one to share) and your choice of either a Toll House cookie or fudge brownie for each sammie.
Delivered for only . . . **9.95 pp**
Limited delivery area
Phone or Fax your order
CALL (312) 664-BRED
FAX (312) 664-2720

*Courtesy of Lettuce Entertain You Enterprises Inc.*

buffet style. Because children are a significant part of the family restaurant's market, a children's menu is usually available. All-you-can-eat buffets represent value and target families and groups (see **Exhibit 2.6**). Quick-service restaurants are softening decor and expanding menu offerings to entire familys.

### *California-Menu Foodservices*

Another type of restaurant is based on the *California-menu* concept. The California menu (so called because it originated there) offers breakfasts, lunches, dinners, snacks, and other foods, all on one menu, so patrons can get almost any meal they want, at any time of the day. Counter service may also be a significant part of this type of operation. These operations may be open 24 hours a day. Prices are as low as possible, and the lower profit margin is offset by high volume and turnover. These operations are usually located in high-traffic areas.

### *Limited-Menu (Quick-service) Restaurants*

About a third of the commercial foodservice segment is made up of *limited-menu,* or quick-service restaurants (QSRs). These operations have only a few items on the menu, and the service is fast. Some specialty units serve only one or two main items. Most use a menu that is permanent and offers no items that change daily, though limited-time special items are offered to bring in new patrons and keep the regulars interested. Volume is emphasized, and prices are low. Fast turnover is required to make a profit.

Quick-service operations try to have a quick seat-turnover rate and serve a simple fare that is quickly prepared, served, and eaten. Many do a large take-out service. Drive-through business has experienced rapid growth.

Many popular items—such as pizza, hamburgers, chicken, sandwiches, salads, soups, snack items, and ice cream desserts—may be offered on the menu along with carbonated beverages. Few limited-menu operations serve alcohol. Most have à la carte items only, but some also offer a limited amount of table d'hôte items, which can be whole meals. A number of these operations may have seated service using tables and booths. No tablecloths or cloth napkins are used. Many items are prepared for carryout.

Quick-service menus are changing; many new units now have menus that emphasize pasta, Mexican foods, and other ethnic dishes. Many items offered contain more healthful foods with less fat and calories. QSRs typically have picked up on trends after white-tablecloth and family operations have taken the lead. By responding sooner, new markets can be drawn to this segment.

Drive-ins are included in the limited-menu category. Pick-up windows may be used for car service, or the operation will be constructed so that patrons can park their cars and receive service from a counter for carryout.

Many limited-menu units are owned by chains or are a part of a franchise operation. The number of limited-menu operations has grown faster than other types of operations and should continue to do so. This is most apt to come from the growth in the number of quick-service operations that offer ethnic or other foods. A challenge in this segment has been the shift in demographics to a more "mature" population.

### *Commercial Cafeterias*

In cafeterias, patrons go to counters where hot and cold foods are offered, select the food, and then take it to a table to eat. A menu board lists the offerings; and items may be marked with their

**Exhibit 2.6** Golden Corral: Menus

## Golden Corral offers an abundant variety of great tasting food daily on the Great Buffet

**On the Hot Choice Buffet you can enjoy all-you-can-eat:**

Fresh Fried Chicken
Rotisserie Chicken
Bourbon Street Chicken
Pot Roast (varieties subject to change)
Meatloaf (varieties subject to change)
Spaghetti and Meat Sauce
Pizza
Macaroni and Cheese
Fresh Made Mashed Potatoes
Green Beans
Baby Carrots
Broccoli, Cauliflower with Cheese Sauce

**From the Cold Choice Buffet you can enjoy:**

Coleslaw
Macaroni Salad
Pasta Salad
Potato Salad
Seafood Salad
Tuna Salad

**From the Soup Choice and Potato Bar enjoy:**

Clam Chowder
Timberline Chili
Fresh Baked Potatoes

**From the Bakery and Dessert Cafe enjoy:**

Specialty Cakes
Carrot Cake
Chocolate Chip Cookies
Fudgy Brownies
Yeast Rolls
Variety Pies
Banana Pudding
Bread Pudding
Cobblers
Sugar Free Bakery Products
Soft Serve Ice Cream

## What you will find on the Sunrise Breakfast at Golden Corral

Scrambled Eggs
Eggs Benedict
Hollandaise Sauce
Spinach Quiche
Bacon & Cheese Quiche
Egg & Sauce Casserole
Hash Brown Casserole
Skillet Hash Browns
Bacon
Corned Beef Hash
Creamed Chipped Beef
Grilled Ham Steaks
Grilled Pork Loin
Sausage Gravy
Sausage Links
Sausage Patties
Grits
Sauteed Onions & Peppers
Whole, Sauteed
Mushrooms
Cheese Sauce
Plain Pancakes
Blueberry and/or Chocolate
Chip Pancakes
Maple Syrup
Sugar-free Syrup
Strawberry Topping
Fruit Topping (Apple, Cherry or
Pineapple)
Whipped Margarine
Whipped Topping
French Toast
Texas Toast
Golden Waffles
Baked Glazed Donut
Cin-a-Gold Cinnamon Roll
Buttermilk Biscuits
Bananas
Fresh, Cut Fruit
Fruit SaladGrapefruit Seed
Orange Juice
Apple Juice
Orange-Guava-Passion Fruit
Milk
Assorted Cereals
Assorted Yogurts
Omelettes
Fresh Fruit Tart
Apple or Cherry Tumo
Banana Nut Bread

*Menu provided courtesy of Golden Corral Corporation*

prices according to where the items are placed. Foods are usually simple and the prices are low, reflecting the lack of seated service. Alcohol is generally not offered. Payment may be made at the counter at the end of the service line or as the patron leaves the operation. In some operations, especially those catering to families and the dinner trade, a worker may be at the end of the counter after all selections are made to carry the patron's tray to a desirable dining spot. This person may also serve water. Condiments and napkins may be on a table in the dining area so patrons can serve themselves. Dining utensils are usually prewrapped in cloth or paper napkins or placed in containers at the serving counter so guests may select their own. Entrees are often foods that hold well in a steam table. If steaks and other grilled items, which do not hold well, are on the menu, employees may cook these to order. When such an order is taken, the patron is given a number or some other type of identification. The patron then chooses a table, and a worker brings the specially ordered item to the table.

Patrons may or may not remove their soiled dishes and trays after eating. Commercial cafeterias usually depend on a high volume and a fast turnover, and are located in high-traffic areas. Cafeterias are often associated with institutional foodservice. The trend seems to be more toward one price or cost-per-ounce buffets.

### *Social Caterers*

Catering enterprises prepare food and beverages in a central kitchen often to take somewhere else for service. They cater to such occasions as picnics, weddings, bar and bat mitzvahs, meetings, banquets, office parties, and other special events. Others cater only small group or individual orders. The catering business has not grown as fast as the rest of the commercial segment, probably because other kinds of foodservice have entered into competition with them. Restaurants, some supermarkets, and even quick-service operations have moved into catering to increase their profits.

Some caterers allow customers to order food by phone and then come in and pick up their orders. These caterers differ from take-out units in that they provide a more complete service.

In some cases, a caterer may be invited to bid for an event, and the patron will select from a group of bidders. This could be for a special occasion or some permanent kind of service, such as catering for a retirement community. In most cases, the client furnishes the space, eating utensils, tables, and chairs. In some instances, the caterer provides everything. Sometimes the caterer may also furnish an orchestra, sound system, or other items. Since the need for service varies, the payroll may be flexible; only a few full-time, permanent employees may be kept on. The volume of business can vary from providing food for a few people to providing a banquet for more than a thousand people in a convention center. Some caterers may restrict their scope to small business parties, teas, or simple buffet dinners. Some caterers work out of their homes, although local health departments may not allow this practice. Some caterers lease commercial space during "down" time.

For catering to be successful, there must be a continuous market for it, and the operation must be properly promoted by advertising. One of the best advertising media for caterers is word-of-mouth by patrons and others who have been to a well-catered event.

The prices for catered events are usually higher than those for regular food and drink offered by a restaurant. The amount of equipment that must be kept in stock, the cost of transportation, and the nature of the business make this necessary. The products offered must be suitable for preparation, holding, packaging, transporting, and serving. The food must be attractive and kept and served hot or cold, according to service and sanitation requirements. Patrons pay for customization and exclusivity.

### Frozen Dessert Units

Frozen dessert units are very popular. Often these are part of a franchise or chain. Menus are restricted generally to the frozen dessert and perhaps some simple beverages, but some menus offer, sandwiches, salads, and other simple foods are obtainable. Usually no alcohol is served. The turnover in seats (if the establishment provides any) is usually fast, and the operation should be located in places where there is a considerable amount of foot traffic. Some do well in large shopping center complexes, Take-out service is almost always provided. Some operations take orders for specially prepared frozen desserts, such as decorated birthday cakes made of ice cream or other special desserts prepared for holidays. Smoothie units, some including simple sandwiches and salads, have entered the market, as have upscale units, some of which also sell fudge and other premium candy.

### Bars and Taverns

This group consists of bars and taverns operated as independent, free-standing units. It does not include bars, lounges, and other alcohol-dispensing units operated in conjunction with another type of foodservice operation. Total sales in 2007 are estimated to be approximately $15.7 billion, about 4.1 percent of the sales of the entire food and beverage industry.

Most bars and taverns are single-owner or family-run units. The trend toward chains and multiple operation formation has not been as great in this segment of the industry as it has been in other segments.

Bars and taverns may serve food—sandwiches, snacks, and other light items—but this is normally limited to merely what is necessary to attract customers. More and more bars are offering unique appetizers to attract patrons. The cost of goods sold is usually lower than that of a foodservice; labor costs are also lower, so the profitability margin may be better for bars than for foodservices. (The average cost of goods sold usually runs from 20 to 25 percent in bars and taverns and from 35 to 45 percent in foodservices.) Some municipalities require a threshold percentage of food that must be sold in order to legally operate. **Exhibit 2.7** illustrates a page from the Anchor Bar menu.

Bars and taverns often build reputations as social gathering centers. Sometimes they provide recreation, such as pool, TV, sports events, videogames, and other attractions. Special drink promotions are common. However, because many states now have *dramshop laws* making bar owners, managers, and servers liable for alcohol-related injuries to third parties, "happy hours" and special promotions that encourage patrons to drink more have all but disappeared. Nonalcoholic and low-alcohol drinks, as well as *designated-driver programs*, are being promoted more often, as is quality-versus-quantity consumption.

Bar and tavern owners and operators are worried about the future because of the growing emphasis on stricter enforcement of *dramshop laws*. Because of very heavy court judgments, insurance rates for operating bars and taverns have risen to almost prohibitive levels. The illustrative menu provided demonstrates a strategy of a shift toward the tavern as a place for family fun. Many U.S. cities and states have put laws into effect in the last several years that enforce smoking bans within their jurisdictions. These laws are frequently challenged by bar and tavern owners associations and other groups that feel the bans negatively effect their business. Business

**Exhibit 2.7**
The Anchor Bar & Restaurant: Home of the Original Buffalo Chicken Wings

FRANK & TERESSA'S ANCHOR BAR
ESTABLISHED 1935

HOME OF THE ORIGINAL

FRANK & TERESSA'S ANCHOR BAR ORIGINAL BUFFALO CHICKEN WINGS

Wings Served with Bleu Cheese & Celery

"THE REAL STORY"...

I would like my customers to know the "real story" behind our chicken wings. One Friday night back in 1964, I was working behind the bar (as I still do now). It was busy that night. My mother, Teressa, was in the kitchen cooking and dad, Frank, was popping in the restaurant greeting customers. It was one of those good Friday nights.

At about 11:30 PM a group of my friends came through the door and they were starving! I told them to wait until midnight so they could have what they wanted. I served another round of drinks.

By midnight, they were really hungry! I ask my mother, Teressa, to fix something for them. About ten minutes later she brought out two plates and set them on the bar. They all asked "What is this?" and to tell you the truth, I also was curious to know what those things were!

They looked like chicken wings, but I was afraid to say so!

To make a long story short, yes, they were chicken wings – She was about to put them in the stock pot (for soup). She looked at them and said, "It's a shame to put such beautiful wings in a stock pot."

The rest is history! Mother Teressa created the famous chicken wing now served all over the country! Wherever you go – when you mention chicken wings you remember about Mother Teressa and that famous night in Buffalo!

Respectfully yours,
"The Rooster"
D. Bellissimo

NEW! ORIGINAL ANCHOR BAR WING SOUP! *Ask your server.*

ANCHOR BAR WORLD FAMOUS WINGS

All of our world famous wing orders are served with traditional celery and bleu cheese, just like Mother served us that famous night in 1964.

MILD, MEDIUM, HOT OR SPICY BARBECUE

Single (10) .................... $8.00
Double (20) .................. $13.00
Bucket (50).................... $26.00

SUICIDAL! IF YOU DARE!

Single (10) .................... $10.00
Double (20).................... $14.00
Bucket (50).................... $28.00

WINGS-TO-GO

Our Authentic Wings are now available FROZEN to take home!
(DON'T FORGET OUR AUTHENTIC SAUCE)

WINGS THAT FLY

WE SHIP Our Wings All Over The Country! Ask Your Server for More Information! or Call: (716) 884-4083



PIZZA

Small Cheese and Pepperoni..$11.00
Any Additional Toppings ........................ $1.00
Large Cheese and Pepperoni $14.00
Any Additional Toppings ........................ $1.50
White Pizza Olive Oil, Chopped Onion, Tomato, Parmesan Cheese and Mozzarella Cheese
............Small $8.00...Large $14.00

*Courtesy of Anchor Bar: Home of the Original Buffalo Chicken Wings*

owners who rely on smoking customers to bring in revenue and health advocates frequently do not agree on the bans. Some feel the bans violate smokers civil liberties. In Madison, Wisconsin a city-wide ban continues to stir controversy as tavern owners claim lost business to other nearby cities that allow smoking. Those who support the ban feel that people who would previously have avoided a smoke-filled environment are now coming to the bars for music and companionship. Others feel it is not the place of government to regulate smokers. A study in El Paso, Texas, performed after a smoking ban was enforced, found that revenue did not drop after the ban. Nevada approved a referendum in 2006 to ban smoking from most indoor places of employment. Establishments that do not serve food are exempt from the law. Some bar and tavern owners in Nevada say they will stop serving food in order to retain the majority of their revenue which comes from gambling. Proponents of no smoking bans feel it takes time for the public to adjust to the bans, but that cleaner air and an improvement in health for workers and customers will eventually lead to bans in most states. This controversial subject will continue to challenge foodservices to find ways to replace revenue potentially lost from smoking bans.

### *Managed Services*

The different types of food contractors can be discussed together, since their modes of operation are similar. A food contractor is paid to take over the foodservice of an operation for a fee, a percentage of sales, a share of the profits, or another financial arrangement. Both commercial and noncommercial organization use contract foodservice companies. Manufacturing and industrial plants are the biggest customers of foodservice contractors. Other users include hospitals, schools, colleges and universities, airlines, and sports centers. Primary and secondary schools are a fast-growing part of this segment, with double-digit growth projected in 2007.

When an operation turns over its foodservice management to a contractor, it usually does so because its major business purpose is not foodservice. Nevertheless, it wants its units professionally run and staffed. Often a professional food contractor with trained employees and experience can do a better job than the owner. Experts in many fields of foodservice operation are on the staff at the central office of the contractor, and they are valuable resources when problems arise; these individuals are not available to the independently run foodservice. The food contractor often has a much better system of production, purchasing practices, and accounting skills than is available to nonexperts.

The amount of subsidization a contractor gets from the contracting company varies. In rare cases, the contracting company may subsidize the entire project and then pay the contractor a fee for operating the unit. This is rare, however, and the practice of subsidization may run from complete to none at all. The frequency of what is done seems to follow a normal curve with some subsidization occurring in most cases.

The practice of companies charging nothing for foods served in their own foodservices is also ending. Companies that once gave free meals to employees have had to end that practice.

One of the unique problems in contract foodservice management is that the unit manager usually has to satisfy two bosses: the home or central office and the client company. This can complicate the management process and divide management's emphasis and loyalty. However, it can also act as a stimulus by ensuring adherence to management procedures as set by the contract manager responsive to the client.

Each different type of operation requires different contractor services. Contracted foodservice is often run as a cafeteria. Industrial or office foodservices may have executive dining rooms with fine seated service, and a cafeteria for employees. Industrial and office contractors also may provide food and drinks for breaks, meals, and special occasions. They may operate over a 24-hour period to serve three shifts. Others may close down their regular foodservice areas and operate from vending units during slower hours. Health facilities must emphasize nutrition and diet. Hospitals often have a cafeteria for employees, a doctors' dining room, a coffee shop for visitors, and, possibly a dining room for nurses.

In some very large offices, factories and health complexes, quick-service contractors such as McDonald's, KFC, and Burger King have been invited in to operate units. Contractors are now even moving into companies and hospitals. They often receive good patronage because the service is fast and the foods are well known.

School foodservices in elementary and high schools have a strong nutritional emphasis, and have to conform to nutritional and other standards established by the federal government if they are participants in the general meal pattern program or the milk program. A school may have separate dining rooms for faculty and staff members, in addition to student cafeterias. Some caterers and chains have entered the school lunch program. Many school districts have had to retrench and reduce subsidization of their foodservice programs. Today a foodservice may be required to break even after paying for lights, janitorial services, and some other expenses. Usually no rent is paid, but in some cases it is. College and university foodservice operations may include residence hall cafeterias, student center coffee shops, and vending operations. Some contractors in the industry will supply janitorial and maintenance services if desired by the client.

Companies contracting with recreational and sports units have a wide variety of requirements. There may be cafeterias, bars, refreshment stands for fans, and clubs for exclusive groups or members of a recreational association. Alcoholic beverages may be provided.

Contractors may also be expected to cater parties, meetings, and other affairs sponsored by a client company. Contractors usually seek out professionally trained managers. Each year, large food contractors (such as Marriott, Sodexo, and Aramark) visit schools where foodservice management is taught to recruit graduates for trainee positions. Many large contractors have well-structured systems for training and a career development plan, in which the trainee may start out as an assistant manager and move up to full management later.

The service of food by contractors for transportation operations, especially airlines, had been a significant business. Organizations such as Marriott may serve thousands of meals a day from one contractor commissary. The meals are packaged so that hot foods can be heated, cold foods kept cold, and the complete meal assembled during service by the flight attendants. Carbonated and alcoholic beverages are usually provided. Special meals are prepared for first-class passengers. On many flights passengers are offered a menu choice. The financial difficulties faced by airlines following September 11, 2001 have resulted in cost cutting affecting on-board meals. Many domestic airlines have reduced or eliminated foodservice from flights of less than three hours in duration or flights which do not take place during a traditional meal period. International and cross-continental flights will generally offer a more traditional foodservice.

### *Lodging Places*

The food and beverage department of a hotel is often a complex system from which a number of different types of foodservices are operated. They may include a coffee shop, banquet facilities, snack bars, bars, a nightclub, fine dining room, specialty restaurants, room service, a cafeteria, and even an employee dining room.

A hotel foodservice department is usually managed by a food and beverage manager responsible for seeing that the operation keeps adequate records and makes a profit. A chef may be responsible to the food and beverage manager, or to the manager or assistant manager of the hotel. A maitre d'hôtel may be in charge of the better dining areas, but a host or hostess might be responsible for service in a coffee shop. A banquet or catering manager handles banquets or special parties. A steward may work with the chef in preparing menus, setting up order requirements, and seeing that linens and silver are properly handled, stored, and accounted for. A steward may also be in charge of storage spaces for food, equipment, and other supplies. The hotel sales department should work closely with the food and beverage department, since this department is usually responsible for making arrangements for special catering and other functions, such as meetings and conferences.

Many hotels operate on the continental system in which an *executive chef* manages operations with a *sous chef* and *chefs du parti* doing the actual food production. The operation is somewhat complex because of the variety of food that must be prepared and the number of different dining areas that must be served. Much food preparation occurs in a central kitchen, which is distributed to the various serving units. Thus, a sandwich grill might get many of its sauces, soups, salads, and other foods from the central kitchen, preparing only sandwiches or grilled items in the grill area.

Many hotel dining rooms stress luxury-type foods and serve alcoholic beverages and wines. Table d'hôte and à la carte menus are often combined. The menu usually features a wide variety of food. Service should be consistent with the menu. Even a hotel's coffee shop can be fairly luxurious, although it may have a different service and offer simpler foods than those served in the main dining room.

Many lodging units cater to conventions, parties, meetings, receptions, and other occasions. This service may be elaborate and may require a special staff. Banquets must be handled with high efficiency and speed. A special kitchen may prepare a large part of the banquet, but the central kitchen can be called upon to supplement many items. Since banquet service may be sporadic, a list of people who can be called on to work as servers, buspersons, and even cooks must be maintained, or the hotel must have a contract with a union or other labor source for its short-term needs. There are many people in the foodservice industry who work only for banquets and are available on call.

Hotels frequently have a problem in that many of the guests who have rooms do not eat at the hotel, especially lunches and dinners. The reason given for this is that customers prefer to eat elsewhere for a change or think the food at restaurants is better or costs less. Thus, many hotels are challenged both to attract their in-house customers and bring in outside trade through creative menus and promotions.

Lodgings generally do more to emphasize their food and beverage business than they did in the past. The reason is that the food and beverage department is now seen as an additional source for revenue and profits. In some hotels and motels, the food and beverage income can be almost equal to that from rooms. Thus, we see many lodging units with elaborate facilities for conventions, meetings, and special events. These can bring in considerable revenue in food and beverage sales, as well as help fill the rooms.

Hotel foodservices have not led the market in acceptance of more modern forms of foods. Some today still cut their meats from carcass or wholesale cuts. However, there is a growing interest in convenience and preportioned foods.

**Exhibit 2.8** is a summary of percentages of income and expenditures that might be considered typical for a full-service hotel.

**Exhibit 2.8** Full-Service Hotels—Statements of Operating Income & Expenses—2005

| | Total U.S. | | | Chain-Affiliated | | | Independent | | |
|---|---|---|---|---|---|---|---|---|---|
| Occupancy (of Sample) | 69.2% | | | 69.8% | | | 62.2% | | |
| Average Size Of Property (Rooms) | 307 | | | 319 | | | 218 | | |
| Average Daily Rate | $142.86 | | | $142.51 | | | $147.25 | | |
| | **Ratio to Sales** | **Per Available Room** | **Per Occupied Room Night** | **Ratio to Sales** | **Per Available Room** | **Per Occupied Room Night** | **Ratio to Sales** | **Per Available Room** | **Per Occupied Room Night** |
| REVENUE | | | | | | | | | |
| Rooms | 62.1% | $35,617 | $142.86 | 63.1% | $35,907 | $142.51 | 52.9% | $32,408 | $147.25 |
| Food | 20.5 | 11,777 | 47.24 | 20.4 | 11,637 | 46.19 | 21.7 | 13,287 | 60.37 |
| Beverage | 5.1 | 2,950 | 11.83 | 5.0 | 2,848 | 11.30 | 6.6 | 4,054 | 18.42 |
| Other Food & Beverage | 4.5 | 2,580 | 10.35 | 4.5 | 2,546 | 10.11 | 4.8 | 2,949 | 13.40 |
| Telecommunications | 1.0 | 562 | 2.26 | 1.0 | 581 | 2.30 | 0.6 | 363 | 1.65 |
| Other Operated Departments | 4.1 | 2,340 | 9.39 | 3.5 | 2,021 | 8.02 | 9.5 | 5,813 | 26.41 |
| Rentals & Other Income | 2.4 | 1,352 | 5.42 | 2.3 | 1,289 | 5.12 | 3.3 | 2,041 | 9.27 |
| Cancellation Fee | 0.3 | 157 | 0.63 | 0.2 | 138 | 0.55 | 0.6 | 360 | 1.63 |
| **Total Revenue** | **100.0%** | **$57,335** | **$229.98** | **100.0%** | **$56,967** | **$226.10** | **100.0%** | **$61,275** | **$278.40** |
| DEPARTMENTAL EXPENSES | | | | | | | | | |
| Rooms | 26.3% | $9,380 | $37.63 | 26.2% | $9,413 | $37.36 | 27.8% | $9,013 | $40.95 |
| Food & Beverage | 74.0 | 12,799 | 51.34 | 73.7 | 12,550 | 49.81 | 76.4 | 15,497 | 70.41 |
| Telecommunications | 94.9 | 534 | 2.14 | 93.3 | 542 | 2.15 | 122.5 | 444 | 2.02 |
| Other Operated Depts & Rentals | 3.5 | 2,022 | 8.11 | 3.2 | 1,795 | 7.13 | 7.3 | 4,486 | 20.38 |
| **Total Departmental Expenses** | **43.1%** | **$24,735** | **$99.22** | **42.7%** | **$24,300** | **$96.45** | **48.0%** | **$29,440** | **$133.76** |
| **Total Departmental Profit** | **56.9%** | **$32,600** | **$130.76** | **57.3%** | **$32,667** | **$129.65** | **52.0%** | **$31,835** | **$144.64** |
| UNDISTRIBUTED OPERATING EXPENSES | | | | | | | | | |
| Administrative & General | 8.3% | $4,730 | $18.97 | 8.1% | $4,621 | $18.34 | 9.6% | $5,911 | $26.86 |
| Marketing | 6.8 | 3,925 | 15.74 | 6.9 | 3,912 | 15.52 | 6.6 | 4,063 | 18.46 |
| Utility Costs | 4.0 | 2,302 | 9.23 | 4.0 | 2,281 | 9.05 | 4.1 | 2,522 | 11.46 |
| Property Operations & Maintenance | 4.8 | 2,755 | 11.05 | 4.8 | 2,711 | 10.76 | 5.3 | 3,226 | 14.66 |
| **Total Undistributed Operating Expenses** | **23.9%** | **$13,712** | **$54.99** | **23.7%** | **$13,525** | **$53.67** | **25.7%** | **$15,722** | **$71.44** |
| GROSS OPERATING PROFIT | 33.0% | $18,888 | $75.77 | 33.6% | $19,142 | $75.98 | 26.3% | $16,113 | $73.20 |
| Franchise Fees (Royalty) | 0.6 | 362 | 1.45 | 0.7 | 391 | 1.55 | 0.1 | 51 | 0.23 |
| Management Fees | 3.3% | $1,894 | $7.60 | 3.4% | $1,928 | $7.65 | 2.5% | $1,524 | $6.92 |
| INCOME BEFORE FIXED CHARGES | 29.0% | $16,632 | $66.72 | 29.5% | $16,823 | $66.78 | 23.7% | $14,538 | $66.05 |
| Selected Fixed Charges | | | | | | | | | |
| Property Taxes | 3.1% | $1,762 | $7.07 | 3.1% | $1,778 | $7.06 | 2.6% | $1,584 | $7.20 |
| Insurance | 1.3 | 741 | 2.97 | 1.3 | 730 | 2.90 | 1.4 | 857 | 3.89 |
| Reserve For Capital Replacement | 2.0 | 1,128 | 4.52 | 2.0 | 1,117 | 4.43 | 2.0 | 1,247 | 5.67 |
| AMOUNT AVAILABLE FOR DEBT SERVICE & OTHER FIXED CHARGES* | 22.6% | $13,001 | $52.16 | 23.1% | $13,198 | $52.39 | 17.7% | $10,850 | $49.29 |
| PAYROLL & RELATED EXPENSES** | | | | | | | | | |
| Rooms | 17.1% | $5,294 | $21.90 | 16.5% | $5,175 | $20.98 | 21.0% | $6,188 | $28.78 |
| Food & Beverage | 44.2 | 6,182 | 25.99 | 43.9 | 5,828 | 23.97 | 46.2 | 8,922 | 41.63 |
| Telecommunications | 88.4 | 406 | 1.62 | 85.0 | 415 | 1.64 | 118.3 | 323 | 1.50 |
| Other Operated Departments | 2.3 | 2,013 | 8.74 | 1.9 | 1,666 | 6.79 | 4.5 | 3,898 | 19.30 |
| Administrative & General | 5.0 | 2,282 | 9.80 | 4.8 | 2,143 | 8.97 | 6.0 | 3,310 | 15.95 |
| Marketing | 3.0 | 1,389 | 5.97 | 3.0 | 1,301 | 5.46 | 3.7 | 2,051 | 9.82 |
| Property Operations & Maintenance | 2.6 | 1,222 | 5.25 | 2.6 | 1,157 | 4.81 | 3.2 | 1,710 | 8.45 |
| **Total Payroll & Related Expenses** | **34.3%** | **$17,412** | **$73.31** | **33.4%** | **$16,516** | **$67.84** | **40.8%** | **$24,136** | **$114.33** |

* Other Fixed Charges include Depreciation and Amortization, Interest, Rent, and Equipment Leases.
** Included in above expenses. Only shown here for additional detail.
Expense ratios to sales for departmental expenses are based on their respective departmental revenues.
All other expenses are based on total revenue. Totals may not add due to rounding.

*Courtesy of The HOST Study, Smith Travel Research*

### *Retail Hosts*

About 133,000 foodservices are operated in retail establishments, such as drug, department, variety, grocery stores, and in conjunction with gas stations.

Department stores emphasize food for shoppers and usually do well with the lunch trade. Many department stores are diversifying their foodservice operations. Some have salad bars and yogurt stands in addition to their regular dining rooms. They may also have an upscale dining room, as well as a moderately priced coffee shop. Often, variety stores have counters where fountain items and light meals are served.

Foodservice operations often are found in gas stations on highly traveled routes. Travelers on these roads find it convenient to get car service and a meal or snack at a single stop. Many interstate highways have such operations, which may be operated by food contractors. Service station units must gear their operations toward families with children, business travelers, and truck drivers.

Grocery stores and convenience stores have become increasingly competitive with foodservice operations, especially since some of the large supermarket chains have found such diversification profitable. Since many groceries already operate bakeries and delicatessens and often sell other prepared foods, it is only a short step to serve these foods to customers. Today one can walk into a supermarket and see a long row or several rows of prepared meats, salads, sandwiches, soups, and desserts attractively arranged and ready for purchase. Often customers supplement their purchase with items picked up off the grocery shelves. As a result, one can get adequate food at a better price than at a foodservice operation. However, well-prepared foods and excellent service continue to be a differentiating factor.

Some supermarket stores operate catering businesses as well. Many convenience stores compete directly with quick-service operations. The advent of megastores hosting complementary foodservice operations has assisted in the rapid growth of this segment, resulting in a projected $25 billion in sales in 2007.

### *Recreational Foodservices*

Recreational foodservice is a $24.3 billion market with a growth rate of approximately 4.8 percent per year. The recreational field consists of sports stadiums, coliseums and arenas, municipal convention centers, concert facilities, movies, bowling alleys, amusement parks, pari-mutuel racetracks, expositions, carnivals and circuses, zoos, botanical gardens, state and national parks, and roller and ice-skating rinks.

The basic menu is a concessions type, which generally is a low-cost, high-profit menu. Foods include hot dogs, popcorn, ice cream, soft drinks, and, in some cases, beer. Some facilities also have full-service dining rooms, cafeterias, and coffee shops.

In 2007, it is projected that recreational facilities will have food and drink sales of $12.6 billion. This segment employs many people part-time during peak seasons. Large industrial caterers may operate by contract to provide for the foodservice needs of convention centers or sports arenas. They usually have an office at the location and operate the business from there. Snack foods, carbonated beverages, hot dogs, peanuts, popcorn, and alcoholic beverages are typical fare. Meal service may also be available in dining areas. Some, such as at racetracks, may offer elaborate meals. Often, service personnel circulate in the stands and sell snack items. Counters where patrons can purchase items are located in selected areas.[1] In smaller stadiums without cooking facilities, quick-service restaurants deliver items such as sandwiches and pizza to be sold.

### *Mobile Caterers*

Prior to World War II, there were few mobile caterers, other than street ice-cream vendors on routes in the summer. During World War II, trucks, vans, and other mobile units began to visit factories and construction sites to sell hot and cold food. Some even cooked food to order. Today, the mobile catering business is a growing part of the foodservice industry. Most of the sales are divided among industrial plants, construction sites, and office buildings and complexes, with the remainder at parks, recreational areas, and special events.

Mobile caterers may be corporations, individually owned, or partnerships. A corporation usually hires people to operate its trucks, but it may lease its trucks to individuals, who then take a percentage of sales. Some also sell the food to them and charge a percentage of sales to operate the route.

The typical mobile catering firm has 17 trucks, and each truck stops at an average of 21 locations a day. Routes are subject to pirating. To hold a business, the route operator must come on a regular basis and provide good service. Some routes may have up to 40 or 50 stops on a 15- to 20-mile run, but many individuals operate shorter routes. Some routes have only a few stops, since enough business may be generated to give a satisfactory return. A route may require 10- to 12-hour workdays, since the truck must be cleaned, serviced, and loaded, in addition to being operated over the route. Some routes may operate every day, but many operate only five days a week, serving workers who are at the stops only on weekdays.

The food must be of acceptable quality and moderately priced. The foods should be those that are satisfactory for such service and hold well. Usually, the same foods that are easily machine vended are sold in a mobile unit, although the mobile unit today may be more flexible and be able to sell a wider variety of foods. Box lunches, cold and hot drinks, sandwiches, hot lunches, coffee, desserts, bagged goods, candy, salads, crackers, and cookies and ice cream are all sold.

Sanitation is important in mobile catering, and many foods will have to be kept under refrigeration or frozen. Hot foods must be kept at or above 135°F (57°C) until they are sold.

### *Vending*

Vending operations continue to grow in popularity. Schools, industrial plants, office complexes, retail stores, and hospitals/nursing homes, as well as colleges and universities, are big users of vending machines. Figures on actual vending sales may be somewhat misleading, since many sales through vending machines are reported as sales of other foodservice operations. Besides servicing vending machines that sell snacks, soft drinks, and hot beverages, about 25 percent of vending companies operate their own commissaries. Some of these companies may even operate cafeterias, concession services, and catering services.

Of the operators offering vended fresh foods, menu items include sandwiches (62 percent), pastries (3 percent), dairy snacks (4 percent), casseroles and platters (10 percent), and salads (13.5 percent). A central commissary should produce sales of not less than 500 to 600 food items per day to be efficient. Otherwise, purchasing fresh food from purveyors, wholesalers, or other vending operators is more cost effective.

More than 20 percent of in-plant feeding operations are serviced completely from vending machines. For in-plant needs, substantial, fresh, high-quality foods that are popular with workers need to be available in the vending machines. As with foods from mobile catering, sanitation is an important factor in selecting, preparing, and holding fresh foods for vending.

Selling fresh food requires a substantial investment in machines and vehicles to transport this food properly. A simple snack machine with a coin mechanism and bill validator will cost $3,000 or more; a machine that dispenses hot beverages, refrigerates fresh food, or holds frozen foods will cost at least two times that. Normally five years is required to recover the cost of a machine. Fresh food costs may be as high as 65 percent, with labor and commissions adding another 21 percent. The presence of an attendant during lunch or dinner at a location will raise labor costs. **Exhibit 2.9** shows the vending sales and expense detail for National Automatic Merchandising Association participants with vending operations. If a vending operation, or any other foodservice operation, transports fresh food across state lines, the commissary where these foods are prepared must meet federal standards for food production facilities and sanitation.

## Institutional Feeding

The institutional segment consists mainly of noncommercial and commercial private and public organizations that operate foodservices in support of the actual purpose of the establishment. They are involved in some of the same types of operations as are enterprises discussed under commercial feeding that *do not* use managed services to run their foodservice operations. Many hospitals and nursing homes manage their own foodservice operations, as do many elementary and secondary school systems, colleges and universities, and prisons. Private clubs and camps are also part of the institutional sector. The trend has been shifting to managed services, particularly in the primary and secondary school sector.

### *Employee Feeding*

Managed services in manufacturing and industrial plants and office buildings do about eleven times as much business as privately owned industrial and commercial units. Public school systems, colleges and universities, hospitals, and other institutions also have employee feeding programs, and many of these will be managed by contract feeding companies. Seagoing ships and inland waterway vessels make up the rest of the employee feeding division.

Employee feeding programs are increasing in popularity as more workers recognize the value and convenience of eating where they work. Employee feeding operations are also in the breakfast trade. Many emphasize more healthful and nutritional foods that are low in fat and calories. More and more, employee feeding trends reflect those of the commercial feeding sector.

Employee feeding programs usually consist of cafeterias, vending operations, and executive dining rooms. Some dining operations may be subsidized by the employer. Attempts are made to keep food costs and prices low to encourage employees to stay on the premises for lunch and for breaks. If meal prices are very low, the foodservice operations may be considered part of the employee's benefit package.

### *Elementary and Secondary Schools*

Nearly 100,000 elementary and secondary schools and more than 2,500 public and private colleges and universities belong to the educational feeding group. Managed services for educational feeding accounted for more than $25 billion in 2006. Primary and secondary managed service feeding was projected to grow faster than any other industry segment in 2006.

Most elementary and high schools belong to the school lunch program. The National School Lunch Program was established by federal law in 1946.[2] The program is administered by the U.S. Department of Agriculture, which distributes funds and surplus foods on the basis of the number and circumstances of the participating children in each state. The program has two objectives: (1) to offer a market for agricultural products; and (2) to serve nutritional foods to school children at

## Exhibit 2.9 Vending Sales and Expense Detail

The table on this page profiles vending sales and expense detail for all participants with vending operations. Firms were categorized on the basis of total company sales volume.

| | Typical NAMA | High Profit NAMA | Sales Under $1 Million | Sales $1–$3 Million | Sales $3–$5 Million | Sales $5–$10 Million | Sales Over $10 Million |
|---|---|---|---|---|---|---|---|
| **Number of Firms with Vending Opns.** | **136** | **33** | **17** | **35** | **20** | **28** | **36** |
| **Typical Vending Sales Volume** | **$3,891,894** | **$4,182,000** | **$350,000** | **$1,885,685** | **$3,410,994** | **$6,307,600** | **$15,976,948** |
| **Vending Sales & Expenses** | | | | | | | |
| **Vending Sales by Category** | | | | | | | |
| Candy, Snacks, Pastry | 38.3% | 38.9% | 48.9% | 40.1% | 35.4% | 37.6% | 35.1% |
| Cold Cup Beverages | 0.8 | 1.0 | 0.0 | 0.7 | 0.5 | 0.8 | 1.3 |
| Hot Cup Beverages | 4.6 | 4.5 | 1.1 | 4.7 | 4.8 | 4.4 | 5.9 |
| Sandwiches, Salads, Entrees, etc. | 11.2 | 10.6 | 3.0 | 10.1 | 13.0 | 11.9 | 13.5 |
| Bottled Drinks | 24.3 | 25.8 | 12.1 | 24.6 | 26.6 | 27.0 | 24.3 |
| Canned Drinks | 18.3 | 17.5 | 28.5 | 16.7 | 17.6 | 17.4 | 18.0 |
| Cigarettes | 0.7 | 0.4 | 0.9 | 0.9 | 1.6 | 0.5 | 0.1 |
| Other Vending Products | 1.8 | 1.3 | 5.5 | 2.2 | 0.5 | 0.4 | 1.8 |
| **Total Vending Net Sales** | **100.0%** | **100.0%** | **100.0%** | **100.0%** | **100.0%** | **100.0%** | **100.0%** |
| Vending Cost of Goods Sold | 48.1 | 47.2 | 49.0 | 48.4 | 47.8 | 47.5 | 47.2 |
| **Vending Gross Margin** | **51.9** | **52.8** | **51.0** | **51.6** | **52.2** | **52.5** | **52.8** |
| Commission Income | 0.0 | 0.0 | 0.0 | 0.0 | 0.0 | 0.0 | 0.2 |
| **Gross Margin + Commission Income** | **51.9** | **52.8** | **51.0** | **51.6** | **52.2** | **52.5** | **53.0** |
| **Vending Payroll Expenses** | | | | | | | |
| Route Salaries & Commissions | 9.3 | 9.3 | 11.1 | 10.0 | 8.8 | 8.3 | 9.0 |
| Location Attendant Salaries & Wages | 0.0 | 0.0 | 0.0 | 0.0 | 0.0 | 0.4 | 0.9 |
| Maintenance & Repair Salaries & Wages | 2.7 | 2.4 | 0.0 | 2.6 | 3.4 | 2.6 | 2.7 |
| Warehouse Salaries & Wages | 1.0 | 1.0 | 0.0 | 1.3 | 1.0 | 1.0 | 1.1 |
| Supervisory Salaries & Bonuses | 3.4 | 3.1 | 5.3 | 3.9 | 3.4 | 3.6 | 2.5 |
| All Other Employee Salaries & Wages | 4.2 | 4.0 | 0.0 | 1.8 | 4.2 | 5.7 | 4.2 |
| Total Salaries, Wages & Bonuses | 20.6 | 19.9 | 16.4 | 19.6 | 20.8 | 21.6 | 20.4 |
| Payroll Taxes (FICA, unemp., workers' comp) | 2.2 | 2.0 | 1.8 | 2.4 | 2.0 | 2.3 | 2.4 |
| Group Insurance (medical, hospitalization, etc.) | 1.6 | 1.8 | 0.5 | 1.4 | 1.8 | 1.7 | 1.9 |
| Employee Benefits | 0.3 | 0.4 | 0.0 | 0.3 | 0.3 | 0.6 | 0.3 |
| **Total Vending Payroll Expenses** | **24.7** | **24.1** | **18.7** | **23.7** | **24.9** | **26.2** | **25.0** |
| **Other Vending Expenses** | | | | | | | |
| Commissions to Locations | 6.6 | 7.0 | 4.8 | 6.3 | 7.5 | 7.0 | 6.5 |
| Vending Equipment Rent & Lease Costs | 0.0 | 0.0 | 0.0 | 0.0 | 0.0 | 0.0 | 0.0 |
| Vending Machine Maintenance & Repair | 1.1 | 1.0 | 1.1 | 1.3 | 1.0 | 1.2 | 1.4 |
| Facilities Expenses | 2.0 | 1.9 | 2.2 | 2.8 | 2.2 | 1.5 | 2.2 |
| Vehicle Expenses | 3.8 | 3.4 | 4.5 | 3.5 | 3.1 | 3.6 | 4.3 |
| Vending Equipment Depreciation | 4.1 | 3.8 | 6.1 | 4.7 | 4.5 | 3.7 | 3.6 |
| All Other Depreciation | 0.1 | 0.1 | 0.0 | 0.1 | 0.5 | 0.2 | 0.4 |
| Sales, Use or Occupational Taxes | 3.7 | 3.9 | 2.9 | 3.3 | 2.9 | 3.9 | 4.1 |
| All Other Taxes (personal & property, etc.) | 0.3 | 0.4 | 0.2 | 0.3 | 0.2 | 0.4 | 0.4 |
| Insurance (business liability & casualty) | 0.6 | 0.7 | 1.1 | 0.7 | 0.7 | 0.7 | 0.5 |
| Over/Short | 0.0 | 0.0 | 0.0 | 0.0 | 0.0 | 0.0 | 0.0 |
| All Other Vending Expenses | 3.4 | 1.9 | 5.1 | 2.9 | 4.1 | 2.5 | 3.3 |
| **Total Other Vending Expenses** | **25.7** | **24.1** | **28.0** | **25.9** | **26.7** | **24.7** | **26.7** |
| **Total Vending Expenses** | **50.4** | **48.2** | **46.7** | **49.6** | **51.6** | **50.9** | **51.7** |
| **Vending Operating Profit** | **1.5** | **4.6** | **4.3** | **2.0** | **0.6** | **1.6** | **1.3** |
| Corporate Overhead Expenses | 0.0 | 0.0 | 0.0 | 0.0 | 0.0 | 0.0 | 0.0 |
| **Vending Profit After Overhead Expenses** | **1.5%** | **4.6%** | **4.3%** | **2.0%** | **0.6%** | **1.6%** | **1.3%** |

*Courtesy of National Automatic Merchandising Association*

low or no cost. In addition, the USDA administers the Special Milk Program, whereby a school gets a cash refund from the government for each half pint of milk it purchases for a child.

To qualify to obtain reimbursements of cash and surplus foods for meals and snacks, a school system must provide each student with foods in compliance with applicable recommendations of the Dietary Guidelines for Americans. Schools are reimbursed based on the number of free, reduced-price and paid meals they serve. More latitude in menu planning is allowed than in the past, but schools must meet nutritional goals of serving approximately 1/3 of the daily nutrition needs for lunches and 1/4 the daily nutrition needs for breakfasts. Standards for the amount of fat, saturated far, protein, Vitamins A and C, iron, calcium and calories are followed in menu planning. Schools may choose to use nutrient analysis menu planning software, approved by USDA, under the Nutrient Standard Menu Planning option, to plan meals that meet nutritional goals through menu items without specific food group requirements. . The foods required under the original law have been changed to reflect current nutrition information. Butter, once required, no longer is; to reduce the amount of fat on the menu, many schools do not serve butter or margarine. Desserts and rich foods are also discouraged. Bread alternatives may include whole-grain rice or pasta. Either flavored or unflavored milk may be offered, but unflavored fluid low-fat milk, unflavored skim milk, or buttermilk must be available. Low-salt or even no-salt foods are encouraged. In 1998, Congress expanded the National School Lunch Program to include reimbursement for snacks served in after school programs to children through age 18.

The foods prepared for the federal program are generally simple and plain. They should have high popularity with children, who have extremely sensitive food tastes and habits.

Today, the school foodservice system is deeply ingrained in our educational system. It is a tremendous force in providing adequate nutrition to many children in this country. In fiscal year 2006, more than 30 million children got their lunch through the NSLP each day. Since the modern program began, more than 187 billion lunches have been served.

### *Colleges and Universities*

Residence halls, student unions, and other college and university foodservices can be huge operations doing millions of dollars of business. Residence halls may be operated on a closed board plan that provides three meals a day to students at an established rate. Students are charged for all meals regardless of whether they eat them. Many schools have modified this program and provide fewer meals at a standard board rate. These schools allow students to select the meals they wish and to pay individually for them. Some others have worked out plans in which only a specific number of meals are purchased during the week. Technology is used to swipe prepaid cards. Innovations in providing students a variety of popular foods in scramble stations has resulted in sales of meal plans to students living off-campus. Alternative venues are sometimes offered to further entice more students to enroll in a meal plan.

Foods must be popular with students. Foodservices must offer natural or vegetarian foods. Pastas, pizza, seafood, and a variety of ethnic food items are well liked. In general, today's college students are more sophisticated and have more varied tastes than students in previous decades. As with school foodservice, the frequent use of casseroles has declined. The food must still be nutritious and provide adequately for dietary needs.

Most college and university foodservices are cafeterias, with students serving themselves beverages, bread, butter or margarine, desserts, salads, condiments, silverware, and napkins. For this reason, many foods may have to be proportioned and dished. Students may also bus their own trays from their tables to dishwashing areas.

In addition to the main cafeteria, a school may operate snack bars, a coffeehouse, a faculty club, and a faculty dining room. These will usually be typical of commercial counterparts, but slanted in price and ambience to suit campus needs. Faculty clubs must provide quality foods for parties, meetings, and receptions. Alcoholic beverages may be sold in some. Faculty clubs frequently resemble regular social clubs. They may even have recreational facilities.

Usually, a central department operates all foodservices on the campus. Registered dietitians may be on the staff to see that menus are balanced and nutritionally adequate. A regular staff of chefs, managers, and other personnel is maintained, along with numerous part-time student employees.

Colleges and universities are finding that they must compete with off-campus foodservice alternatives more directly than they did in the past, by combining marketing and merchandising.

## Transportation Feeding

Airline feeding is the largest transportation foodservice, doing close to $2 billion in sales. Ships, buses, and railroads make up a smaller percentage of sales in this group.

### *Airline Feeding*

Airline feeding on domestic flights of less than three hours or during time periods not associated with a meal have nearly ceased entirely. Many flights on domestic carriers, particularly economy class, are likely to offer a sandwich, hearty salad, or snack for sale in limited quantities. In many respects, a large kitchen producing airline foods is like a central commissary, but with some very significant differences. Timing demands are tremendous. Foods must be ready as needed. Thus, quick production, plus rapid transportation to the planes, is required. Often, an incoming plane needs to be ready for takeoff in about 20 to 30 minutes. This means that a rapid system for unloading used food items and loading on new must be worked out. The delay of a plane because food is late is a serious problem. However, foods cannot be produced too far in advance, or quality and safety may be lost. There is a need to develop maximum efficiency in work methods, and many units now use automation and labor- and time-saving methods.

A production center may produce foods for one or several airlines. A local foodservice company may set up a central production unit to sell meals and operate the airport foodservices. As an alternative, a large airline may operate its own production center and provide foods for other airlines.

Often an airline uses a cycle menu—a menu that repeats at regular cycles. A six-day cycle prevents frequent passengers from getting the same foods if they travel a certain day of the week. Each airline has its own standards and own menus, with some now having meals (often snacks) that never change. Food standards and food specifications are usually tightly written directives that leave the producing unit little leeway to use a different item. The menus vary depending on the length and type of the flight.

Several menus may be required for the same plane: for example, a choice for first-class passengers, a menu for coach class, plus any special dietary items ordered by passengers. Thus, if there are a large number of planes to be loaded in one day, there may be a set number of foods to produce at one time. Usually production centers are advised 14 to 30 days in advance that a plane must be loaded, but the actual passenger count is not given until several hours before flight time. Often, though, there is a preliminary count available approximately 24 hours or more before. Special meals must be ordered 24 hours in advance, but some planes are able to accommodate requests just before loading because they carry special foods for this purpose. This is possible due to the limited menus now served on most flights.

Each plane may also require its own type of dishes with accompanying equipment. It is not unusual for a large center to produce 6,000 to 8,000 meals a day with only a limited number of common items used for all planes. In addition, alcoholic and nonalcoholic beverages, ice, water, tea, coffee, milk, napkins, nuts, and other items are provided.

Because of the large inventory of different kinds of dishes, tableware, equipment, and other items, the production center must have considerable storage space for holding items for each airline. Transportation of foods and equipment to and from the center to planes must also be well organized. The operation of a food production center for airlines is quite complex.

Good flexibility must be established. Changes can occur rapidly, and the center must meet them. For example, a plane may be routed to a different airport and have to be loaded there.

The federal Public Health Service has jurisdiction over sanitation on public transport carriers moving across interstate lines, and the rules are stringent. No food or beverages returned on one plane can be used on another, unless they are packaged items, such as canned beverages, bagged pretzels or nuts, or bottled alcoholic beverages. Sanitation standards for the facilities, equipment, workers, and work methods are quite high. Besides meeting federal standards, an operation often must meet those of the local authorities.

With the advent of domestic flights' curtailment of complimentary foodservice, the opportunity may exist for airport terminal operators to offer boxed meals and snacks. This may initiate a new avenue for sales to this market.

### *Railroad and Bus Foodservices*

Railroad feeding today is very different from the past, when passenger trains served meals in dining cars that competed in quality and service with the best restaurants, and terminal stations provided foodservice that ranged from quick-service to fine-dining units. Today there are fewer passenger trains, and many on short runs serve only sandwiches or other fast foods. However, others serve meals that are reminiscent of the past, with gourmet offerings, fine service, and excellent decor. Most of these trains make longer trips and also have a snack bar offering lighter foods.

*Luxury ride* trains with accompanying fine dining cars stop to pick up diners and move much slower than commuter trains. Obviously, this type of foodservice operation must maximize sales to pay for itself.

Normally, railroad foodservices are not very profitable, but government railroad subsidies have been substantial and can enable some operations to break even or lose only a small amount.

Bus terminals are busier places and normally have quick-service operations. Some also have coffeehouses. Bus travel has not slowed as much as railroad travel, and so has been able to support foodservices better. Only rarely will any kind of food or beverage be sold on board the bus, except at stops where vendors may board the buses and sell snack foods and drinks.

## Health Services Feeding

Within the foodservice industry, the health service category includes nursing homes; homes for the aged, blind, mentally and physically disabled; and homes for orphans. Hospitals, nursing homes, and long-term-care facilities all provide some dietary foodservice. Hospitals do more than $12.4 billion worth of business a year, while nursing homes and others do more than $6.4 billion. Together these health service foodservice operations do about 4.6 percent of the total sales in the foodservice industry.

### *Hospitals*

A hospital's foodservice is usually directed by a professional with a college degree in foodservice management or an administrative dietitian with a degree in dietetics: Dietitians are registered by the American Dietetic Association and must achieve education and professional requirements set by the ADA. Some hospitals use a professional foodservice manager to administer the different foodservices, and dietitians to handle dietary concerns. A chef or food production manager works under the director of the hospital's foodservice. Hospital food is generally simple, but good quality. The nutritional value of the food and sanitation are vital considerations in production.

Most hospitals prepare a general menu from which patients on nonrestricted diets select the foods they desire. Modified diets, such as low-sodium, are adapted from the general menu. For this reason, a general menu must be planned with a view toward its being used for modified diets. However, food choice should not be dictated by this consideration alone.

Most hospitals use a cycle menu. Since patients' stays are usually short, most hospitals find they can operate with a rather short cycle. The food must be nourishing, simple, and not too highly seasoned.

Some hospitals use a cook-chill method in which food is cooked in batches enough for two or more meals. After cooking, meals are packaged and rapidly chilled, then stored to be withdrawn as needed at a future date. The food is then reheated and served. This method reduces labor and waste, and, when done properly, gives adequate-quality food.

The method of delivering and serving the food to patients may vary. Some hospitals have *centralized services*, where foods are dished up in the central production area and then sent to patients. Various methods of keeping the food hot or cold are used. Some hospitals have trayveyor systems that deliver food to the various floors in a short time. Others use insulated carts in which hot or cold food is loaded and sent to floors for delivery. A hot *pellet* (a round piece of metal weighing about a pound) may be put under dished food to keep it at a desirable temperature. Also available is an insulated server in which hot or cold foods can be kept with little or no loss of temperature for several hours. Thermalization units can be used that heat foods quickly for service. Microwave ovens also can be used to heat foods.

With the increasing use and availability of cook-chill, quick-frozen, convenience, and preportioned foods, many newer hospitals are developing decentralized foodservice systems. Only a limited amount of bulk preparation occurs in a central area. This is then sent to the various pantries on individual floors, where these foods are combined with others to make patient meals. Thus, a foodservice may purchase all of its salad greens ready-to-use. Frozen entrees may also be purchased. These foods will then be sent in the proper amount to the smaller pantries where final preparation occurs. Some pantry staffs wash the dishes and do a great deal of the cleanup. Foods in these pantries may be heated by microwave, quartz (infrared), or conventional ovens.

Hospitals serve patients who are ill and, consequently, under considerable emotional stress. While the food may be of good quality, patients may be dissatisfied with it because of their own physical or emotional condition. The food becomes a dog that the patient "kicks" in order to take out his or her fears and frustrations. Because of this, many hospitals now recognize the need to "sell" patients on the food, just as commercial units do. Food should be attractive, properly garnished, served at the right temperature, and have good form, color, texture, and flavor. While patients on modified diets may have little choice as to their food, every attempt should be made to cater to patients' personal tastes. For this reason, more and more hospitals today run special menus and promotions. *Room service* is an emerging hospital trend whereby patients are offered a greater degree of service, menu choices and control within their prescribed diet. Hospital foodservice may be contracted to a managed service provider or done in-house. Some municipalities have gone to using the hospital kitchen as the central commissary for the provision of elementary and secondary school lunches.

### *Long-term Health Facilities*

Long-term care facility foodservice operations are much like those in hospitals, but may not have the wide number of modified diets, although many patients in nursing homes will be on some special type of diet. The food must be nutritious and fairly simple. Many older people and people with long-term illnesses have problems eating, so much of the food must be soft and suitable for easy chewing and swallowing. However, special soft and bland diets should be prepared only for those who must eat them, while others can have more normal, textured food.

Most long-term facilities operate on low budgets and, as a result, the foodservice department must watch its costs. A Certified Dietary Manager, foodservice manager or supervisor will be responsible for direct operations management. Registered dietitians may act as consultants, calling once a week or sometimes as seldom as once a month. Thus, in some facilities, responsibility for management may rest on an individual who has little foodservice management or dietary training.

The equipment and layout of the foodservice unit may not be designed for maximum productivity. Because of this, menus and the foods that can be produced must necessarily be restricted. Although the food should be simple and close to what patients are used to, it should not lack variety. With a little care, even low-cost foods can be highly varied, interesting, desirable, and nourishing. With the population of persons over age 65 increasing, this field is predicted to experience career opportunity growth. Not all clients are of reduced means, and as a result, a rising number of facilities are catering to the desires of wealthy residents.

## Clubs

Club foodservices usually emphasize luxury-type foods. Clubs serve members who use the club for enjoyment, relaxation, or to entertain guests. However, some also offer quick-service food in addition. Food quality must be high. The food and service must give pride to members and impress guests who eat there. The market is usually restricted to members and guests.

In addition to parties, social clubs may serve breakfast, lunch, and dinner. Many such clubs exist in urban areas, and they may provide rooms for resident members. Serving meals and catering to events are usually big sources of revenue. Many clubs sell alcoholic beverages.

Country clubs are social clubs, but usually have recreational and sports facilities, such as pools, tennis courts, and golf courses, in addition to the foodservices. Club foodservices may vary from fine dining to snack bars.

Most clubs have a wide fluctuation in the amount of business done. At times they will be extremely busy, with every foodservice operating at full or even greater capacity, while at other times, little or no business will be done. This makes it difficult to schedule production and labor.

The menu must include food items such as steak or chicken that can be quickly prepared to meet a sudden surge in business. Some convenience items must be available so they can be used quickly if the demand arises.

Club foodservices must endeavor to attract members. A meal featuring ethnic foods, a fashion show with a luncheon, dances, and golf tournaments may be used to draw in members. Clubs frequently structure fees so that members must pay for a set dollar amount of foodservice, whether they choose to partake or not in order to support facilities. This may cause members to partake of services en masse at the end of a billing cycle.

## Military Feeding

The value of food served to military personnel at posts and in other military units is not included by the National Restaurant Association in its classifications. The only sales that are included are those in officers' and noncommissioned officers' (NCO) clubs and in military exchanges.

Military-feeding operations have gone through major changes and will continue to change. These changes reflect the changes in the troops in the post-Vietnam and Cold War era. There are more married enlistees than ever, and military commanders consider the stability of those families critical to reenlistment. The trend in military clubs in moving away from operating them as drinking places and toward using them as family centers and fitness clubs.

*Post exchanges* (large variety stores on a military base or post) continue to offer quick-service options, snack bars, and sometimes coffee shops.

Feeding the troops is still a priority, and in this area the changes are consistent with changes in the commercial feeding segment. The food is still substantial but more healthy than in the past. More selections are offered. Also, quick-service units are springing up on military bases and posts, offering even more variety.

## Central Commissaries

The production of central commissaries in our classification is included in the different types of units served by them. However, since these commissaries are specialized kinds of food production operations, they merit separate discussion.

Chain and large foodservice operations use central commissaries to produce much of the food they sell. School systems are currently using them. Some large foodservices have a substantial quantity of their foods produced in them. These foods are shipped to satellite units, where they are reprocessed and served.

Some central commissaries may receive only large shipments and then reship the food items to individual units in smaller lots. Michigan State University's central commissary acts as a warehouse and breakdown center where fresh vegetables, groceries, meats, and other supplies are received, broken down, and reshipped to campus units. Another central commissary in the city of Los Angeles school district does this, but prepares basic dry bakery mixes, pudding mixes, and sauce mixes, in addition to reshipping other items. The advantage in such a system is that lower prices can be obtained, since the central commissary can qualify for brokerage rather than wholesale prices.

Other central commissaries may prepare entrees, meats, salads, vegetables, desserts, bakery items, and soups ready-for-use. These may be shipped portioned in bulk lots to be processed by the satellite. The foods may be shipped frozen, chilled, or cooked and ready-to-eat.

A central commissary is not a kitchen as much as a factory in which foods are quickly mass-produced on assembly lines. Special equipment must be used. Some foods are cooked in huge steam-jacketed kettles holding 2,000 gallons or more of food. The food is also cooled down in these kettles by shutting off the steam and running refrigerated glycol or water into the jacket. Huge electric stirrers are used to move the food while cooking and cooling. Special pumps and other units must be used to put the food into the kettles and remove it. Highly sophisticated packaging machines must also be used.

Some foodservices are becoming disenchanted with central commissary production because cost savings are not substantial. Some chains have shut down their units. However, the use of commissaries to prepare food for school systems and other institutions continues.

# Forces of Change on the Foodservice Industry

The foodservice industry operates in an environment facing economic and social changes. The foodservice industry is also dynamic, and forces within the industry will generate change. **Exhibit 2.10** lists some of the broad predictions made in the National Restaurant Association's study, *Foodservice Manager.*

## The Economy

The foodservice industry sales are equal to 4 percent of the U.S. gross domestic product. The world economy continues to grow in influence on the national economy. Competition among producing nations has become fierce. Although the foodservice industry will face tremendous economic and competitive challenges, it will continue to grow. The biggest gains will most likely take place in managed services in schools, health care, and retail hosts.

The commercial segment of the foodservice industry—by far the greatest part of the industry—is very dependent on the amount of disposable income consumers have to spend. Whenever consumers have less to spend, foodservices must compete for the available discretionary income. Health care costs represent a significant expense to commercial foodservices and continue to rise.

## The Social Pattern

The Bureau of Census predicts that while the U.S. population will increase, there will be a change in the makeup of its age groups. Birth rates have slowed. The percentage in the total population of people under 20 will decrease, as well as the percentage of people 20 to 44. The percentage of those between 45 and 65 will increase substantially, and there will be a significant increase in the percentage of people over 65. We have about 34 million people in the population today who are over 65. By the year 2020 there will be over 55 million.

**Exhibit 2.10** Highlights of the 2006 Restaurant Industry Forecast

- Greater diversity on the menu, among restaurant customers and in the restaurant workforce.
- More emphasis on technology to boost efficiency and productivity, enhance the customer experience, provide online training and upgrade marketing through the Internet, e-mail, and broaden and improve frequent diner and gift card programs.
- An unabated demand for convenience provided by drive-through, curbside service, delivery, and take-out embraced by all levels. Full-service restaurants are likely to become aggressive with these untapped opportunities.
- Heightened focus on health and nutrition, while taste remains paramount. Restaurateurs will need to work to wow nutrition-conscious consumers with big taste and provide options for Americans with every dietary preference.
- Recruitment and retention issues, along with talent development and career ladder progress, are expected to remain a number-one concern among a significant number of operators.
- A continued pivotal role for tourism in restaurant industry sales. Travelers and visitors account for an estimated 40 percent of sales in fine dining and 25 percent of sales at casual and family-dining operations. The expected growth of domestic and international travelers is welcomed.
- Increased emphasis on value. This may result in more experimentation with limited-time menu offerings in quick service and expanded use of frequent diner programs.
- Increased attention to energy efficiency as utility expenses eat more of the operators' bottom line.
- Restaurants as homes away from home with amenities such as televisions, wireless Internet access, and softened ambiance at quick-service establishments.
- A bright future as sales have moved from $43 billion in 1970 to surpassing the half-trillion mark in 2006. Americans have embraced eating away from home as an integral part of their lifestyles. The restaurant industry truly has become an economic juggernaut.



The makeup of our social living structures affects the industry's economy. Family-style units profit in a stable population group. Couples and singles usually patronize a different type of unit and eat a different kind of food. The number of married couples with children has fallen from 40 percent in 1970 to 23 percent in 2003. Married couples without children decreased only slightly, from 30 percent in 1970 to 28 percent in 2003, according to the U.S. Census Bureau. The number of single people with children has also increased. The number of people living alone has gone from 17 percent in 1970 to nearly 26.5 percent in 2003. These figures indicate a decrease in the number of births per year. Family size has been shrinking. The fact that people live longer and the number of senior citizens has increased, as noted above, also influences the way population growth affects the industry.

This change in our population will mean that the kinds of operations and kinds of foods needed to satisfy patrons will change. **Exhibit 2.11** compares family compositions of 1970 to those of 2003.

Ethnic differences and the rapid rise of Hispanics as the second largest percentage of the population will also play an important part in dictating styles of operation and the kinds of foods

**Exhibit 2.11** Households by Type: 1970–2003

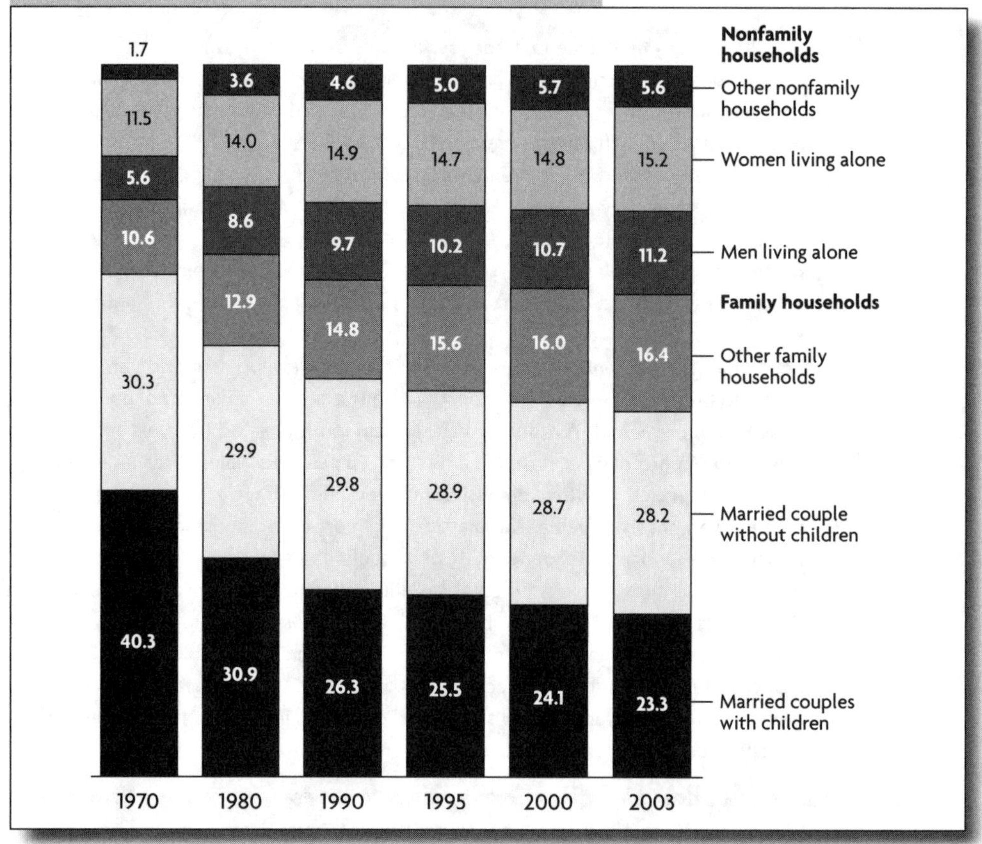


*U.S. Census Bureau, Current Population Survey, March and Annual Social and Economic Supplements: 1970 to 2003*

served. Because of the emphasis on healthful foods, certain foods that seem to be healthful will become more popular.

## The Labor Force

The industry workforce will change because of the way our population is growing. The white population will increase at a rather slow pace, while the African-American population will grow quickly, increasing from 13 percent in 2005 to 14.6 percent by 2050. The pace of growth of Asian and Hispanic groups will be quite accelerated, estimated to reach one third of the total U.S. population by 2050 at 8 percent and 24.4 percent, respectively. The National Restaurant Association and others predict that the percentage of non-Hispanic whites working in our industry will drop substantially, to be replaced largely by African-Americans, but more substantially by Asians and Hispanics. Women will make up over 55 percent of the foodservice workforce. Challenges in handling a multicultural group of employees will be multiplied, and the need to introduce more training programs will continue.

The predictions are also that with the increasing number of immigrants working in the industry, training needs will increase. A large number of walk-in workers have little or no work experience and must receive their training on the job. This will be an ongoing challenge to foodservice employers.

Hospitality schools will also have to fine tune their programs to produce supervisors, managers, and others who direct operations who are knowledgeable about marketing, human resources management, and computer systems. The skills involved in managing diversity will have to be taught.

The Department of Labor has identified hospitality as a high-growth industry. Further, it has cited the need to counter negative stereotypes, expanding the youth labor pool and targeting untapped labor pools as issues. Reducing turnover and addressing language skills were seen as needs, as well as developing consistent training models, improving "soft skills" and skill certifications.

## Healthful Foods

Consumers today are much more aware of the relationship between their state of health and what they eat. This has had a great impact on the foodservice industry. Patrons today want more fresh vegetables and less fatty meats and foods; fish and poultry are more in demand. Businesses are working to remove trans-fats from foods that they produce and serve. Many menus now offer special items to meet the demand for healthful foods. Culinology®, the blending of culinary arts and food science, is a fast-growing discipline that applies both the chef's and the food scientist's approach to food product and menu development. This field offers the opportunity to produce the flavorful, healthful products that consumers are seeking out in order to enhance health without sacrificing taste. The Research Chefs Association (RCA) promotes Culinology® and the partnership between chefs and food scientists because they offer the potential for improved food products, including products that are more healthful and cater to consumer trends. For example, the practice of Culinology® may increase the flavor and appeal of low-fat, low-carb, or low-sugar foods. Another example would be the addition of almond skins (typically discarded) to tortilla production in order to reduce the sodium and increase the fiber, improving the overall nutritional value of the final product.

## Employee Recruitment and Retainment

As noted, the foodservice industry has more people on its payroll than any other, employing nearly 12.8 million employees in 2007. It is estimated, by the National Restaurant Association that by 2017, the industry will employ 14.8 million people. At one time or another, 40 percent of all Americans will work in the industry. Thirty-two percent will find their first job within the industry. Employment time is shorter than in many other industries because it is often a means of livelihood while one prepares for a career in another area. This is a contributing factor to the high labor turnover. In some operations, the turnover is over 100 percent per year. Another reason for the turnover is that the industry pays low wages; some employees are paid below the minimum

wage because tips are considered a part of the total wage. Other reasons for the high labor turnover are the long hours and irregular schedules demanded of some employees.
Turnover is costly; it costs money to hire replacements and train them. Often the replacement employee, new to the job, lacks the proficiency in doing the job of the former employee. For instance, if it costs $300 to replace a worker and the operation makes a 5 percent profit, $6,000 in sales are required to make up for this cost.

The foodservice industry realizes its problem and has taken steps to try to reduce employee turnover. The advantages of industry careers are stressed and many try to compete with other industries to pay wages that attract good employees. However, according to predictions, the labor recruitment and retainment problem will remain. Overcoming negative stereotypes about the foodservice industry is a challenge that may be met, in part, by publication of the opportunities available. Over 60 percent of foodservice managers have incomes of more than $50,000. Our industry employs more minority managers than any other industry. **Exhibit 2.12** highlights this growth.

## Government Regulations

For many years the foodservice industry was largely ignored by governments at both federal and local levels. Laws and regulations that did affect the industry were those made in general for all industries. Laws such as those that established minimum wage rules, child labor laws, and the Occupational Safety and Health Act (OSHA), and laws regarding the employment of people with disabilities, income tax, and unemployment benefits are examples of this detached regulation. However, during the past few years, state and federal governments have singled out the foodservice industry for special regulation.

The majority of the laws that affect foodservice operators directly are written at the state and local levels. These include truth-in-menu laws, laws requiring nonsmoking sections in public dining areas, sanitary regulations, third-party liability (dramshop) laws, and liquor laws.

The federal government has also become interested in foodservice industry practices. Tip-reporting regulations under the tax code requirements predominantly affect foodservice employers and employees. Further government regulation has resulted from changes in U.S. immigration policies.

Certainly, in the future we will see further direct government regulation of the foodservice industry. The National Restaurant Association, state restaurant associations, and other groups representing the foodservice industry are aware of this and are active in political and legal areas. Among the issues that will continue to affect the foodservice industry are continued heavy regulation of liquor sales and service, including tighter dramshop laws and truth-in-menu regulations regarding additives and preservatives, and possibly laws restricting their use. The industry has lobbied to obtain relief from lawsuits such as those regarding obesity-related issues.

## Foodservice Industry Trends

Trends within the foodservice industry itself are important to watch if one is concerned about growth. These include changes in patron preferences, menu trends, and the availability of new technology in food equipment and supplies.

**Exhibit 2.12** Restaurant Industry Employment Growth by Occupation 2006–2016

**Ten-Year Restaurant-Industry Job-Growth Projections Show Demand Increasing Across Range of Restaurant Occupations**

As the restaurant workforce grows from 12.5 million in 2006 to an expected 14.4 million by 2016, that growth will be spread across a range of occupations.

In terms of proportional job growth, the strongest gains are projected among positions that combine both food preparation and service—particularly at quickservice restaurants. These positions are expected to number more than 2.7 million by 2016, an approximately 23 percent increase over their 2006 level.

The number of chefs and head cooks is projected to increase roughly 16 percent between 2006 and 2016, with the number of positions increasing to 168,000. Waiters and waitresses also are among the fastest-growth occupations, adding a projected 403,000 jobs—or more than 17 percent—between 2006 and 2016.

Management positions also should register steady growth during the next 10 years. The restaurant industry is projected to add 47,000 foodservice managers between 2006 and 2016, an increase of more than 11 percent. During the same period, the number of first-line supervisors and managers of food-preparation-and-serving workers is expected to increase by 119,000, or nearly 16 percent.

| | | | —Employment Change, 2006–2016— | | |
|---|---|---|---|---|---|
| | 2006 | 2016 | Number of Jobs | Total % Change | Avg. Annual % Change |
| Total Restaurant-Industry Employment | 12,489,000 | 14,406,000 | 1,917,000 | 15.3% | 1.4% |
| Foodservice managers | 410,000 | 457,000 | 47,000 | 11.5 | 1.1 |
| Food preparation and serving related occupations | 11,064,000 | 12,822,000 | 1,758,000 | 15.9 | 1.5 |
| Supervisors, food preparation and serving workers | 905,000 | 1,047,000 | 142,000 | 15.7 | 1.5 |
| Chefs and head cooks | 145,000 | 168,000 | 23,000 | 15.9 | 1.5 |
| First-line supervisors/managers of food preparation and serving workers | 760,000 | 879,000 | 119,000 | 15.7 | 1.5 |
| Cooks and food preparation workers | 3,027,000 | 3,403,000 | 376,000 | 12.4 | 1.2 |
| Cooks | 2,087,000 | 2,273,000 | 186,000 | 8.9 | 0.9 |
| Cooks, fast food | 610,000 | 640,000 | 30,000 | 4.9 | 0.5 |
| Cooks, institution and cafeteria | 440,000 | 449,000 | 9,000 | 2.0 | 0.2 |
| Cooks, private household | 8,000 | 8,000 | 0 | 0.0 | 0.0 |
| Cooks, restaurant | 790,000 | 916,000 | 126,000 | 15.9 | 1.5 |
| Cooks, short order | 239,000 | 260,000 | 21,000 | 8.8 | 0.8 |
| Food preparation workers | 940,000 | 1,130,000 | 190,000 | 20.2 | 1.9 |
| Food and beverage serving workers | 5,714,000 | 6,770,000 | 1,056,000s | 18.5 | 1.7 |
| Bartenders | 484,000 | 526,000 | 42,000 | 8.7 | 0.8 |
| Fast food and counter workers | 2,725,000 | 3,315,000 | 590,000 | 21.7 | 2.0 |
| Combined food preparation and serving workers, including fast food | 2,220,000 | 2,726,000 | 506,000 | 22.8 | 2.1 |
| Counter attendants, cafeteria, food concession, and coffee shop | 505,000 | 589,000 | 84,000 | 16.6 | 1.6 |
| Waiters and waitresses | 2,300,000 | 2,703,000 | 403,000 | 17.5 | 1.6 |
| Food servers, nonrestaurant | 205,000 | 226,000 | 21,000 | 10.2 | 1.0 |
| Other food preparation and serving related workers | 1,418,000 | 1,602,000 | 184,000 | 13.0 | 1.2 |
| Dining room and cafeteria attendants and bartender helpers | 437,000 | 502,000 | 65,000 | 14.9 | 1.4 |
| Dishwashers | 532,000 | 580,000 | 48,000 | 9.0 | 0.9 |
| Hosts and hostesses, restaurant, lounge and coffee shop | 323,000 | 376,000 | 53,000 | 16.4 | 1.5 |
| All other food preparation and serving related workers | 126,000 | 144,000 | 18,000 | 14.3 | 1.3 |
| Other eating and drinking place occupations* | 1,015,000 | 1,127,000 | 112,000 | 11.0 | 1.1 |

* Includes operational, business, financial, entertainment, sales, administrative and transportation occupations

*Source:* National Restaurant Association, based on Bureau of Labor Statistics data

Among the menu items that have grown in popularity are healthful foods, entrees, salads, Tex-Mex, Asian, wraps, tea, and bottled water. Operators who are aware of the differences between menu trends and menu fads will consider national trends in light of local tastes and eating patterns.

Growing consumer interest in convenience has resulted in the segue of retail operations into the foodservice field. This is one of the fastest-growing segments of the foodservice industry, and growth is expected to continue.

The need for the industry to reduce costs—due to increased energy costs and expected increases in labor—has brought about such changes as increased availability of self-service, take-out, curbside service, and menu items not easily prepared at home.

The industry has recognized consumer demand for more healthful foods and has taken steps to meet the demand. Many foodservices are trying to emphasize health in menu items. Some are turning to the emerging field of Culinology®.

One specific area of substantial growth—take-out, curbside, and delivery—can be a major part of almost any type of foodservice operation. The National Restaurant Association says that the increase in working women, singles, childless couples, and two-income households has led to an increase in foodservice take-out sales, and that this growth is likely to continue. Working women are responsible for much of the growth in the take-out business. Snack foods, partial meals, and gourmet dishes are all popular. The market is divided into consumers who buy foods to take to eat during the work-day and consumers who purchase food to take home. Each of these groups wants different foods. The first group of buyers wants snacks, light lunches, and other items that provide nourishment during work hours. The second group buys a complete or partial meal, and may be more interested in gourmet meals, especially items they would not fix at home.

## Technology

Technology will continue to change the industry, not only because of the continued production of convenience foods that reduce labor in the operation, but also because of more modern equipment, facilities, and modes of operation. The availability and cost of energy will also influence the foodservice industry and will help shape its future.

Looking back 50 years and comparing production and service then and now, we recognize that vast changes have occurred. Foods 50 years ago were neither as fresh nor as sanitary as they are today. Our grading standards and our ability to hold foods safely were not as good. Many foods now come to operations processed to such a degree that less labor is needed to get them ready for service. Many different kinds of cooked foods, ready for service, are on the market. Frozen foods, as we use them today, are less than 50 years old.

More than 70 years ago, foodservices began to use mechanical dishwashers and mixers. Coal, wood, and fuel-oil ranges and ovens were common in comparison with the thermostatically controlled units in almost universal use today. There were no electric meat saws, cubers, or grinders. Steam-jacketed kettles and pressure cookers were just appearing. The modern deep-fat fryer was unknown. Potatoes and other vegetables and fruits were prepared by hand. Convection, combination convection/conventional, thermal conduction burners, and microwaves can cook foods more quickly while retaining their character and quality. Technology has taken us a long way.

Foodservice operators are now demanding more energy-efficient appliances and more convenience foods preportioned for quick production and service. Improvements in packaging should help increase the shelf life of many food items. The *aquaculture* or farm-type raising of fish and shellfish is improving the prices and availability of such items. *Hydroponic* growing (growing plants in water rather than soil) of vegetables has become more common.

Perhaps one of the biggest changes is occurring in the way we are packaging our foods, and this will continue. Today we can use *controlled atmosphere packaging,* which seals in atmosphere around the fresh fruit or vegetable and keeps freshness for a much longer time. New labeling laws will make it possible for consumers to know more about what they are buying so they can better evaluate healthful qualities. New fat substitutes in foods give flavor without the calories and cholesterol. New noncaloric sweeteners have been produced, which are even more sweet and leave less aftertaste than early noncaloric sweeteners containing phenylalanine.

The computer has potential for application in all areas of food production and service. In the near future, few operations will be without some sort of computer application. Not only has technology become a factor in simplifying internal operations, but its expanded use by other organizations, such as banks and suppliers, helps to network a number of operations, including bank deposits, reorders, and credit checks. Internet marketing is a newer arena. Forty-five percent of 25- to 34-year-olds have used the Internet to find out information about a restaurant they haven't patronized before. Websites include menus and often allow customers to make reservations online.

In the next 50 years, we can look forward to even greater technological progress because we have far more technology available to improve food production than ever before.

## SUMMARY

The foodservice industry is marked by wide diversity in operating units and by yearly sales that make it the largest private-sector employer in the nation. The three major divisions are:

(1) commercial restaurant services that account for 91.5 percent of total foodservice sales;

(2) noncommercial restaurant services contributing 8.15 percent; and (3) the military restaurant segment, which accounts for 0.35 percent of sales. Eating places do most of the business in the first group, consisting of restaurants, lunchrooms, limited-menu operations, commercial cafeterias, social caterers, and frozen dessert stands. Bars and taverns that are not a part of another foodservice operation do just over 4 percent of the total sales in the commercial feeding category. Managed services in the commercial group do about 6.7 percent of all foodservice business. Lodging places do about 5 percent. Foodservices in retail units, recreation and sports facilities, mobile caterers, and vending units account for the rest of the sales in the eating places group, about 9.1 percent.

Employee feeding provided in-house has shown steady decline. Slighter but clear decreases in school and university feeding done in-house are also in evidence. Hospital and nursing care feeding shows a slight increase, perhaps more due to the aging population than a rise in in-house foodservice. Clubs, sporting, and recreation camps is the only sizable group showing significant increases. Smaller growth is seen in the categories of transportation and community centers. The military feeding facilities consist of officers' and NCOs' clubs and military exchanges.

Restaurants provided more than $70 billion in meal and snack occasions in 2005. This amounts to 47 percent of what Americans spend on food. Typical restaurant food costs are about 30 to 35 percent of sales. Payroll and employee benefits cost around 30 to 35 percent of sales and are rising. The average industry profit is generally small, from 3 to 7 percent.

The type of ownership in the foodservice industry has had a gradual change. At one time the industry was dominated by family-owned single units, but this started to change after World

War II. Today, while independently owned units still predominate, the percentage has dropped considerably. More than 70 percent of restaurants are single-unit operators. Quick-service chains and franchises within a corporate structure have begun to take over the sales of the industry and promise to grow at a rapid rate in the future.

The foodservice industry is the biggest employer in this country, outside of government, including both full- and part-time employees. Turnover is very high at all levels. Training of these employees and others already in the industry will be necessary to improve the quality and quantity of the labor supply.

Trends in the U.S. social structure and economy influence the foodservice industry. Perhaps one of the largest factors has been the growth in the number of women working outside the home, who eat out more and have more discretionary income to spend when they do. Another factor in our social structure is the growth in the number of single people, and the fact that people are marrying later and having fewer children. The size of households is smaller, with more and more families headed by single women. These factors bring more people out to eat and affect the type of foodservice operations these groups patronize. The growing number of retirees is also a factor, having implications for the foodservice industry not only in menu planning, but also for the labor supply. Retirees represent a readily available and, many times, more reliable labor supply than teenagers, whose populations have fallen.

Evidence of industry change that can influence the foodservice industry includes economic conditions; government control; changing social patterns; the interest of people in nutrition, value, and diet; the high employee turnover rate in the industry; the advent of the computer; and the popularity of convenience foods. Technology will continue to affect the future of the foodservice industry. New methods of packaging, holding, and portioning foods will help keep costs low and preserve quality for consumers. Computers are the major technological boon to the foodservice industry. They can be used in some way in almost every operation.

## Questions

1. How are foodservice labor requirements likely to be changed by technological advances?
2. How are changes in the population likely to affect the foodservice industry in terms of the menu? Of labor?
3. Describe some of the trends in consumer preferences that will affect menu planning.
4. Contrast the major differences in commercial and noncommercial/institutional feeding services in terms of goals, menus, and people served.
5. How do government regulations currently affect the foodservice industry? How might they affect it in the future?
6. What labor requirements have the potential to be taken over by technology? Which labor requirements don't? Why?

# Mini-Cases

2.1 A county hospital is debating the possibility of running its own foodservice. The foodservice will serve multiple groups including patients, employees, the executive (physicians) dining room, the visitor and volunteer cafeteria, and the café/coffee shop. Options include expanding as a central commissary to the county school district and other area hospitals, which are smaller in size. The alternative would be going to a contract foodservice provider. Weigh the positives and negatives of each option. Which option would you prefer to operate as a manager, and why?

2.2 You've inherited a good number of vending machines from a distant relative, including coffee/hot beverage, soda/cold beverage, snack, cold potential hazardous food (sandwich/salad), hot potentially hazardous food (french fries), and a unit that dispenses frozen food and contains a microwave for patrons to heat and eat. As you have received the machines from a distant location, you now must find a new a place to put them. Where in the community would you seek to locate each of the variety of machines you have received? Why would vending machines be more suitable than traditional foodservice venues in the locations you have chosen?

2.3 Develop a quick-service restaurant concept that you believe could be trendsetting. Describe the market for which this menu meets an unmet need and how you plan to carry out the central theme.

# 3

# PLANNING A MENU

## Outline



## Objectives

After reading this chapter, you should be able to:

1. Identify and characterize various menus used in the foodservice industry and explain the needs met by each variety.
2. Describe what is meant by *meal plan* and explain how menus are developed for them.
3. Explain how menus are organized and structured traditionally and the process by which they are derived.
4. Describe the various tools used to plan menus.
5. Compare and contrast institutional and commercial menus.

## Introduction

For foodservice consumers, a menu is a list, often presented with some fanfare, showing the food and drink offered by a restaurant, cafeteria, club, or hotel. For the manager of a foodservice establishment, however, the menu represents something significantly more: It is a strategic document that defines the purpose of the foodservice establishment and every phase of its operation.

In considering the menu, we may think of it generally in two ways: first, as a working document used by managers to plan, organize, operate, and control back-of-the-house operations, and second, as a published announcement of what is offered to patrons in the front of the house. In its first model it serves a variety of functions: as a guide to purchasing; as a work order to the kitchen, bar, or pantry; and as a service schedule for organizing job duties and charting staff requirements in all departments. In the second case it is a product listing, a price schedule, and the primary means of advertising the food, beverages, and service available to guests.

A good menu should lead patrons to food and beverage selections that satisfy both their dining preferences and the merchandising necessities of the operator. It can serve as public notice of days and hours of operation, inform patrons of special services available, narrate the history of the establishment and significant material concerning its locale; and even be used to inform patrons of new ways to enjoy the dining experience, including descriptions of unusual or exciting dishes, drinks, or food techniques. The importance of this selling opportunity makes it critical to design a menu that sells.

Selling is a goal not only of commercial operations but of noncommercial and semicommercial institutions as well. For instance, a hospital menu must offer items and present them in a manner that pleases patients and leads to favorable impressions. The same applies to other institutional menus. All menus should be an invitation to select something that pleases.

## What Is a Menu?

As far as we know, the first coffeehouses and restaurants did not use written menus. Instead, waiters or waitresses simply recited what was available from memory. Some Parisian operations had a signboard posted near the entrance, describing the menu for the day. The maitre d'hôtel stood near the sign and, as guests arrived, described the various offerings to them and took their orders.

Eventually, some restaurants in Paris originated the custom of writing a list of foods on a small signboard. Waiters hung this on their belts to refresh their memories. As menus became more complex, and more items were offered for sale, this method became too confusing—both to the patron and to the servers—and written menus entered into general use. Grimod de La Reyniere, an eighteenth-century gourmet who outlived both Louis XVI and the French Revolution, noted that many restaurants handed out bills of fare for patrons to take home with them, both as souvenirs and as advertisements to bring in more business. It was at the time of Napoleon that the restaurants in Paris began to reproduce private gourmet dinner menus for use in the public eating rooms. The elaborate menus written traditionally for the great banquets of the nobility thus found their way into the hands of the emerging middle class.

The word *menu* comes from the French and means "a detailed list." The term is derived from the Latin *minutes,* meaning "diminished," from which we get our word *minute.* Based on this, perhaps, we can say that a menu is "a small, detailed list."

Instead of *menu,* some use the term *bill of fare.* A *bill* is an itemized list, while *fare* means food, so we can say the term means "an itemized list of foods." This seems to be just another way of saying the same thing—*menu.*

## The Purpose of the Menu

The job of a menu is basically to inform—inform patrons of what is available at what price, and also to inform workers of what is to be produced. But it is much more than that. The menu is the central management document around which the whole foodservice operation revolves.

"Start with the menu" is a familiar byword of the foodservice trade. The menu should be known at the initial stage when planning a foodservice enterprise because it describes the very nature of the undertaking and the scope of the investment.

Management professionals have known for many years that in order for a company to succeed, it needs to have a clear idea of where the business is headed and how it plans to get where it wants to go. This process is known as long-range or *strategic planning.* In addition to developing strategic plans, management must also create short-range or *tactical plans* that define how the various parts of the organization must function in order to achieve strategic plans.

One of the first activities a manager must perform is to create a *mission statement* or statement of purpose for the organization. The mission for a school district's foodservice might be to provide wholesome, nutritional meals to students from a variety of economic circumstances and faculty during the school year. A commercial operation's mission might be to serve unique Mexican food at moderate prices to customers of all ages. Once the mission statement is developed, the organization must develop objectives or goals that the establishment wishes to attain. These may be expressed as *profitability objectives* (written in specific numbers), *growth objectives* (addition of units), market share objectives, or any number of other objectives. **Exhibit 3.1** shows a list of common organizational objectives.

Organizational objectives are turned into specific strategic plans that answer questions concerning exactly how the company intends to fulfill its mission and objectives. Managers often create these plans after studying environmental conditions. The business environment a foodservice establishment finds itself in may be relatively stable and risk-free. More often, though, the environment is one of frequent change and high risk. Patrons' desires change rapidly, while vigorous competition is a constant challenge. Working within these environmental constraints, the manager must develop a strategy to accomplish the objectives found in the mission statement. An example may be the task of converting a productivity objective into a strategy for increasing worker efficiency, coupled with a strategy for increasing equipment efficiency.

The next stage in planning is developing the financial plan. This is a complicated activity, whose details are beyond the scope of this book. Successful managers usually consult with a Certified Public Accountant, reputable hospitality consultant, or other knowledgeable financial advisor when formulating these plans. A numerical road map of the future is helpful in many ways, so this important area of planning should not be overlooked. Without long-range financial direction, managers are without a way of measuring operational success.

**Exhibit 3.1** Common Organizational Objectives

| Type of Objective | Sample Objective |
|---|---|
| Diversification | Operate table service, catering, and gourmet food store units |
| Efficiency | Operate using skilled workers and automated processes where possible, maximizing productivity. |
| Employee welfare | Provide a workplace free of sexual harassment, drug use, and alcoholism that encourages maximum growth. |
| Financial stability | Maintain key financial ratios in accordance with internal standards. |
| Growth | Increase sales at a rate of 10 percent per year. |
| Management development | Train, develop, and provide opportunity for committed employees to become senior corporate managers within ten years. |
| Market share | Enter and compete in markets where the company is likely to have the major market share. |
| Multinational expansion | Operate throughout North America and Japan. |
| Product quality and service | Serve wholesome, nutritious foods in a friendly, homelike environment. |
| Profitability | Operate with an average profitability of 30 percent over total sales. |
| Social responsibility | Hire and train economically disadvantaged people in the community. |
| Ecological | Recycle and use recycled products. Use organic meats and produce. |
| Ethical | Provide wholesome, healthful, appealing menu choices to patrons. |

Related to the financial plan is the establishment of operating budgets. Successful managers perform this task regularly. Budgets are usually created on an annual basis, and are usually expressed in terms of monthly increments.

Remember that no manager is so clever or so knowledgeable that he or she can do all of this work alone. A management team, composed of people highly qualified in many skills, will usually be more effective than one manager working alone. Although the foodservice industry is one of the few remaining industries where individual achievement is both possible and highly prized, most successful individuals learn to listen to suggestions from others before making the final decisions on their own. More and more restaurant companies are allowing management team members to buy into ownership.

## Who Prepares the Menu?

Because the menu is the essential document for successful operation of the establishment, menu planners must be highly skilled in a number of areas. They must know both the operation and the potential market. They should know a great deal about foods; how they are combined in recipes, their origin, seasonal preparation, presentation, and description. They must also understand how various recipes can be combined, which menu items go together, and which do not. Planners must also be aware of how operational constraints, such as costs, equipment availability, and the skills of the available labor force, affect the final menu selection. They must be able to visualize how the menu will appear graphically, what styles will look good, and which may be inappropriate for the particular operation. Finally, planners must be skilled at communicating successfully with patrons through the menu.

If these requirements seem daunting, perhaps there is a way to compensate. One key skill in managing is the ability to work in groups. Perhaps a management team is the answer (in even a small establishment, the cook and host can meet). While no one person in the organization may possess all the skills required to write the perfect menu, it is likely that a group can be formed whose membership combines all of the skills mentioned. Group preparation of the menu has one additional advantage, in that it gets the various members of the operational staff into a receptive frame of mind, anticipating the changes about to be implemented. Front-of-house personnel understand who the guests are and what has historically appealed to them. Back-of-house personnel know the strengths and limitations of the kitchen and are knowledgeable and enthusiastic about creating selections.

Menu planning is a time-consuming and detailed task, and should not be done quickly or haphazardly. Treat menu planning for what it is—the most critical step in defining the operation.

## Tools Needed for Menu Planning

What are the tools needed to prepare a menu? First, a quiet room where one can work without disturbance. A large desk or table is needed so materials can be spread out. These include a file of historical records on the performance of past menus, a menu reminder list, a file of menu ideas, and sales mix data indicating which items may draw patrons away from specials the operation wants to sell. A list of special-occasion and holiday menus should be on hand. Costs and the seasonality of possible menu items should also be available.

## Market Research

Market research is the link between the consuming public and the seller. This link consists of information concerning the public's buying preferences and how well the seller's business meets those preferences *as viewed by the potential buyer.* Formal market research is a scientific process of collecting data, analyzing the results, and communicating the findings and their implications to the seller. Anyone planning a menu is well advised to use as many forms of market research as possible. The more specific the research is (e.g., research concerning a specific group of potential patrons in a specific geographic location), the more reliable will be the results. At a minimum, the menu planner should take advantage of information available from marketing research firms. The federal government, financial newspapers and magazines, stock and investment firms, and many other large companies all provide the results of their market research. Trade associations, such as the National Restaurant Association and state associations, constantly produce market research reports that can be very useful in determining market trends.

Market studies have always been expensive. Traditionally, therefore, only the largest hospitality companies and chains could afford to conduct their own studies. However, it is beginning to be more feasible for smaller organizations to conduct market studies. An operator might, for example, commission a local college or business school to conduct market research. Advertising agencies can sometimes conduct a useful study for just a few thousand dollars. Convention and

Visitors Bureaus and Chambers of Commerce often collect data about visitors and residents and these figures are available for little or no cost. Managers or owners of a small operation can also conduct studies personally.

A key element of a market study is to determine exactly what kind of information is needed. On the one hand, if a menu planner wants to create a menu that will increase sales from existing customers, then the market data should be gathered only from that group. On the other hand, if the menu planner is attempting to draw in new customers, then that group should be targeted.

The next key element is to decide how the research will be conducted. Methods include conducting interviews, mailing out questionnaires, observing customers, and holding focus groups. Each is useful in eliciting specific kinds of information.

The questions asked of research subjects can be either open-ended or closed. An example of a closed question, which requires only a one-word answer, is "Do you prefer seafood or steak?" An open-ended question, which requires that the respondent supply a longer answer, might be "What are the most important factors that help you decide where to eat out?"

The information obtained through market research must coincide with other menu planning factors, such as purchasing costs and staff capabilities. Rich Melman of Lettuce Entertain You says, "Listen to the customer. Know what they want before they are aware of it."

# Menu Planning Factors

When planning a new menu, there are several important factors to consider. A new menu should be planned sufficiently in advance of actual production and service to allow time for the delivery of items, to schedule the required labor and to print the menu. Some menus must be planned several months before use. Operational needs will dicate how far in advance of use new menus should be prepared. Menu design and printing are covered in detail in Chapter 7.

A number of menus may be needed by a single operation. A large hotel might need several types of menus for different dining areas and specialty events. Some operations might need even more. Many operations change their menus seasonally. Atmosphere, theme, patrons, pricing structure, and type of service are all factors in determining what sorts and how many types of menus an operation will need.

# Types of Menus

## À La Carte Menu

An *à la carte* menu offers food items separately at a separate price. All entrees, dishes, salads, and desserts are ordered separately; the patron thus "builds" a meal completely to his or her liking. À la carte menus often contain a large selection of food items and, consequently, often lead to increased check averages. Commercial operations often find this menu type highly profitable, provided they can keep food spoilage—which can result from the large number of offerings—under control. Having variety within a manageable range of products is desirable. See **Exhibit 3.2.**

**Exhibit 3.2**
Brasserie Jo:
À la Carte Menu

TELEPHONE 312.595.0800
FACSIMILE 312.595.0808
EMAIL LEYE63@LEYE.COM
WWW.BRASSERIEJO.COM

# BRASSERIE JO

59 WEST HUBBARD STREET

Brasserie - Brewery: Lively Place That Is Both Restaurant and Cafe

## Sélection De Bières
BEERS

•HOPLA!, *Biere de la Brasserie* 5.00 •Two Brothers French Country Ale, *Warrenville, IL* 5.75
• Allagash White, *Portland, ME* 6.50 •Great Lakes Eliot Ness Vienna Lager, *Cleveland, OH* 5.75 •Rogue Dead Guy Ale, *Newport, OR* 5.75
•Anchor Steam, *San Francisco, CA* 5.75 •Sierra Nevada Pale Ale, *Chico, CA* 5.75 •Dogfish Head 60 Minute IPA, *Milton, DE* 6.00
•Staropramen Czech Pils, *Czech Republic* 6.00 •Erdinger Hefe-Weizen, *Germany* 7.50 •Unibroue Maudite, *Canada* 7.75 •Maredsous "8", *Belgium* 8.75
•Monks Café Flemish Sour Red Ale, *Belgium* 8.75 •Tripel Karmeliet, *Belgium* 8.75 •De Koninck, *Belgium* 7.50 •Saison Dupont, *Belgium* 9.00

## HORS D'OEUVRES

### WARM

- Soup du Jour . . . . 5.95
- **Onion Soup Gratinée,** Baked With Gruyere . . . . 6.95
- Lobster Bisque á la JO . . . . 8.95
- Baked Brie, Apricot, Almonds, Haricot Verts . . . . 8.95
- Crêpe of Mushroom, Ham, Swiss Chard, and Cheese . . 7.95
- **Escargots en Cocotte,** Garlic Butter . . . . 8.95
- Alsace Style House Made Gnocchi . . . . 8.95
- **Onion Tarte Uncle Hansi** . . . . 7.95
- Frog Legs Garlic Provençal . . . . 9.95
- Grilled Shrimp and Spicy Merguez . . . . 9.95
- Steamed Mussels, Alsace Riesling Wine . . . . 8.95
- Ratatouille and Goat Cheese Tart . . . . 8.95

### COLD

- 1/2 Dozen Oysters du Jour . . . . 11.95
- Shrimp Cocktail á la Française . . . . 8.95
- Duck Rillette, Country Toast. . . . . 8.95
- Charcuterie, Paté and Cheese Platter. . . . . 11.95
- **Chicken Liver Mousse and Country Paté** . . . . 8.95
- **Smoked Salmon,** Crispy Potatoes, Horseradish . . . . 12.95
- Cured Ham Prosciutto Style, Arugula . . . . 9.95
- **Assiette of Cheese,** Spiced Fruit Compote . . . . 9.95

## TARTES FLAMBÉES
CLASSIC ALSACE STYLE PIZZA

- Tarte Flambée Classic, Fromage Blanc, Onion, Bacon . . . . 8.95
- Tarte Flambée Pear, Bleu Cheese, Walnut, Garlic Chips. . . . 9.95
- Tarte Flambée Smoked Salmon, Capers, Onion . . . . 12.95
- Tarte Flambée Escargot, Leek . . . . 10.95
- Tarte Flambée Merguez, Garlic Aioli . . . . 10.95

## SALADES

- Boston Bibb á la Française, Vinaigrette . . . . 7.95
- Frisée Salade Lyonnaise, Bacon, Poached Egg . . . . 8.95
- Belgian Endive, Walnuts, Bleu Cheese . . . . 8.95

## Les Plats du Jour
DAILY SPECIALS

MONDAY
Julia Child's Beef Bourguignonne
22.95

TUESDAY
Veal Wiener Schnitzel
24.95

WEDNESDAY
Moroccan Style Couscous
22.95

THURSDAY
Cassoulet Toulousain
19.95

FRIDAY
Lobster & Seafood Bouillabaisse
24.95

SATURDAY
Beef Medallion Bordelaise
29.95

SUNDAY
Slowly Roasted Pork,
18.95

## PLATS PRINCIPAUX

### POISSONS

- **The Famous Shrimp Bag, Lobster Sauce, Herb Rice** . . . . 22.95
- Sautéed Walleye Pike, Warm Mustard Vinaigrette, Frisée Salade . . . . 22.95
- Filet of Trout Grenobloise, Capers, Lemon . . . . 19.95
- Salade Niçoise, Ahi Tuna in Olive Oil . . . . 18.95
- Steamed Mussels, Alsace Riesling Wine, Pommes Frites . . . . 18.95
- Herb Crusted Salmon, French Lentils . . . . 21.95
- Striped Sea Bass, Braised White Beans, Pepper Compote . . . . 23.95
- **Sautéed Skate Wing, Caper-Brown Butter, Pommes Purée** . . . . 19.95

### STEAKS

- **Classic Steak Frites,** Mustard Butter, Pommes Frites . . . . 19.95
- Steak Frites, Bordelaise Butter or Roquefort Butter, Pommes Frites . . . . 19.95
- Hanger Steak, Sauce Bearnaise, Pommes Pailles . . . . 19.95
- 10 oz. New York Strip Steak Au Poivre . . . . 29.95
- **Steak Tartare, Always Cold, Pommes Frites** . . . . 18.95

### SPECIALITEES

- Roasted 1/2 Chicken, Herbes de Provence, Pommes Frites . . . . 17.95
- **Red Wine Chicken Coq au Vin, Alsace Kneffla** . . . . 21.95
- Roasted Rack of Lamb, Parsley-Mustard Crust . . . . 26.95
- Papa Jo Beef Tongue, Pommes Purée, Mustard Sauce . . . . 18.95
- **Choucroute Alsacienne, Cabbage Alsace-Style, Smoked Meats** . . . . 24.95
- Roasted Duck Breast, Braised Red Cabbage, Pommes Purée . . . . 21.95
- Grilled Lemon Chicken Paillard, Petite Salade Parmesan . . . . 18.95
- Roasted Pork Loin, Swiss Chard, Rosemary Roasted Potatoes . . . . 19.95
- Sauteed Calf's Liver, Braised Onion and Madeira . . . . 19.95
- Wild Forest Mushrooms, Hand Made Pasta . . . . 16.95
- The 20 Vegetable Bag, Roasted Pepper Sauce . . . . 16.95

### Vegetables and Potatoes

- Braised White Beans . . . . 5.95
- Ratatouille . . . . 6.95
- Haricot Verts . . . . 6.95
- Pommes Frites . . . . 4.95
- Pommes Purée . . . . 4.95
- Alsace Kneffla . . . . 4.95

**Please alert your server if you have special dietary restrictions due to a food allergy or intolerance.**

CIGARS, PIPES: NON!
CIGARETTES: OUI!

**Reserve one of our intimate wine rooms or the Salon Privé for your next soirée from 8 to 300 guests.**

SERVICE NON COMPRIS
A service charge of eighteen percent is included for parties of five or more.

*Courtesy of Lettuce Entertain You Enterprises Inc.*

## Table d'Hôte Menu

The *table d'hôte* menu groups several food items together at a single price. This often can be a combination, such as a complete meal of several or more courses. Often there is a choice between some items, such as between a soup or salad or various kinds of desserts. This type of menu often appeals to patrons who are unfamiliar with the cuisine offered by the establishment. It is an excellent way to introduce the new patron to fine dining as perceived by the establishment. Another advantage of this type of menu is the limited number of entrees that must be produced. These menus can combine wine selections with each course, further enhancing the dining experience, especially for those hesitant to order wine. Very often à la carte and table d'hôte menus are combined. Even quick-service restaurants use *value menus* that bundle components.

## Du Jour Menu

A *du jour* menu is a group of food items served only for that day (*du jour* means "of the day"). See **Exhibit 3.3.** The term is most often associated with the daily special, the *soup du jour* being one example. Again, these items often are combined on à la carte and table d'hôte menus. One profit-boosting technique is to offer daily specials that use foods purchased at a reduced price or to use surplus goods, or goods whose expiration date is in sight.

**Exhibit 3.3**
Eiffel Tower: Sample Menu

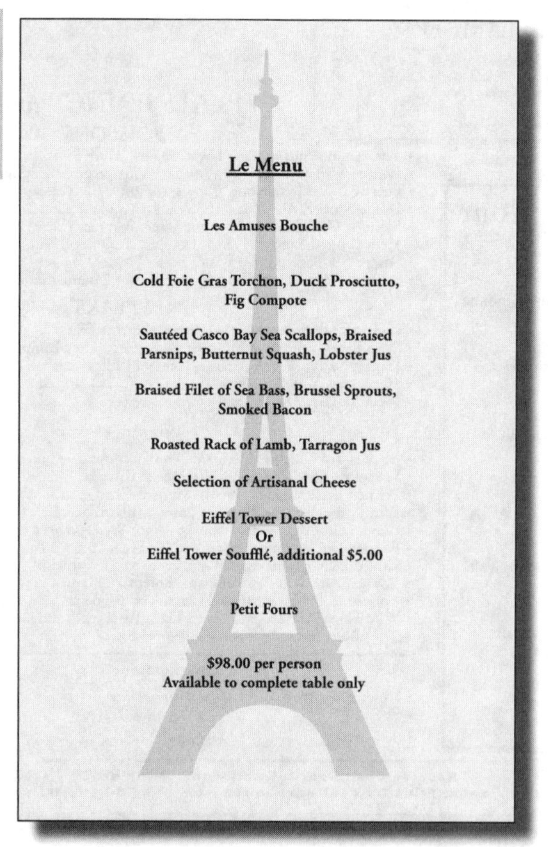

Le Menu

Les Amuses Bouche

Cold Foie Gras Torchon, Duck Prosciutto, Fig Compote

Sautéed Casco Bay Sea Scallops, Braised Parsnips, Butternut Squash, Lobster Jus

Braised Filet of Sea Bass, Brussel Sprouts, Smoked Bacon

Roasted Rack of Lamb, Tarragon Jus

Selection of Artisanal Cheese

Eiffel Tower Dessert
Or
Eiffel Tower Soufflé, additional $5.00

Petit Fours

$98.00 per person
Available to complete table only

*Courtesy of Lettuce Entertain You Enterprises Inc.*

## Limited Menu

A *limited* menu is simply one on which selections are limited in some way. Often associated with quick-service operations or cafes, the limited menu allows the manager to concentrate his or her efforts in training, planning, and calculating food cost or other menu analyses. Cost control is one very important benefit available to operators using limited menus. These menus do not look different from those previously mentioned with the exception of offering fewer choices. This is not entirely a new concept; a restaurant in Paris operated from 1729 to the 1800s offering only one menu item, chicken, cooked in several ways.

**Exhibit 3.4**
WOW BOA:
Limited Menu

WOW
MENU
BAO
BAO

**bao** .................... $1.29
kung pao chicken
chicken teriyaki
thai curry chicken
green vegetable
spicy mongolian beef
bbq pork

**party bao** *also sold *frozen* to take home
six pack* .................... $7.49
dozen .................... $14.79

**hot rice bowls** .......... $4.19
chicken teriyaki
kung pao chicken

**napa bowls** .......... $4.19
chicken teriyaki
kung pao chicken low-carb

**combos** ................ $4.69
1) 2 bao & eight veg salad
2) 2 bao & hot & sour broth
3) 2 bao & pad thai salad

**salads** ................ $3.79
pad thai chicken *or* eight vegetable

**soup/broth**
thai hot & sour broth ............ $2.19
thai chicken noodle soup ......... $3.29

**beverages**
fresh ginger, ginger ale .......... $2.19
hibiscus iced tea ................ $1.69
lemonade ....................... $1.69
bottled water .................... $1.69
soft drinks ...................... $1.49
frozen thai coffee or mocha ...... $3.19
hot sweet, ginger tea ............ $2.19
hot tea ......................... $1.69

water tower place, chicago, 312-642-5888

*Courtesy of Lettuce Entertain You Enterprises Inc.*

Today's menu might feature pizza, burgers, or sushi or any other food. See **Exhibit 3.4.**

## Cycle Menu

A *cycle* menu refers to several menus that are offered in rotation. A cruise ship, for example, may have seven menus it uses for its seven-day cruise. At the end of the seven-day cycle, the menu is repeated. The key idea here is to inject variety into an operation catering to a "captive" patronage.

Some hospitals might use only a three-day cycle. When patron stays are fairly long, such as in nursing homes, prisons, or on long cruise journeys, the cycle must be longer. Some operations of this type may have four-cycle menus for a whole year that change with the seasons of the year to allow for seasonal foods as well as menu variety. **Exhibit 3.5** shows an elementary school's cycle menu for one month. It is important when using a cycle menu over a long period to rotate Sunday and holiday selections, since people seem to remember what they had to eat on Sundays and holidays. Holidays require special meals to mark the occasion. Where fresh food is used versus convenience food, cycle menus can create efficiency through planned product utilization. Tuesday's meatloaf can appear again in Thursday's marinara sauce.

## California Menu

The *California* menu, called that because it originated there, offers breakfast, snack, lunch, fountain, and dinner items that are available at any time of the day. (See **Exhibit 3.6.**) Thus, if one patron wants hot cakes and sausage at 6:00 P.M. and another patron wants a steak, french fries, and a salad for breakfast, each is accommodated. This menu is typically printed on heavy, laminated paper so it does not soil easily. Many hotel room-service menus are based on this design as well.

**Exhibit 3.5**
Cycle Menu: Cherry Hills Elementary School

# February 2006

## American Heart Month

| MONDAY | TUESDAY | WEDNESDAY | THURSDAY | FRIDAY |
|---|---|---|---|---|
| **Spike reminds us, "Climb new heights with a strong heart and a healthy lunch!"** | | 1 Breakfast<br>Pancakes & Syrup<br>Fruit or Juice & Milk<br><br>Lunch<br>Breaded Chicken<br>Dinner Roll<br>Green Beans<br>Fruit, Juice & Milk | 2 Breakfast<br>Pancakes & Syrup<br>Fruit or Juice & Milk<br><br>Lunch<br>Hot Dog<br>Baked Beans<br><br>Fruit, Juice & Milk | 3 Breakfast<br>Pancakes & Syrup<br>Fruit or Juice & Milk<br><br>Lunch<br>*Domino's Pizza*<br>Tossed Salad<br>with dressing<br>Fruit, Juice & Milk |
| 6 Breakfast<br>Hot Pocket<br>Fruit or Juice & Milk<br>***Super Bowl Monday***<br>Lunch<br>***Breakfast 4 Lunch***<br>French Toast &<br>Sausage<br>Peach Cup<br>Juice & Milk | 7 Breakfast<br>Hot Pocket<br>Fruit or Juice & Milk<br><br>Lunch<br>Cheeseburger<br>Served with<br>French fries<br><br>Fruit, Juice & Milk | 8 Breakfast<br>Hot Pocket<br>Fruit or Juice & Milk<br><br>Lunch<br>Grilled Cheese<br>Served with<br>Soup<br><br>Fruit, Juice & Milk | 9 Breakfast<br>Hot Pocket<br>Fruit or Juice & Milk<br><br>Lunch<br>Spaghetti with<br>Mozz Sticks<br>Pudding Cup<br><br>Fruit, Juice & Milk | 10 Breakfast<br>Hot Pocket<br>Fruit or Juice & Milk<br><br>Lunch<br>Schwan's Pizza<br>Served with<br>Veggie Sticks<br><br>Fruit, Juice & Milk |
| 13 Breakfast<br>Pancakes & Syrup<br>Fruit or Juice & Milk<br><br>Lunch<br>Chicken Nuggets<br>Dinner Roll<br>Green Beans<br><br>Fruit, Juice & Milk | 14 Breakfast<br>Pancakes & Syrup<br>Fruit or Juice & Milk<br><br>*Happy*<br>***Taco Tuesday***<br>Hard Taco, Meat,<br>Cheese, lettuce & Salsa<br>Mexi Corn<br>Fruit, Juice & Milk | 15 Breakfast<br>Pancakes & Syrup<br>Fruit or Juice & Milk<br><br>Lunch<br>Breaded Chicken<br>Dinner Roll<br>Pudding Cup<br><br>Fruit, Juice & Milk | 16 Breakfast<br>Pancakes & Syrup<br>Fruit or Juice & Milk<br><br>Lunch<br>*Domino's Pizza*<br>Tossed Salad<br>with dressing<br><br>Fruit, Juice & Milk | 17<br><br>No School |
| 20<br><br>President's Day<br>No School | 21 Breakfast<br>Cheese Omelet<br>Fruit or Juice & Milk<br>***President's Day Treat***<br>Lunch<br>Cheeseburger<br>Served with<br>Pudding Cup<br><br>Fruit, Juice & Milk | 22 Breakfast<br>Cheese Omelet<br>Fruit or Juice & Milk<br><br>Lunch<br>Chicken Patty<br>Sandwich<br>Potato Rounds<br><br>Fruit, Juice & Milk | 23 Breakfast<br>Cheese Omelet<br>Fruit or Juice & Milk<br><br>Lunch<br>Breaded Chicken<br>Dinner roll<br>Green Beans<br><br>Fruit, Juice & Milk | 24 Breakfast<br>Cheese Omelet<br>Fruit or Juice & Milk<br><br>Lunch<br>Pizza Bagels<br>Served with<br>Tossed Salad<br><br>Fruit, Juice & Milk |
| 27 Breakfast<br>Hot Pocket<br>Fruit or Juice & Milk<br><br>Lunch<br>Chicken Nuggets<br>Dinner Roll<br>Green Beans<br><br>Fruit, Juice & Milk | 28 Breakfast<br>Hot Pocket<br>Fruit or Juice & Milk<br><br>Lunch<br>***Taco Tuesday***<br>Soft Taco, Meat,<br>Cheese, lettuce & Salsa<br>Mexi Corn<br>Fruit, Juice & Milk | **MORE INFO:**<br>**Help Speed Up the Lunch Lines!**<br>1- Practice your child's number with them at home. Especially if they don't purchase everyday!<br>2- Prepay by Check. Avoid cash handling in line. | | |

**SERVED DAILY**

***Daily Breakfast Alternates***

Blueberry Muffin

Super Donut

Bagel & Cream Cheese

Cereal & Toast

*Breakfast is served with fruit or juice & low-fat or skim milk*

***Daily Lunch Alternates***

Bagel Box Lunch (Bagel, cream cheese, yogurt & cheese stick)

Peanut Butter & Jelly Sandwich on white or wheat bread

Beef Hot Dog

***Weekly Lunch Alternates:***

Jan 30$^{th}$ – Feb 3$^{rd}$
Turkey Hoagie

Feb 6$^{th}$ – 10$^{th}$
Garden Salad With Cheese

Feb 13$^{th}$ – 17$^{th}$
Chicken Caesar Wrap

Feb 20$^{th}$ – 24$^{th}$
Cheese Cubes, Wheat Crackers & Peaches

Feb 27$^{th}$ – Mar 3$^{rd}$
Garden Salad With Cheese

*Lunch is served with fruit, juice & low-fat or skim milk*

**Cherry Hill Elementary School Breakfast & Lunch Menu**

**Special News...**

**Breakfast Price \$1.00 Lunch Price \$2.00**
(Eligible students may receive free or reduced price meals-
Call 761-6200 for information)

Menus are subject to change without notice.

*Courtesy of Aramark School Support Services*

**Exhibit 3.6** Waffle House: California Menu

WAFFLE HOUSE®
www.wafflehouse.com

AMERICA'S PLACE TO WORK, AMERICA'S PLACE TO EAT®

# BREAKFAST

*Taste why we're the world's leading server of Waffles, Omelets, Cheese 'N Eggs...*

## WAFFLES

*From the originator of the world's most perfect waffle...*

|  |  |
|---|---|
| WAFFLE | 2.45 |
| Pecan WAFFLE | add .45 |
| DOUBLE WAFFLE | add .99 |
| Buttermilk WAFFLE *NEW* | 2.45 |
| Chocolate Chip WAFFLE *NEW* | 2.90 |
| Butterscotch WAFFLE *NEW* | 2.90 |

## EGGS

Served with Toast & Jelly; and choice of Grits, Hashbrowns or Tomatoes

|  |  |
|---|---|
| 3 Eggs | 3.10 |
| 2 Eggs | 2.75 |
| Cheese 'N Eggs<br>Served with Raisin Toast & Apple Butter, | 3.40 |
| Steak & Eggs, 5 oz. NY Strip | 7.35 |
| Country Ham & Eggs | 5.80 |
| Egg whites only ("Hold the Yolks") | add .50 |

## OMELETS

|  |  |
|---|---|
| Cheesesteak | 5.50 |
| Chili Cheese | 4.85 |
| Fiesta | 5.30 |
| Ham & Cheese | 4.75 |
| Turkey & Cheese | 5.50 |

Get both GRITS & HASHBROWNS ONLY 50¢ MORE

add a side of BOTH BACON & SAUSAGE ONLY $3.00 or EITHER for 2.00

## EGG SANDWICHES/WRAPS/MELTS

Texas Melts served on Texas Toast with DOUBLE Cheese

| | Sandwich | TEXAS Melt | Wrap |
|---|---|---|---|
| Bacon, Egg & Cheese | 3.10 | 3.75 | 3.80 |
| Sausage, Egg & Cheese | 3.10 | 3.75 | 3.80 |
| Sausage Egg Cheese Raisin | 3.10 | | |

Served on our delicious, signature Raisin Toast

*Add Hashbrowns for only $1.30*

## SIDES

Jimmy Dean

|  |  |
|---|---|
| Bacon *(3 slices)* | 2.00 |
| Sausage *(2 patties)* | 2.00 |
| City Ham | 2.25 |
| Country Ham | 3.50 |
| Bowl of Grits | 1.30 & 1.80 |
| Cheese Grits | add .35 |
| Texas Toast | 1.60 |
| Raisin Toast | 1.45 |
| Toast *(White or Wheat)* | 1.30 |

## HASHBROWNS

*"World's Leading Server of REAL Hashbrowns"*

|  |  |
|---|---|
| REGULAR | 1.30 |
| DOUBLE | 1.80 |
| TRIPLE | 2.05 |
| SCATTERED on the grill AND... | |
| SMOTHERED – *ONIONS* | add .25 |
| COVERED – American *CHEESE* | add .35 |
| CHUNKED – "Hickory Smoked" *HAM* | add .50 |
| DICED – *TOMATOES* | add .25 |
| PEPPERED – Jalapeño *PEPPERS* | add .25 |
| CAPPED – *MUSHROOMS* | add .70 |
| TOPPED – Bert's *CHILI*® | add .60 |
| "SCATTERED ALL THE WAY" | ~~4.20~~ *SPECIAL* 3.99 |

## VALUE MENU

*For Jrs. & Srs. and Everyone else*

|  |  |
|---|---|
| Waffle | 2.45 |
| Cereal *(Regular/Large)* | 1.70 & 2.30 |
| 1 Egg Breakfast<br>Served with Toast & Jelly; and choice of Grits, Hashbrowns or Tomatoes | 2.30 |
| Egg Sandwich | 1.75 |
| Bacon *(2 slices)* | 1.60 |
| Sausage *(1 patty)* | 1.40 |
| "Original" Hamburger | 1.25 |
| "Original" Cheeseburger | 1.60 |
| "Original" Double Cheeseburger | 1.99 |
| "Original" Meal<br>(2 "Originals" & Hashbrowns) | 2.99 |
| *add Lettuce and Tomato for .35 each "Original"* | |
| Grilled Cheese | 1.70 |
| Pie (Single Slice) | .99 |

## DESSERTS

Try our Aunt Maggie's® Pie — Sara Lee

|  |  |
|---|---|
| Pie – Southern Pecan, Triple Chocolate | |
| Single Slice | .99 |
| Double Slice | 1.80 |
| Assorted Pastries | Ask your server |

**Public Health Advisory:** Eggs, hamburgers and steaks may be cooked to order. However, the consumption of raw or undercooked food such as meat, chicken and eggs which may contain harmful bacteria, may cause serious illness or death, especially if you have certain medical conditions.

*Thank You! You had a choice and you chose Waffle House. We appreciate your business and hope you choose to eat with us often. Please send any comments regarding our food and service to : "Let Us Know" Dept., P.O. Box 6450, Norcross, GA 30091 / 1-877-9WAFFLE (992-3353)."We cannot accept any checks – Thank you for your cooperation – A 10% Service Charge is added to all 'To-go' orders for the servers who prepare them"*

*Courtesy of WH Capital, L.L.C. and Waffle House, Inc.*

**Exhibit 3.6** *(Continued)*

*GOOD FOOD FAST & Friendly™ OPEN 24 HOURS*

# LUNCH/DINNER

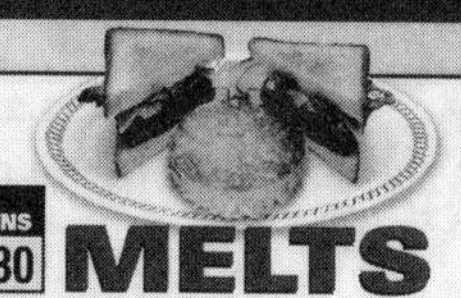

ADD HASHBROWNS ONLY $1.30

## MELTS

Wheat Bread – Grilled Onions – Double Cheese

TEXAS Cheesesteak™ Melt ........ 3.70
TEXAS DOUBLE Patty Melt ........ 4.10
TEXAS Turkey Melt *NEW* ........ 3.70
Lib's Patty Melt™ (¼ lb.*) ........ 3.05
Chicken Melt ........ 3.50
Sausage Melt ........ 3.05

*Any Melt on Texas Toast add .30 Bacon add 1.00*

## WRAPS & SANDWICHES

Enjoy any Sandwich or Hamburger on Texas Toast add .30 or as a Wrap in a Grilled Flour Tortilla add .70

Sandwich only

Bacon Chicken Cheese ........ 4.80
TEXAS Bacon Turkey *NEW* ........ 4.70
Chicken Sandwich ........ 3.45
Ham & Cheese (Lettuce & Tomato) ........ 3.20
Bacon Lover's B.L.T. ........ 3.30

*World's leading server of T-Bone Steaks, Pork Chops, Melts...* USDA CHOICE

CHOICE CUTS

## STEAKS

All Steaks and Big Deal Dinners come with a regular Garden Salad, Hashbrowns and Texas Toast

T-Bone Steak, 10 oz. ........ 9.85
Sirloin, 5 oz. NY Strip ........ 8.15
Chop Steak ........ 6.20

## BIG DEAL DINNERS

Papa Joe's® Pork Chops ........ 6.75
Grilled Chicken ........ 6.00

*with Mushrooms add .70*

## CHILI & SOUP

Bert's Chili™ *(Regular/Large)* ........ 2.45
Bert's Best Bowl of Chili ........ 3.80
Large bowl of Bert's Chili™, Smothered, Covered, Chunked & Peppered
Walt's Soup™ *(Regular/Large)* ........ 2.45
Our own Chicken Noodle Soup *(seasonal offering)*

## WAFFLE HOUSE COFFEE

Top off the ultimate dining experience with a hot cup of our freshly ground Waffle House Coffee. The best coffee for your meal.

Regular or Decaf ........ 1.15
To Go ........ 1.15 & 1.55

FREE Refill on Coffee, Iced Tea, Soft Drinks and Hot Tea

## HAMBURGERS

USDA CHOICE

with Lettuce, Tomato, Onion and Pickles

DOUBLE ¼ lb.* Cheeseburger ........ 3.50
¼ lb.* Cheeseburger ........ 2.75
¼ lb.* Hamburger ........ 2.40
"Original" Hamburger ........ 1.25
"Original" Cheeseburger ........ 1.60

*Mushrooms add .70 Bacon add 1.00*
*Pre-cooked weight.

## GARDEN FRESH SALADS

Grilled Chicken Salad ........ 4.10
Chef's Salad ........ 4.10
Garden Salad *(Regular/Large)* ........ 2.10 & 2.40

*with Bacon add 1.00*

## BEVERAGES

Alice's Iced Tea™ *(Freshly Brewed)* ........ 1.25
Hot Tea ........ 1.25
Swiss Miss Hot Chocolate ........ 1.25
Ice Cold Milk (2%) ........ 1.25 & 1.45
To Go ........ 1.25 & 1.70
Chocolate Milk ........ 1.75
Minute Maid Orange Juice *(Regular/Large)* ........ 1.25
To Go ........ 1.25 & 1.70
Soft Drinks ........ 1.25

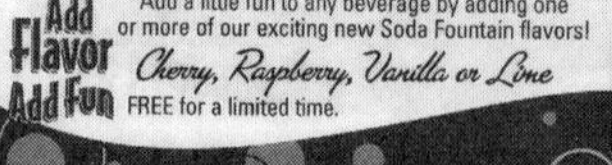

Add Flavor Add Fun

Add a little fun to any beverage by adding one or more of our exciting new Soda Fountain flavors!

*Cherry, Raspberry, Vanilla or Lime*

FREE for a limited time.

SODA Fountain

THE FUN IS IN Heinz THE FLAVOR!

# The Meal Plan and the Menu

Menu planners must keep in mind that there are two things a patron decides when selecting a meal from a menu. The first is what the sequence of foods will be, and the second is which items will be selected. The meal planner must do the same: (1) decide on courses and their sequence, and (2) establish the specific foods in each course.

These two factors make up the *meal plan,* or the manner in which foods are grouped for particular meals throughout a day. The typical American meal plan is three meals a day, with about a fourth of the day's calories eaten at breakfast, a third at lunch, and the balance at dinner, although some may reverse the lunch and dinner percentages. For a table d'hôte menu, some knowledge of the meal plan is required, even for partial meals. On an à la carte menu, a meal plan is not needed since the patron establishes one by selecting the items.

In this country, food patterns have been changing. Some menus may be written for a four- or five-meal-a-day plan. This gives a greater division of calories but the same total amount in a day. Most meal plans call for all the food to be consumed within a 10-hour period, but if a snack is served at night, the fasting period may be less than 14 hours.

Staying too strictly with a meal plan can become monotonous, so it is recommended that a plan be varied occasionally. Thus, a supper of vegetable soup with crackers, large fruit plate with assorted cheeses, bran muffins and butter, brownie, and a beverage may be a relief from the often-consumed meat, potatoes, and hot vegetable dinner. **Exhibit 3.7** shows a breakfast meal plan and a menu based on it.

**Exhibit 3.7** Breakfast Meal Plan and Menu

| Meal Plan | Menu |
|---|---|
| Fruit or juice | Orange, grapefruit or tomato juice |
| Cereal | Granola or oatmeal, milk |
| Entree | Eggs any style or breakfast burrito |
| Bread | Bran muffins or cornbread |
| Beverage | Tea, coffee, or cocoa |

## Menu Organization

Menus usually group foods in the order in which they are intended to be eaten. Typical menus begin with appetizers and end with desserts. Several sample courses are shown in **Exhibit 3.8.**

Within the entree category, it is usual to split the various food offerings into the following categories: seafood and fish, meat (beef, lamb, and pork), poultry (chicken, duck, etc.), and others such as pasta or meatless entrees.

Each menu category should offer choice. It is important to note that the foods within a major

**Exhibit 3.8** Sample Menu Organization

| Coffee Shop | French Restaurant | Hospital | Family Dining | Steakhouse |
|---|---|---|---|---|
| Appetizers and side dishes | Hors-d'oeuvre | Appetizers and soups | Appetizers | Appetizers |
| Salads | Potages (soups) | Salads | Soups | Soups and salads |
| Sandwiches | Salad | Entrees | Salads | Entrees |
| Hot entrees | Sorbet | Vegetables | Entrees | Side dishes |
| Fountain items | Entrees | Desserts | Side dishes | Desserts |
| Desserts | Plateau de fromage (cheese platter) | | Desserts | |
| | Entremets (small desserts) | | | |

category must differ in style of preparation. For example, every entree should not be fried. Various preparation methods should be used such as poaching, roasting, grilling, frying, and baking. While it may not be feasible to include all methods of preparation, a menu heavily weighted toward one method or another will be unbalanced. Only in an operation specializing in one preparation style is repetition recommended.

It is often said that variety is the spice of life. It certainly is the key to creating a menu that sparks patrons' interest and encourages them to come back again and again. The variation of cooking methods is a subtle form of variety. The variations in tastes and textures are less subtle. It is important that foods vary from spicy or hot to bland or mild. Even the most adventurous tongue needs a rest, which is why some dinner restaurants are serving an intermezzo course (sometimes a fruit sorbet) between courses. Both spicy and hot dishes can appear on a menu. (Hot dishes cause burning on the tongue, while spicy items incorporate complex flavors.)

Various cooking methods will affect menu items' texture as much as their ingredients.

Variety extends to the visual as well. There is a saying that people "eat with their eyes." Chefs are concerned with balancing color as well as texture. Whether an ingredient is cut, ground, minced, cubed, sliced, pared, or kept whole affects its visual appeal as well as its "mouth feel." Even the use and kind of garnish can distinguish items and make an otherwise common dish seem exotic.

## How Many Menu Items?

The number of menu items found on menus varies greatly, from the simplest limited-menu operations (one operation in London has only one choice each for appetizers, soup, salad, entree, and dessert) to Chinese menus offering many choices. Management usually wants to limit the number of offerings in order to reduce costs and maintain good control. The patron may desire fewer items also. Some studies have shown that people who are confronted with a large number of choices tend to fall back on choices they have made before.

The key word in choosing the items for a menu is *balance*. There must be a balanced selection that allows for different tastes. It is necessary to balance selections within food groups. Thus, a lunch menu might offer a minimum number of appetizers, say two juices (one fruit and one vegetable), a fruit, and a seafood cocktail. This minimum offering could be expanded on the dinner menu into choices of both hot and cold appetizers. Gourmets may substitute a hot seafood appetizer for the classical fish course, served before the entree.

The number and kinds of soups also depend on the meal period. Lunch patrons may wish to have a bowl of hearty soup such as vegetable or black bean or a combination cup of soup and half a sandwich. Cold soups or lighter soups are popular. However, the dinner patron may want soup as a beginning for a large meal. This patron may wish to have a light appetite-stimulating soup like a consommé or miso. The balance with the amount of food offered with the rest of the meal must be struck between light clear soups, purees, and bisques, and the heavier chowders and stewlike soups.

The typical menu has at least five or six entrees. Theme operations, such as a seafood restaurant, naturally will specialize in certain entree categories. They must still offer some alternative dish; for example, a seafood restaurant may offer a nonseafood item for those patrons who will not (or cannot) eat seafood but who are with a party that does. Nonmeat items (e.g., pasta or vegetable dishes) are increasingly necessary on menus to provide choices for those who choose to limit or eliminate meat from their diets. If balance is required, the entrees should be divided among beef (more than 50 percent of the meat consumed in this country is beef), pork and ham, poultry, lamb and veal, shellfish, fish, egg dishes, cheese dishes, and non-meat main dishes. If only two food items appear on the menu they may both be beef (one roasted and one a steak dish) or one beef and one of the other meat categories. If two fish or shellfish items appear, they should be quite different, such as a lean, white-fleshed fish like a sole or flounder, and a fatter one such as salmon.

If one is limiting the entrees to five or less, it is best to remember that quantity is no substitute for quality. A few items prepared and served with absolute perfection are preferable to a dozen mediocre items.

The ideal menu must also provide for balance and variety in the vegetable, sauce, and starch dishes as well. The vegetables must be chosen to complement the entree choices. Thus, on a menu in which there is ham, chicken, or turkey, an offering of sweet potatoes or yellow winter squash will complement the meal. Potatoes, rice, wild rice, polenta, pasta, and other starches are also an important part of a menu.

When selecting vegetables, as with all menu items, you must be especially aware of their flavors, colors, and texture contrasts. Popular vegetables like asparagus, peas, string beans, and carrots should be offered, but for variety's sake it may be wise to include less familiar vegetables as well. Variety can also be achieved by using different cooking, methods, or serving some vegetables raw. Vegetables can be steamed, grilled, roasted, pureed, diced, julienned, made into fritters, or served with a variety of sauces like mustard, tomato, or cheese. Classical egg- and fat-based sauces are less popular than they once were.

Again, variety must not be carried to extremes; three to five starches, and five or so vegetable dishes are more than sufficient for most menus.

The number of salads offered as meal accompaniments (as opposed to salads as main dishes) has decreased. For many years table d'hôte meals included a choice of several salads. Now the typical menu includes only one or two salads, usually a tossed green salad, or a special salad like a Caesar salad. However, as more people strive for a healthy lifestyle, many menus are offering several generous-portioned salads as main courses; grilled meat or seafood are commonly added to larger-sized specialty salads. Variety is achieved, not only in ingredients—arugula, endive, spinach, artichokes, hearts of palm, nuts, seeds, patés, and edible flowers are just some of the creative ingredients being used—but also by offering a variety of salad dressings, featuring flavored vinegars and oils, honey, cheese, and mustards. The self-serve salad bar trend has crested as more operators become concerned with the labor and food costs associated with salad bars, as well as heightened fears concerning the potential health risks associated with self-serve environments, and fear of sabotage.

Another trend in menus has been to offer fewer dessert choices but to make those items offered more elaborate. Some operators see dessert as a low-profit area and confine the choices to an ice cream or cookie. Others see dessert as a selling opportunity, where sweets or pastries can be coupled with unique after-dinner drinks to generate sales. The dessert course must be treated as seriously by the manager as any of the other food courses. One way to free up tables for other diners is to offer dessert in a separate room. Several operations have turned this into a very profitable way to keep diners on site and spending additional dollars on after-dinner drinks. Cheese courses and post-dessert chocolates are seen in fine dining.

We look at breakfast menus last. This important meal is characterized by high volumes and low profits. The public seems to have a clear idea of what it is willing to pay for breakfast, and operators emphasizing this meal period are often caught in the price trap. While there is no study to support the idea, it may be conjectured that the reason for this price problem is that breakfast, more than any of the other meals, consists of food items that patrons regularly prepare at home. This knowledge of the costs and skills required might cause a patron to develop firm attitudes concerning what is a fair price to pay for breakfast.

One way to boost check averages is to offer breakfast items that the public cannot relate to their at-home cost. Elaborate breakfasts or unusual breakfast combinations may be the answer to this dilemma. An analysis of 50 breakfast menus from hotels around the country showed that the average breakfast menu offers about 100 choices. There are usually five to ten juices, five or more fruit dishes, a hot cereal, and a dozen or more cold ones offered with and without a variety of toppings. One may find eggs cooked in perhaps four to six different ways (there are about 100 ways to cook an egg in formal French cooking). Meat dishes may include hams, steaks, bacon, sausages, and other meats. French toast, pancakes, toasts, and rolls may add another five to ten items to the menu, with side dishes increasing the menu size even further. Finally, there are the requisite beverages from coffee and tea to smoothies, herbal teas, and fresh squeezed juice combinations.

It is possible to limit the breakfast menu to perhaps 30 or 40 items, but many menu planners feel the need to offer more.

## Menus for Various Meals and Occasions

Many operations have one menu that covers choices for the three meals of the day, with the times of availability included. Other operations have separate menus for each meal, and still others have separate menus for special occasions and parties. The types and number of menus needed will vary from one operation to another.

Each operation must rely on patron preference, costs, and operational goals to dictate what items should be included on each menu. The following section should serve as a guide to meeting general menu requirements.

## Breakfast Menus

Some breakfast menus are printed on the regular menu, while others appear as separate menus. An example of a breakfast menu is shown in **Exhibit 3.9.** A California menu that offers breakfast during all hours of operation may list the breakfast on a side panel or list food items in chronological order. Some have breakfast items on the back of a placemat. A children's breakfast menu may be offered in some units.

Both à la carte items and table d'hôte breakfasts should be on a breakfast menu. Table d'hôte offerings should list a *continental breakfast* that includes a juice (usually orange), a bread item (usually a sweet roll, muffin or bagel, but a croissant is common), and a beverage (usually coffee or tea). It should also list heavier breakfasts. A juice or fruit may or may not be included with these. Meat, eggs, or other main dishes will be accompanied by toast, hot breads, biscuits or rolls, and perhaps hashed brown potatoes or grits (in the South). Hot cakes, French toast, or waffles are served sometimes with bacon, ham, sausage, or even an egg. A beverage is often included with these breakfasts.

The offerings of table d'hôte breakfasts should be balanced. A familiar one includes eggs fixed any style and priced for one or two eggs. Omelets or eggs with bacon, ham, or sausage are other offerings. Egg-white or Egg Beaters omelets are common. Meats alone may also be offered. A pancake and a waffle breakfast offering is usual. Sometimes a hash main dish with or without poached eggs, steak plain or with eggs, or other main dishes may be on table d'hôte listings. Items such as Huevos Rancheros, Eggs Florentine, breakfast burritos, or other occasion foods can be included, depending on the type of operation. Specials, such as a low-calorie breakfast, a steak-and-egg breakfast, a high-protein breakfast, or a child's breakfast may be offered. It is important to also have something low-calorie, low-fat, low-cholesterol, or low-salt that guests might want. Tomatoes or fruit may substitute for starch.

Table d'hôte and similar breakfasts that bring in a higher check average should be in the most prominent place on the menu. These higher-income items should also be given as effective a presentation as possible, with large type, bracketing, and effective description.

Numbering breakfasts on a large menu makes them easier to order, both for the patron and the server. Specialties, such as a variety of syrups, jellies, and jams, may serve to encourage choices of desirable menu items.

The menu order for breakfast items is usually as follows:

1. Fruits and juices
2. Breakfast grains
3. Eggs alone or combined with something else
4. Omelets
5. Meat and other main dishes
6. Pancakes, waffles, and French toast
7. Toast, rolls, and hot breads
8. Beverages

**Exhibit 3.9**
Breakfast Menu: Heartland Café

Delivered 11 am - 1 pm M-F • 11 am - 2 pm Sa, Su, Holidays

## Fresh Juices (8 oz & 12oz)

Orange Juice, Grapefruit, Apple Cider, Pink Lemonade .......... **3.50/4.00**
Carrot Juice, Carrot, Beet & Celery **4.00/4.50**

## Bottled Juice

Black Cherry, Cranberry Nectar, Grape, Mango Nectar (organic), Morning Blend, Papaya Nectar, Peach (organic), Pear (organic), Pineapple, Pineapple Coconut, Pomegranate, Very Veggie (organic) ....... **3.50/4.00**

## Milk (8 oz & 12oz)

**Soy Milk** vanilla/plain ......... **3.00/3.50**
**Rice Dream** vanilla/plain ....... **3.00/3.50**
**Cow's Milk** ............... **2.50/3.00**

## Fresh Fruits & Yogurt

**Bowl of Sliced Bananas** ......... **3.25**
with yogurt ...................... 4.25
**Bowl of Yogurt** .............. **3.25**
bananas, raisins & seeds ........... 4.00
bananas and granola .............. 5.00
mixed fruit ...................... 5.50

## Breakfast Grains

**Heartland Granola** with milk ......... **5.00**
bananas ........................ 5.50
yogurt .......................... 6.00
bananas or mixed fruit with yogurt ..... 6.50
**Steel Cut Oats** with milk .......... **4.50**
bananas ......................... 5.00
**Hot Grits** butter & maple syrup ...... **4.50**
bananas ........................ 5.00

## Breads

we use organic cornmeal, wheat & white flour

Our famous Cornbread .............. 1.50
Wheat, White or Pita Bread ............ 1.50
Grilled Cornbread, Muffins or English Muffin 1.75
Whole Wheat Biscuits (two, breakfast only) . 2.00
Big Cinnamon Roll .................. 2.25
Corn Tortillas (4) or Flour Tortilla (2) .... 1.25

## Eggs & Omelettes

Served with choice of wheat, white, cornbread, biscuit or English muffin & choice of rice, grits or potatoes

**Two Farm Fresh Eggs**
Cooked to order and served with breakfast potatoes, rice, grits, or oatmeal ..... **5.00**
turkey ham, bacon or sausage ........ 7.00
**Plain Omelet** ............. **5.00**
**Cheese Omelet** with choice of Monterrey Jack, Cheddar, Feta, Swiss, Cream or Soy **6.00**
**Sautéed Mushroom & Cheese Omelet** .......... **6.50**
**Honeyed Apples, Raisins & Cheese Omelet** .......... **7.00**
**Peach & Cheddar Omelet** ..... **7.00**
**Kate's Omelet** with sautéed spinach and cream cheese ............ **7.00**
**Mexican Omelet** with sautéed mushrooms, green peppers, onions, cheese and salsa ............. **7.50**
**Chili, Avocado & Cheese Omelet 8.00**
**Everything Omelet** with choice of bacon, ham or sausage with broccoli, mushrooms, green and red pepper, onions, spinach, tomato and cheese ... **9.00**
**Veggie Omelet** with spinach, mushroom, green and red peppers, broccoli, onions, tomato and cheese ............ **7.50**
**Make Your Own Omelet** egg only. **5.00**
add cheese ..................... 1.00
add avacado .................... 2.00
any vegetable ...................... .50

## Sat & Sun Only

**Eggs Benedict** Two fresh eggs, Hollandaise sauce & turkey ham on an English muffin with breakfast potatoes ............ **9.00**
**Eggs Florentine** Two fresh eggs, Hollandaise sauce & spinach on an English muffin with breakfast potatoes ...... **9.00**

## Pancakes & French Toast

Served with Wisconsin pure maple syrup

**Whole Wheat or Buckwheat (vegan) Pancakes** .................... 2 @ 5.00 • 3 @ 6.00
**Blueberry, Peach, Banana, Strawberry or Mixed Berry Cakes** 2 @ 6.00 • 3 @ 7.00
**Nutty Cakes** .................. 2 @ 6.25 • 3 @ 7.25
**Fruity Nutty Cakes** .. 2 @ 6.50 • 3 @ 7.50
**French Toast** organic wheat or white .. 5.50
**Fruity or Nutty French Toast** ..... 6.50

## Breakfast Specialties

**Heartland Buffalo Breakfast Plate**
Grilled slices of Buffalo Roast, grilled peppers & onions, eggs or scrambled tofu, potatoes or beans and choice of bread .......... **9.00**
**Heartland Panhandler**
Choice of eggs or tofu (or both) scrambled with potatoes, mushrooms and green onions, bread ....... **6.75**
cheese .................... 7.75
meat add .................. 2.00
**Veggie Panhandler** Choice of eggs or tofu (or both) scrambled with spinach, broccoli, potatoes, mushrooms & green onions, bread .............. **7.50**
cheese .................... 8.50
meat add .................. 2.00
**Huevos Rancheros** Two eggs cooked any style served with beans of the day, a scoop of brown rice, corn tortillas and a side of salsa .............. **6.00**
with cheese ................ 7.00
**Breakfast Burrito** Eggs, refried beans, avocado, jack cheese, lettuce & tomato, warm tortilla, potatoes or rice .... **7.00**
turkey ham, bacon, sausage ..... 9.00
**Scrambled Organic Tofu** Seasoned with tamari and sesame seeds, sautéed with green onions, served with a scoop of brown rice and cornbread ...... **7.00**
**Heartland Breakfast Combo** One pancake, two eggs any style and choice of turkey ham, bacon or sausage ... **7.00**
**Heartland Breakfast Sandwich**
Fried egg, turkey ham, bacon or sausage & melted cheese on an English muffin ............ **5.00**
plus potatoes or rice .......... 6.00
**Hot Biscuits, Gravy & Grits**
Heartland wheat biscuits, grits, and choice of gravy .......... **6.00**

Menu and prices subject to change

## Breakfast Side Orders

Cup of Gomashio (sesame/salt) by request

|  |  |
|---|---|
| Turkey or Veggie Sausage Gravy | 2.00 |
| Mushroom Gravy | 1.00 |
| Egg a la carte | 1.25 |
| Grilled Buffalo Roast | 4.00 |
| Veggie or Turkey Sausage | 2.25 |
| Turkey Bacon or Turkey Ham | 2.25 |
| Scrambled or Cubed Tofu | 3.25 |
| Brown Rice or Beans of the Day | 2.00 |
| Home Fried Potatoes | 2.00 |
| Cheese | 1.00 |
| Avocado Slices | 3.00 |
| Cup of Salsa | 1.00 |
| Yogurt, cup | 2.00 |
| Cup of Fruit | 3.00 |

**VIEW OUR MENU ONLINE AT**
**www.heartlandcafe.com**

**GIFTS & BASICS IN THE HEARTLAND GENERAL STORE ARE AVAILABLE BY DELIVERY TOO!**

*Courtesy of Heartland Café, Chicago, IL*

Side orders must be placed in available space. On the à la carte menu, items should be grouped together in the same order.

Breakfast menus should cover less space than the other meal menus and usually should have larger type, because people are not yet awake. Do not list items only as "juices" or "cereals," but list each offering separately. Also list essential information, such as how long breakfast is served and special breakfast facilities.

Special breakfasts may have to be catered. A wedding breakfast may start with champagne, gin fizzes, a fruit punch, or juices. A fresh fruit cup is often served as the first course. If not, then a fruit salad is appropriate. A typical main dish is served with a high-quality sweet roll. Eggs Benedict would be suitable for the early party-type breakfast, or eggs with sausage, ham, or bacon. An omelet of some type would also be a good choice for this type of breakfast. If the affair is held late in the morning, the main dish can be creamed chicken or seafood crepes or smoked salmon. A beverage choice is offered. If wine is served at the table, it is usually a chilled white or a semi-dry blush.

Buffet breakfasts are popular. Some may be offered to allow guests to quickly obtain what they want and leave. Many people at conventions or meetings are in a hurry and may patronize the foodservice operation if they feel they can get what they want quickly. Some who usually skip breakfast may still be enticed to come in and get a quickly served buffet continental breakfast.

A buffet breakfast for more leisurely dining may be much more elaborate and feature a wide choice of juices and fruits; cold items, such as cheese, sliced baked ham, lox and bagels, even salads; and hot dishes, such as scrambled eggs, assorted breakfast meats, pepper steaks, hash, and different kinds of omelets. Omelet, crepe, and waffle *action stations* produce customized food and serve to entertain patrons. Side dishes, such as hashed brown potatoes or grits, can be included. Assorted hot breads and sweet rolls are offered, along with a beverage choice. The drink is often poured at the table by servers, the guests selecting the other foods they wish at the buffet. On the most elaborate buffets, a dessert may be offered. Depending on the occasion, champagne or alcoholic beverages may be provided. Wine service may be offered with a white wine (not completely dry) or a blush.

A *hunt breakfast* is an elaborate buffet that may include broiled lamb chops, steaks, roast beef, grilled pork chops, pheasant, hare, venison, or other items. A hunt breakfast originated as a meal before or after a hunt and was intended for hearty eaters leading a vigorous life—they had to eat that way.

*A chuck wagon breakfast* should feature sourdough hot cakes, steaks, eggs, hashed brown potatoes, and perhaps freshly cooked doughnuts. Grits or biscuits can be substituted for the breads. (A chuck wagon is the meal wagon that was used to feed cowboys when they were away from the ranch.)

*A family-style breakfast* is one in which the food is brought to the table and guests serve themselves. Camps or bed-and-breakfasts might use this style to create a casual, warm atmosphere for patrons. This method is also used by restaurants that appeal to families and groups.

Group breakfasts should be planned carefully. Eggs and other breakfast items cool rapidly and can go "green" if held too long. Do not attempt difficult egg preparations, such as omelets or Eggs Benedict, unless they can be prepared and served properly. Toast is difficult to serve because it gets cold and chewy quickly. Hot breads, such as biscuits, muffins, and cinnamon rolls, are easier to handle. American service is usually used, but Russian service is sometimes also used (see Chapter 12).

## Brunch Menus

A brunch menu should combine items usually found on breakfast and lunch menus and provide for substantial meals. (See **Exhibit 3.10.**) A fruit juice or fruit should be offered. The main dish should be substantial—omelets; crepes; a soufflé; a small steak with hashed brown potatoes; chicken livers and bacon; or a mixed grill of lamb chop; sausage, bacon, grilled tomato slice, and potato. Hot breads and a beverage choice should be offered. A fruit salad or grilled vegetables are popular. Shrimp and grits, breakfast burritos, pasta, and regional specialties may also appear.

## Luncheon Menus

Luncheon menus may contain a wide assortment of foods, from complete table d'hôte meals to snacks. Offer a wide number of à la carte items with combinations, such as a sandwich and a beverage, or a cup of soup, salad, and dessert with beverage. A few casserole dishes can be offered. Items such as sandwiches, salads, soups, and fountain products can be stressed. A lunch menu can more easily feature economical purchases than the dinner menu.

Many units have modestly priced items and attempt to cover costs with volume and fast turnover. Occasion foods—specialty items created to enhance a special occasion—may be profitable, if especially slanted to the trade. Executives, expense account patrons, and tourists may wish more elaborate menus. Thus, a club, better hotel, or fine restaurant that they patronize may feature a higher-priced list for lunch. Alcoholic beverages may or may not be offered.

Lunch menus should have permanent à la carte offerings on the cover, but may also present daily offerings. The permanent menu will offer sandwiches, salads, fountain items, and desserts. Flexible menus are more typical of lunch than any other meal. Inserts or table displays may be used to call attention to specials. Specials frequently are created to make use of food before its expiration date.

Lunches for groups usually are complete meals. Clubs or organizations may meet at lunchtime, and a main dish with vegetables, salad, dessert, and beverage will be included. A first course may be added for a more elaborate luncheon. Party or occasion foods may be offered, but since most diners have little time, the menu must allow them to eat quickly, have time for scheduled events, and get back to work. However, if the luncheon is to last for a longer period and features an important speaker, the group may want to have more luxury-type food. The foods should fit the occasion and the particular group.

## Afternoon Menus/Coffee and Tea House

Many operations have little or no business in the afternoon following lunch. To bring in customers, a different menu can be designed with specials to catch the afternoon and shopper trade. These menus should appeal to people most likely to eat out in the late afternoon—retirees, shoppers, entrepreneurs, tourists, students, and homemakers. The foods should be snack-type with considerable occasion appeal, different from the usual menu. Thus, after the lunch-hour rush, an operation might put on a special menu with small sandwiches, desserts, fountain items, smooth-

**Exhibit 3.10**
Brunch Menu: Deluxe Restaurant

Sunday Brunch

**Classic Caesar Salad**
crisp romaine lettuce, herb croutons, shaved parmesan & classic Caesar dressing Sm.5.5 Lg.7.5

**Deluxe Salad**
mixed greens with Italian blue cheese, d'Anjou pear, and toasted chili pecans, in roasted shallot vinaigrette Sm. 5.5 Lg. 7.5

**Mediterranean Salad**
mixed greens tossed with kalamata olives, feta cheese, shaved red onion, tomato, cucumber & pine nuts in Aegean dressing 8
Add Grilled Free-Range Chicken to Any Salad 6

**Maine Lobster Omelet**
three local brown eggs filled with select lobster meat, wild mushrooms & French brie* 14

**Tomato, Sweet Basil & Feta Cheese Omelet**
fresh tomato, chiffonade of fresh basil and feta crumbles in a three egg omelet* 8.5

**Lump Crab and Chevre Cheese Omelet**
fresh lump crabmeat, asparagus tips, local goat cheese with herb aioli* 11.5

**Eggs Benedict**
two poached all natural eggs on a toasted English muffin with Canadian bacon and hollandaise* 8.5

**Huevos Rancheros**
grilled flour tortillas topped with black beans, salsa fresca, poached all natural eggs, and Vermont white cheddar* 8.5

**Eggs Florentine**
two poached eggs on a toasted English muffin with sautéed spinach and hollandaise* 8.5

**Hickory Smoked Salmon**
atop a split country biscuit with Vermont white cheddar, poached eggs & grain mustard cream sauce* 9

**Creamy Blue Crab Artichoke Sauce**
over two poached all natural eggs on a toasted English muffin* 9.5

**Pecan Swirl French toast**
vanilla cinnamon scented sweet pecan swirl bread, served with pure Canadian maple syrup* 8.5

**Smokehouse Omelet**
apple-wood smoked bacon, smoked cheddar cheese, folded in 3 egg omelet * 8.5

**Blueberry Cornmeal Pancakes**
blueberry enriched sweet pancakes served with pure Canadian maple syrup* 8.5

**Vegetable Omelet**
ribbons of carrot, zucchini and red onion; Vermont white cheddar, and herb aioli* 8

**Fresh Catch**
today's selection served with garden-herb aioli, scrambled egg, your choice of side and fresh fruit 14

**Kobé Beef Burger**
served with apple-wood smoked bacon, smoked cheddar and hand cut gold potato frittes 12

**Red Corn Tacos**
filled with beef picadillo, shredded lettuce, onions, smoked cheddar, zesty black beans and salsa fresca 10

**Buttermilk Battered Crispy Grouper Nuggets**
accompanied by scrambled eggs, cheese grits and dill remoulade 12

**Granola, Fresh Fruit & Yogurt**
Deluxe granola + vanilla flavored yogurt with seasonal fruit 6

**served with* either **Stone Ground Grits** or **Herb Roasted Potatoes, Fresh Fruit, and fresh baked muffin**
*add 1.5 for egg white omelets ***Sunday Brunch Served from 10:30am until 2:00pm***

*Courtesy of Deluxe Restaurant, Wilmington, NC; Chef Keith Rhodes*

ies, fruit plates, cookies, and pastries. Gourmet coffees, teas, and fresh-squeezed juices often are featured with extensive choices.

Many people can be induced to eat dinner early if a lower-priced menu is offered. A variety of items should be included, along with one or two light desserts. *Early-bird* menus are often smaller-portioned, abbreviated versions of the regular menu.

## Dinner Menus

The dinner menu usually has more specialty items than others and attempts to feature more occasion foods. The menu must be carefully directed to patrons.

The typical American meal plan of a salad, main dish, potato, vegetable, dessert, and beverage is probably most appropriate for the family market, though some variation of it likely suits all operations. (See **Exhibits 3.11** and **3.12.**)

This traditional family dinner is as well liked as it once was, but families often cannot eat together, so meals like pizza or hamburgers might be in order. However, for the usual family dinner meal, a soup, salad, or other appetizer may be served as a first course. Meats such as steaks or roasts are popular, but other kinds of meat, chicken and other poultry, fish, and shellfish should also appear. Game and other meat, pastas, and some specialty foods may also be offered.

Families who often dine out know exactly what they want from any menu. Comfort food, such as mashed potatoes, macaroni and cheese, and brownies, are often appealing.

Menus are also needed that feature value-priced items for those who want convenience rather than luxury. Budget-conscious diners may not have much to spend but will want enough to eat. Some operations find this trade a satisfactory source of revenue. Value menus or bundling is a feature of many quick-service restaurants. When creating these menus, the restaurant puts several items together to make a meal and prices the total lower than the cost of those items purchased separately.

In attempting to develop memorable dinner menus, it is important to be aware that service and decor are as essential as food. Complete follow-through on *all* details of an idea is a requirement. Too many menus attempt to create a food *atmosphere* on the menu only to have the rest of the performance a dismal failure, or vice versa. The operation must be constantly watched to see that there is complete follow-through. If a menu features Mexican, Greek, or other ethnic foods, the cuisine should be authentic. Research is necessary to verify authenticity, yet modification may also be needed to suit the nonauthentic palate. A very hot Indian curry served with Bombay duck and all the side dishes may be delicious to someone who knows this food, but it could be a disappointment to someone who isn't familiar with it.

International and ethnic fusion foods have become popular and help give menus interest and variety. If offered, they need to be made correctly with high-quality ingredients. An important consideration in featuring many ethnic foods is that they usually do not require the most expensive ingredients. Thus, they may be more profitable to serve than typical American foods. The fact that these foods are not normally a part of our diet makes them good occasion foods. Novelty can also be achieved by unique service.

It is becoming more and more common on table d'hôte meals to omit appetizers and desserts and have these selected from a special menu. **Exhibit 3.13** shows an attractive dessert menu. Furthermore, many operations today serve the salad as a first course, making it fairly substantial, and then omit the salad with the main course.

**Exhibit 3.11**
Dinner Menu: Shaw's Crab House

All Oysters
— Shucked —
To Order

CRAB HOUSE

Fresh Seafood
—in the Heart—
of Chicago!

## TODAY'S HALF SHELL OYSTERS

| | doz | ½ doz |
|---|---|---|
| Sunny sides (*Crassostrea virginica*), Prince Edward Island | 23.95 | 11.95 |
| Fanny Bay (*Crassostrea gigas*), Bayne Sound, British Columbia | 23.95 | 11.95 |
| Tatamagouche (*Crassostrea virginica*), Tatamagouche Bay, Nova Scotia | 23.95 | 11.95 |
| Caraquet (*Crassostrea virginica*), Caraquet Bay, New Brunswick | 23.95 | 11.95 |
| Wellfleet (*Crassostrea virginica*), Cape Cod, Massachusetts | 23.95 | 11.95 |
| Raspberry Point (*Crassostrea virginica*), Prince Edward Island | 23.95 | 11.95 |
| **Oyster Sampler** | 23.95 | 11.95 |

## APPETIZERS

Cajun Popcorn Shrimp .......... 8.95
Coconut Shrimp Orange Marmalade .......... 8.95
Crispy Calamari Cocktail Sauce .......... 8.95
Baked Clams Casino .......... 8.95
Shrimp Cocktail .......... 10.95
New Zealand Green Lip Mussels .......... 11.95
Oysters Rockefeller .......... 11.95

## SALADS

Cole Slaw .......... 2.95
Shaw's Caesar Salad .......... 4.95
House Mixed Greens .......... 4.95
Double Wedge Blue Cheese Dressing .......... 6.95
Chopped Harvest Salad .......... 7.95
Greens, Dried Cherries, Apple, Pear & Sunflower Seeds

## SOUPS

| | CUP | BOWL |
|---|---|---|
| Seafood Gumbo | 3.95 | 4.95 |
| New England Clam Chowder | 3.95 | 4.95 |
| Lobster Bisque | 3.95 | 4.95 |

## APPETIZER PLATTERS

Hot Appetizer Combination .... (Per Person) 11.95
Crab Cake, Crispy Calamari & Popcorn Shrimp
Cold Appetizer Combination ... (Per Person) 14.95
Oysters, Shrimp & King Crab

Platters Serve 4-5 People

Grand Sushi Platter .......... 64.95
Lobster, Spicy Tuna & Mediterranean Maki; BBQ Eel & Shrimp Nigiri; Tuna, Salmon & Yellowtail Sashimi
Grand Hot Shellfish Platter .......... 69.95
Crab Cakes, Calamari, Popcorn Shrimp, Coconut Shrimp, Oysters Casino & Clams Casino
Grand Cold Shellfish Platter .......... 79.95
Maine Lobster, Oysters, King Crab & Shrimp

## SEAFOOD AT SHAWS

Commitment to quality is a source of pride at Shaw's. We fly in seasonal seafood daily from the Atlantic, Gulf and Pacific Coasts. To preserve flavor, fish is filleted on premise as needed, never in advance. Shaw's actively supports the management of America's fisheries to preserve fish and water quality for future generations.

## CRAB APPETIZERS

Crab and Artichoke Dip .......... 10.95
Maryland Style Crab Cake .......... 12.95
Chilled Alaskan King Crab Bites .......... 16.95

## SUSHI BAR

King Crab California UN-ROLLED .......... 11.95
Tuna and Avocado Tartare Yuzu Juice .......... 12.95
Shaw's Charred Sashimi Tuna .......... 13.95
Tuna Sampler Nigiri, Maki & Sashimi .......... 24.95

## NIGIRI - SASHIMI

Per Piece

Shrimp (Ebi) .......... 2.00
Salmon (Sake) .......... 2.50
BBQ Eel (Unagi) .......... 3.00
Tuna (Maguro) .......... 3.50
Yellowtail (Hamachi) .......... 3.50
King Crab (Kani) .......... 4.00
Fatty Tuna (Toro) .......... 9.95
Nigiri Combination (5 pieces) .......... 12.95
Sashimi Combo Tuna, Salmon & Yellowtail .......... 17.95

## MAKI SUSHI

Spicy Salmon .......... 7.95
Philadelphia Roll .......... 7.95
Spicy Tuna .......... 8.95
BBQ Eel & Avocado .......... 8.95
Shrimp Tempura .......... 8.95
Salmon & Snow Crab .......... 8.95
Lobster, Avocado & Cucumber .......... 9.95
Blue Crab & Shrimp .......... 9.95
Acapulco Roll .......... 9.95
Futomaki .......... 9.95
King Crab California .......... 10.95
Spicy Shrimp & King Crab .......... 12.95
Rainbow Roll .......... 13.95
Mediterranean Roll .......... 14.95
Maki Combination (serves 2) .......... 16.95

18% Gratuity will be
— Added —
To Parties of 6 or Larger

Florida Stone Crab
—Now In—
Season

*Courtesy of Lettuce Entertain You Enterprises Inc.*

**Exhibit 3.11**
*(Continued)*

All Seafood is
— Subject to Season, Weather —
And Fishing Conditions

CRAB HOUSE

Shaw's is Proud
— To Serve —
Trident King Crab Legs

## FRESH FISH SPECIALS

**Pan Seared Lake Superior Whitefish** Mixed Greens, Mashed Potatoes ........ 19.95
**Sauteed Lake Erie Yellow Perch** Lemon Butter, Cole Slaw ........ 19.95
**Parmesan Crusted George's Bank Haddock** Sauteed Spinach, Lemon Butter ........ 22.95
**Sauteed Lake Erie Walleye** Horseradish Crust, Green Beans, Herb Tomato Butter ........ 22.95
**Grilled North Carolina Swordfish** Bacon & Garlic Crust, Mushroom Sauce, Sauteed Oyster Mushrooms .. 24.95
**Sauteed Florida Grouper** Pumpkin Seed Crust, Delicata Squash Puree, Lemon Butter Sauce ..... 27.95
**Roasted Alaskan King Salmon** Parsnip Puree, Autumn Vegetable Saute ........ 28.95
**Pan Seared Alaskan Halibut** Walnut and Baby Beet Salad, Orange Dressing ........ 28.95
**Grilled Gulf Yellowfin Tuna** Crispy Noodles, Peanuts, Ginger Soy ........ 29.95
**Sauteed Alaskan Halibut Steak** Herb Butter, Matchstick Fries ........ 30.95

## SHAW'S SEASONAL CRAB

**ALASKAN RED KING CRAB LEGS – Our Specialty – (1 1/2 lb)** ........ 48.95
**Maryland Style Crab Cakes** Blue Crab Meat Hand Picked Exclusively for Shaw's ........ 23.95
**Jumbo Lump Crab Cake** Hand Picked Jumbo Lump Blue Crab ........ 26.95
**Chilled WholeWashington Dungeness Crab** Mustard Mayonnaise ........ 39.95
***Fresh* Florida Stone Crab Claws (1lb)** ........ **Large (48.95) Jumbo (58.95)**
**Seasonal Crab Combination** Alaskan Red King Crab Legs, Crab Cake, Dungeness Crab ...... 52.95

## LOBSTER

**WHOLE MAINE LOBSTER – Live From Our Tank** ........ **1.5 & 2 lbs.**
**Steamed or Broiled Maine Lobster** Drawn Butter ........ **(per lb)** 25.95
**Blue Crab Stuffed Maine Lobster** ........ **(per lb)** 33.95
**Lobster, Brie & Penne Pasta Au Gratin** ........ 23.95
**Twin Australian Lobster Tails (2) 6 oz** Drawn Butter ........ 51.95

## PRIME STEAKS

**Filet Mignon (8oz)** Bearnaise Sauce or Horseradish Cream ........ 29.95
**Filet Mignon (12oz)** Bearnaise Sauce or Horseradish Cream ........ 39.95
**Horseradish Crusted Filet Mignon (8oz)** ........ 32.95
**Blue Cheese Crusted Filet Mignon (8oz)** ........ 33.95
**Oscar Style Filet Mignon (8oz)** King Crab, Asparagus, Bearnaise Sauce ........ 37.95
**New York Strip (14oz)** Horseradish Cream ........ 39.95

## SURF & SURF ▪ COMBINATIONS ▪ SURF & TURF

**Shaw's Seafood Platter** Garlic Shrimp, Sea Scallops, Crab Cake ........ 23.95
**Club Room Combination** 10oz New York Strip, Sea Scallops, Garlic Shrimp ........ 29.95
**Shaw's Surf & Turf** 8oz Filet Mignon & Australian Lobster Tail ........ 51.95
**The Signature** 8oz Filet Mignon & Alaskan Red King Crab Legs ........ 54.95
**Shaw's Surf & Surf** 6oz Australian Lobster Tail & Alaskan Red King Crab Legs ........ 55.95

## CHICKEN, SHRIMP & SCALLOPS

**Parmesan Crusted Chicken** Sauteed Spinach, Lemon Butter ........ 14.95
**French Fried Shrimp** Hand Breaded, French Fries, Cocktail Sauce ........ 19.95
**Griddled Garlic Shrimp** Garlic Butter ........ 19.95
**Sauteed Bay of Fundy Sea Scallops** Shaw's Signature Rice, Sauteed Spinach, Lemon or Garlic Butter . 24.95

## POTATOES

**Baked Potato** ........ 5.95
**Mashed Potatoes** ........ 6.95
**Hashed Browns with or without Onions** .... 6.95
**Crab House Fried Potatoes** ........ 6.95
**Potatoes Au Gratin** ........ 6.95

## VEGETABLES

**Sauteed Spinach with Garlic** ........ 4.95
**Shaw's Creamed Spinach** ........ 5.95
**Steamed Green Beans, Lemon Butter & Tomatoes** .. 5.95
**SteamedBroccoli** ........ 5.95
**Steamed Asparagus** ........ 6.95

Alert Server for
— Dietary Restrictions from—
Food Allergy or Intolerance

Blue Crab Meat
— Hand Picked—
From Louisiana

Courtesy of Lettuce Entertain You Enterprises Inc.

**Exhibit 3.12** Dinner Menu: Ambria

*Ambria*

**First Plates**

Ibérico Plate of Pata Negra Salchichon and Chorizo
26.00

Duck Country Pâté with House-Cured Cornichons
Roma Apple Sauce, Missouri Black Walnuts, Country Toast
16.00

Atlantic Blue Fin Tuna Tartare, Porcelain Garlic Aioli
Salad of Haricot Vert and Breakfast Radishes, Pimentón d'Espelette
15.00

Pemaquid Oysters—*Casino Style*
Melted Leeks, Spanish Pimentón, Applewood Smoked Bacon
16.00

Tasmanian Red Sweet Crab Salad, Green Apple Emulsion
Chufa Nut Infusion, Szechuan Pepper, Saffron Rice Crisp
17.00

Crispy Veal Sweetbreads, Black Trumpet Mushrooms, Quail Egg
Lentils du Pays, House Cured Pork Belly, Roasted Veal Jus
17.00

Dayboat Sea Scallops 'à *la Plancha*', Butternut Squash Puree
Piperada, Herb Salad, Jamon Serrano Vinaigrette
18.00

Nantucket Bay Scallops, Chanterelle Mushrooms
Manchego Cheese, Frisée, Red Mustard Greens
22.00

**Salad**

Organic Mesclun Salad, Garden Herbs, Autumn Vegetable Crudités
Garlic Croutons, Dijon Mustard Vinaigrette
12.00

**Soups**

Cépe Mushroom Velouté with Butternut Squash and Duck Confit, Chestnut Cream
10.00

*Please alert your server if you have special dietary restrictions due to a food allergy or intolerance.*
Out of courtesy for all our guests, please refrain from cellular phone use in the dining room.

*Courtesy of Lettuce Entertain You Enterprises Inc.*

**Exhibit 3.12** *(Continued)*

**Entrées**

John Dory *'a la Plancha,' Avalanche Farms* Celery Root
Baby Leeks, Apple Cider Gastrique with Pine Nuts
41.00

Slow Roasted Tasmanian Salmon, Crushed French Fingerling Potatoes
Chanterelle Mushrooms, Salsa Verde
38.00

Atlantic Blue Fin Tuna, Marcona Almond Mojo, Quinoa
Malabar Spinach, Leeks, Smoked Date Purée
39.00

Maine Lobster Roasted in the Shell, Black Trumpet Mushrooms
Judión White Beans, Tomato Confit, Shellfish Emulsion
41.00

Loin of Lamb, Chickpea Crêpe with Braised Lamb Shank
Baby Carrots, Basil Purée, Roasted Piquillo Peppers, Lamb Jus
43.00

Roasted Center Cut Milk-Fed Veal Chop, Chanterelle Mushroom
Marmalade *Snug Haven* Spinach, Boudin Blanc, Veal Madeira Sauce
45.00

Muscovy Duck Breast, Basmati, Black Forbidden and Wild Rice
Duck Leg Confit, Butternut Squash, Quince, Duck Jus
38.00

Pecan Crusted Venison Loin, *Tipi Farms* Red Cabbage,
Cardamon Carrot Purée, Baby Turnips, Sauce Grand Veneur
40.00

*For Two to Share, Carved Tableside*
Bone-In 30 oz Prime Beef Ribeye, French Horn Mushrooms, Local Spinach
Piquillo Pepper, Soufflé Potatoes, Rioja Red Wine Sauce
92.00

36 Hour Braised Milk Fed Veal Shank, Gremolata, Chanterelle Mushrooms
Autumn Compote of Orchard Fruits, Roasted Parsnip Purée, Veal Braising Jus
76.00

*Chef/Proprietor: Gabino Sotelino* *Chef de Cuisine: Christian Eckmann*
*Ambria supports the Green City Market and sustainable farming*

**Exhibit 3.13**
Dessert Menu: Everest

*LES DESSERTS*

***CHESTNUT***

*Milk Chocolate, Chestnut Fondant Vermicelli,*
*Mandarin Coulis*
** Pappy Van Winkle's 20 Year Bourbon 28*

***CARROT***

*Carrot Cake, Fromage Blanc Glacé,*
*Walnut Nougatine, Carrot Coulis*
** NV Ceretto Moscato d'Asti 12*

***PUMPKIN***

*Truffelbert Farm Hazelnut Tart,*
*Pumpkin Parfait and Parsnip Cream*
** 2003 Oremus Late Harvest Tokaji 16*

***CLASSIC ALSACE VACHERIN***

*Pear and Chocolate Glacés, Vanilla Sauce*
** Merryvale Antigua Dessert Wine 20*

** Recommendation for Dessert Wine by the Glass*

*LES DESSERTS*

***APPLE BEIGNET***

*Granny Smith Apple Beignet*
*Rose Hip Coulis, Kirsch Cheese Ice Cream*
** 1998 Klipfel SGN Gewürztraminer 22*

***CHOCOLATE***

*Fantasy of Chocolate, Selection of Five Chocolate Tastes,*
** Sandeman Founders Reserve Port 14*

***CHICORY***

*Chocolate Caramel Gateau,*
*Chicoree Leroux Ice Cream, Caramel Jus*
** Grahams 20 Year Tawny Port 20*

***BANANA***

*Terrine of Chocolate Crèpe and*
*Banana Brulée, Maple Cap Mushroom Syrup*
** 2000 Château Rayne-Vigneau Sauternes 22*

*Desserts 18*
*Complimented with Petits Fours and Mignardises Service*

*Courtesy of Lettuce Entertain You Enterprises Inc.*

## Formal Dinner Menus

Very few formal dinners are served today that follow the traditional French style of three settings with a progression of courses for each setting. People now find it difficult to eat this much food. Even a more simple formal meal can be exhausting unless properly planned. Portions for a formal dinner should be adequate but restrained. The food should be selected to give a progression of flavor sensations, avoiding any heavy sweetness until the very end of the meal to "silence the appetite."

A formal meal should give time for the guests to appreciate the food and service and to converse. The most formal meal today usually does not have more than eight courses.

The first course of a formal meal can be oysters or clams on the half shell, a seafood cocktail of some type, a canapé, or fruit, such as melon or mango. Some may omit this course if cocktails and appetizers are served before the meal. Soup is the next course, and this should be quite light, such as a consommé, miso broth, or a light cream soup. On all courses, garnishes are important and may be the only item accompanying the food.

The fish course comes after the soup, followed by the poultry course. The roast or joint, with perhaps potato and vegetable, is the next course. The salad course is next, followed in turn by cheese and then the dessert course. In some formal meals an ice cream or sherbet (sorbet) may come right after the roast course. The meal may end with demitasse coffee, nuts, mints, and bonbons.

Wines and alcoholic beverages for this type of meal should be selected carefully. Cocktails made from spirits, such as martinis, manhattans, or scotch and sodas, blunt the taste buds. Apertif wines are usually better before meals. Nevertheless, many Americans prefer the former. The trend, however, is toward moderation. Sweet wines and drinks should be avoided, except at the end of the meal, where a sweet dessert wine may be served with the dessert or sweet liqueurs after the dessert. A wine such as a dry white may be served with the first course and with the soup. A light, white dinner wine would typically follow with the fish and poultry. A full-bodied dry red wine is served with the main course. A very good dry red wine may be served with the cheese, followed by a light-bodied but refreshing sweet dessert wine. Port is sometimes served in lieu of dessert or with dark chocolate. The tendency in formal meals is to limit the number of wines. As few as one or two wines may suffice.

A properly planned formal dinner should compare with a symphony or concerto. It should have different movements and themes in the foods and wines. The progression of themes should lead to a taste climax and then gradually recede in intensity as other foods follow. There are places in the meal where the foods should be modest, quiet, and subtle in flavor; and in other places the food and flavor should stand out in a loud crescendo. Each course should lead into the next, and the meal should be a continuity of related taste themes just as a good symphony has continuity of musical themes.

Lightness and delicacy should be the motif of the appetizer and soup. Both should be refreshing and somewhat zestful to whet the appetite. They should be a good introduction to the foods to follow. The fish course should be bland but sufficiently pronounced to give contrast with the courses that preceded it and with the more pronounced flavors in the poultry course that follows. The poultry course should be light and delicate. The roast can be a red meat, and should be the peak of the meal. At this point, flavor values are more pronounced, but excessive richness, sweetness, or sharpness in flavor should be avoided.

The salad should be a relieving course. Its cool crispness and slightly tart flavor should be delicate and possess a distinctive quality, giving a respite from the heavier foods that have gone before. The Italians say the salad should "clean the palate and the teeth." Any oiliness or sharpness in flavor should be avoided. The tangy cheese course that may follow the salad should renew the jaded palate and set it properly for the concluding course—a sweet dessert. This should be light and delicate, and not too sweet. It should end the meal with some finality, but not on a heavy note.

The most formal meals are served without bread or butter. Salt and pepper are not on the table. In fine commercial dining rooms, bread, butter, salt and pepper are often found on the table.

A less formal dinner may have only three to five courses. The first course can be a fruit, juice or seafood cocktail, melon, canapé, oysters, or clams, followed by a soup and then a fish course.

Or there can be either an appetizer or a soup and no fish course. The main course is next, accompanied by a starch and vegetable and, often, a salad. A dessert follows. Rolls and butter are usually also served.

## Evening Menus

Some places operate after dinner hours. This may be to catch those who are out at the theater, a film, an athletic event, or some other attraction, or it may be that there is a good trade potential for night workers, and students. Some operations may attempt to attract business with entertainment that may either be quite elaborate, such as a night club, or very basic, such as a jukebox. Still others may be simple venues that function more as social gathering places with limited menus. It is possible to combine night customers, with separate areas for those who want entertainment and those who only want food.

Many operations remove specials and table d'hôte dinners from the à la carte dinner menu and offer customers a selection of à la carte items on their evening menu. Some may pull the whole dinner menu and have another completely different evening menu emphasizing snack foods, such as sandwiches, pizza slices, ice cream, and pastries. Such a menu may be desirable to reduce the need for a large kitchen staff after dinner hours. In this case, most menu items are available up front where servers can get them. If an operation is near a theater or similar entertainment area, the menu should emphasize good desserts, specialty coffees, and perhaps light supper items. A *graveyard shift* menu generally includes breakfast offerings for workers coming from or going to work the night shift.

A good analysis of the desires of the market and its potential should be made if one hopes to draw a trade that seeks entertainment as well as food and drink. The menu should then be designed to meet the demand for the appropriate kind of entertainment rather than attempting to create a demand. The menu should be supported by appropriate emphasis on props, such as servers' uniforms, dishes and glassware, table decorations (candles, vases, flowers), lighting, and entertainment. **Exhibit 3.14a** illustrates a light-dinner section of a menu. **Exhibit 3.14b** illustrates an entire menu devoted to light offerings.

## Special Occasion Menus

Quite a few hotels, clubs, and restaurants are able to develop a substantial party and catering business. This can be profitable and considerably supplement overall income.

The special occasion menu must be planned in detail to suit the occasion. While there is a general pattern to events such as wedding receptions, buffet dinners, and cocktail parties, each should be characterized by its own special arrangements. Thus, some menus may have to be planned for a special occasion and not used again. Some events will be quite formal, while others will be very informal. The menu should carry out the spirit of the theme. Creativity and ingenuity are required. Color, decor, props, foods, and service should be combined to give correctness and novelty to the occasion.

A menu for a party should be carefully analyzed to see that it does not make excessive demands on servers, dishware, or equipment. The foods selected should be easy to make and to serve. They need to be able to withstand delays while retaining quality and remaining sani-

**Exhibit 3.14a** Lite Dinners Included in Petterino's Theatre District Menu

# Petterino's™

## Appetizers

Chicken Liver Paté en Ramekin, FRESH APPLE SLICES, TOASTED BRIOCHE . . . . . 5.95
Crispy Coconut Jumbo Shrimp, COCKTAIL AND POLYNESIAN SAUCES . . . . . 9.95
Luscious Louie Cocktail, KING CRAB AND JUMBO SHRIMP, AVOCADO–TOMATO SALAD . . . . . 12.95
Asparagus Milanese, CRUNCHY PARMESAN CRUST, LEMON AIOLI . . . . . 6.95
Wild Mushroom Fricassée, TOASTED BRIOCHE . . . . . 7.95
French Fried Calamari, TARTAR & COCKTAIL SAUCES . . . . . 9.95
Shrimp de Jonghe, GARLIC CRUSTED, SHERRY BUTTER SAUCE . . . . . 9.95
Iced Jumbo Gulf Shrimp, COCKTAIL SAUCE OR SAUCE PIQUANTE (HOT!) . . . . . 12.95
Artichoke and Crabmeat Gratin, FRITZEL'S TOASTED CHIPS . . . . . 12.95
All-Lump Crabcake, AVOCADO–TOMATO SALAD, TARTAR SAUCE . . . . . 9.95

## Soups

| | Cup | Bowl |
|---|---|---|
| Tomato Bisque, CROUTON | 2.50 | 3.95 |
| Midwest Corn Chowder, BACON | 2.50 | 3.95 |
| Red Snapper Soup, AU SHERRY | 2.95 | 4.50 |

## Salads

Petterino's Salad, MIXED GREENS, CARROTS, RADISHES, CUCUMBERS . . . . . 5.95
Caesar Salad, REGGIANO PARMESAN . . . . . 6.95
Judge Rizzi Wedge, TOMATO, MAYTAG BLUE CHEESE, BACON, HEARTS OF PALM . . . . . 6.95
Vine Ripened Tomato and Sweet Onion Salad, BASIL, ARTURO'S VINAIGRETTE . . . . . 6.95
The "Kup" Chopped Salad, CORN, AVOCADO, BACON, EGG, OLIVE, BLUE CHEESE AND MORE . . . . . 6.95
Pump Room Spinach Salad, BACON, CHOPPED EGG, RADISHES, GREEN ONIONS, MUSTARD DRESSING . . . . . 6.95

*Caesar • Roquefort • Green Goddess • Dijon Vinaigrette • Lemon-Soy Vinaigrette • Italian Vinaigrette*

## Lite Dinners

Petterino's Chicken Hash, SHIRRED EGGS . . . . . 9.95
Grilled Chicken Sandwich, MAYONNAISE, FRENCH FRIES . . . . . 9.95
Cheeseburger, CHEDDAR, SWISS OR MAYTAG BLUE CHEESE, FRENCH FRIES . . . . . 9.95
Irv Kupcinet Chicken Chopped Salad, CORN, AVOCADO, BACON, EGG, OLIVE, BLUE CHEESE AND MORE 12.95
Chopped Steak, "3 of a Kind," ONIONS, PEPPERS, MUSHROOM SAUCE, FRENCH FRIES . . . . . 14.95
Prime Club Steak, HAND SLICED, TOASTED BRIOCHE, AU JUS, FRENCH FRIES . . . . . 17.95
Pot Roast, ONIONS, CARROTS, MASHED POTATOES . . . . . 16.95

A LETTUCE ENTERTAIN YOU® RESTAURANT

• *For your gift giving* – THE WHOLE YEAR 'ROUND – LETTUCE ENTERTAIN YOU GIFT CARDS •

*18% GRATUITY WILL BE ADDED TO PARTIES OF SIX OR MORE*
FOR YOUR CONVENIENCE, WE ACCEPT ALL MAJOR CREDIT CARDS

DESSERTS BAKED DAILY IN OUR OWN BAKESHOP

*Courtesy of Lettuce Entertain You Enterprises Inc.*

**Exhibit 3.14b**
Lite Fare Menu: Eiffel Tower

# Eiffel Tower Lunch Menu

**Starters**

**Cream of Wild Mushroom**
7.95

**Bib Lettuce à la Française**
8.95

**Salads**

**Tomato, Buffalo Mozzarella, Basil and Pinenuts**
16.95

**Sautéed Shrimp, Mango, Tahitian Style**
19.95

**Grilled French Sandwiches**
-served with small salad and fingerling chips-

**Brie, Golden Delicious Apple, Toasted Walnuts**
14.95

**Classic Ham and Cheese Croque Monsieur**
14.95

**Roasted Chicken, Herbed Cheese, Tomato**
14.95

**Steak, Roquefort, Red Onion Compote**
14.95

**Plats Chauds**

**Maine Lobster, Musnroom Tart**
19.95

**Braised Center of Wild Halibut, Fine Herbs**
19.95

**Filet of Trout, Compote of Tomato, Tarragon**
18.95

**Slow Braised Salmon, Lemon, Horseradish Olive Oil,**
18.95

**Chicken, Asparagus, Potato Lyonnaise Salad**
19.95

**Aged Parmesan Crusted Chicken, Gratin, Légumes**
18.95

**Spicy Moroccan Style Lamb Hamburger**
18.95

**Skirt Steak Roulade, Potato Brick, Bordelaise Sauce**
22.95

*Courtesy of Lettuce Entertain You Enterprises Inc.*

**Exhibit 3.15a** Special Occasion Menu/New Years Eve Pre-Theatre: Everest

*2005*

***New Year's Eve Pre-Theatre Menu***

***Les Preludes***

~

***Maine Lobster, Field Mache Mango and Walnuts***

***Sautéed New York State Foie Gras, Winter Fruit Compote,***
***Alsace Marc de Gewurztraminer***

***Marbre of Pheasant and Partridge, Truffle Coulis***

***Sautéed Casco Bay Sea Scallops, Celeriac Puree, Jus de Poulet***

***Shirred Farm Egg, Osetra Caviar, Alsace Marc de Gewurztraminer***
***(Additional $39.00)***

~

***Salmon Soufflé Auberge de L'll Paul Haeberlin***

***Sautéed Veal Medallion, Wild Mushrooms,***
***Galette of White Corn, Winter Vegetables***

***Beef Tenderloin sautéed Artichoke and Salsifis, Sauce Béarnaise***

~

***Le Pre Dessert***

~

***Composition of Five Everest Chocolate Desserts***

***Classic Madagascar Vanilla Crème Brulee***

***Slow Braised Roasted Pineapple, Fromage Blanc Glace***

***Assortment of Ice Creams and Sorbets***

~

***Petits Fours and Mignardises***

***Pre Theatre Wine Pairing Additional $69 Per Person***

*Courtesy of Lettuce Entertain You Enterprises Inc.*

**Exhibit 3.15b** Special Occasion Menu/ Pre-Theatre-Petterino's

**Pre-Theatre Menu**

**(Three Course Price-Fixed Menu for $27.95)**

**Soup or Salad**
Tomato Bisque
Midwest Corn Chowder
Petterino's Salad MIXED GREENS
Traditional Caesar Salad PARMESAN

**Entrees**
Char-Broiled Whitefish HERB-MARINATED, LEMON
Garlic Crusted Sea Scallops SHERRY BUTTER
Chicken Picatta LEMON-CAPER BUTTER SAUCE
"Classic" Steak Diane FILET MEDALLIONS, DIANE SAUCE ADD $2.00

**Mini Desserts**
Real Chocolate Pudding
Chocolate "Applause" Cake
Petterino's Famous Cheesecake

*Courtesy of Lettuce Entertain You Enterprises Inc.*

tary. They should be suitable for the occasion and the clientele. Most operations that specialize in these functions have a list of offerings that meet these requirements. **Exhibit 3.15a** and **b** illustrates.

## Party Menus

A wide variety of parties may be served by a catering department. These may be as simple as punch for a dance or coffee for a meeting. Others may provide a simple dessert before a fashion show. Others may cater formal luncheons or dinners.

Party menus will vary according to the event, and thus there may be constant change. However, some operations preplan menus and have these available for clients when they call to discuss an event. This helps to standardize production and allows an accurate calculation of costs. If a special menu has to be developed, the individual planning the menu and establishing the price should know the complete costs so that an adequate price can be charged.

Most operations doing party business use standard printed forms for details. (See **Exhibit 3.16.**) The number of copies distributed may vary, and each operation will have its own special procedure. It is important that all departments affected have a coordinated plan and that the system work smoothly. There are many details in planning parties. Party business can be profitable, but it is very demanding of personnel. Good use of forms and procedures and a smooth, efficient system can do much to add to profits and make party business less demanding.

Some foodservices have catering departments that do considerable amounts of business and set up a sales office to handle booking arrangements. They often have preplanned menus with prices for patrons. It may be necessary to quote a tentative price at first and then later, when menu and service costs can be more accurately determined, set a firmer price. Catering menus can be developed for breakfasts, continental breakfasts, buffet breakfasts, luncheons, luncheon buffets, dinners, buffet dinners, receptions, holiday celebrations, cocktail parties, birthdays, weddings, liquor (spirits), and wine lists; prices can be established for shows and for entertainment. **Exhibit 3.17** is an example of a catering menu with choices.

## Tea Menus

There are both high and low teas. A low tea might be simply tea, served perhaps with lemon, milk, and sugar. A more elaborate tea would add a simple dessert item or cookies, mints, bonbons, and nuts; a still more elaborate one might add a frozen dessert and fancy sandwiches. Often a tea is served from a table with the beverages served by an honored host or hostess at one end of the table and coffee served at the opposite end by another honored host. These hosts are usually changed at set times. The placement of the tea table should be considered carefully because traffic can be a problem.

**Exhibit 3.16** Typical Catering Form

**BANQUET PROSPECTUS**

Name of Organization
Address Phone
Nature of Function
Day Date Time
Room Rent
Name of Engager
Address Phone
Responsibility of Party
Price per Person $ Gratuity: Minimum Number Guaranteed: Maximum Attendance:

**FOOD MENU**

**BEVERAGE MENU**
Cash: Charge:
Corkage:
Room for Bar:
Open: Close:
Bartender Charge:
Minimum Charges (4 hrs):
Types of Beverages:

**Staff:**

Table Arrangements: Set Up By: Head:
Flowers: Centerpiece: Checking:
Ticket Table: Blackboard: P.A.: Stage:
Lectern: Screen: Piano: Music:
Cigars: Cigarettes: Platform: Dance Floor:

**REMARKS:**

**COMMENTS BY PARTY:**

**Copies to:**
- ❑ Customer
- ❑ Catering Office
- ❑ Maitre D'
- ❑ Chef
- ❑ Houseman
- ❑ Bar
- ❑ Accounting

**Date Booked:**

BY

A high tea is somewhat like a meal. It is served late in the afternoon and is more substantial than a low tea. It is often used as a light meal between lunch and a late evening dinner. It is most frequently served in homes, and in some luxury hotels.

A tea may be used as a reception event for a speaker, followed by a talk in a room nearby. It can also be used to honor individuals or an event.

With the popularity of coffeehouses, as well as variety teas as menu elements, coffee and tea house menus are gaining in appeal. Styles range from traditional to Bohemian.

**Exhibit 3.17**
Catering Menu: Wildfire Party Menu

# Wildfire's Club Supper Menu

Perfect for large gatherings...
Everyone in your party can enjoy a variety of items
when you order the Club Supper Menu.
Minimum party size of 6.
Our friendly staff can make suggestions to accommodate one and all. Enjoy!

## Starters

Choose Two
Spinach & Artichoke Fondue
Horseradish Crusted Cherrystone Clams · Tomato, Basil & Three Cheese Pizza
Four Cheese Crusted Portobello Mushrooms
Wood Oven Baked Goat Cheese · Wild Mushroom Pizza
Wildfire's "Specialty" Wood Roasted Mussels (seasonal availability)
Grilled Chicken and Portobello Mushroom Skewers
Wood Oven Roasted Lump Crab Cakes ($2 per person extra)
Jumbo Shrimp Cocktail ($2 per person extra)
Roasted Sea Scallop Skewers ($2 per person extra)

## Salads

Choose Two
House Salad Bowl with 3 dressings: House vinaigrette, Ranch, 1000 Island
Chopped Tomato and Red Onion Salad · Caesar Salad
Spinach Salad · Wildfire Chopped Salad

## Main Courses

Choose Two
Spit Roasted Herb Chicken · Barbecued Chicken · Baby Back Ribs
Lemon Pepper Chicken Breast · Swordfish "London Broil"
Cedar Planked Salmon · Roumanian Skirt Steak
Basil Hayden's® Bourbon Tenderloin Tips · Mushroom Crusted Pork Chop
Fresh Fish of the Day ($3 per person extra)
Roasted Prime Rib of Beef ($3 per person extra)
Filet Medallions Oscar ($4 per person extra)
New York Strip Steak ($4 per person extra)
Filet Mignon ($4 per person extra)
Horseradish Crusted Filet Mignon ($5 per person extra)

## Potatoes and Vegetables

Choose Two
Redskin Mashed Potatoes · French Fries
Fresh Vegetable of the Day · Wild Rice
Creamed Spinach · Steamed Broccoli with Herb Butter
Wood Roasted Mushroom Caps · Wildfire Cheddar Double Stuffed Potato

## Desserts

Choose Two
Triple Layer Chocolate Cake · Wildfire Ice Cream Sandwich
Seasonal Berry Crisp with Ice Cream · Homemade Key Lime Pie
Fresh Baked Skillet Pie with Ice Cream · Chocolate Peanut Butter Pie · Seasonal Pie of the Month
Classic N.Y. Style Cheesecake choose 1 topping: Mixed Berries, Hot Fudge or Snickers

28.95
per person

All Club Suppers are served family style. Menu prices are subject to change.



*Courtesy of Lettuce Entertain You Enterprises Inc.*

## Reception Menus

Receptions resemble teas but may feature alcoholic beverages. A wedding reception usually has a punch made of fruit juice, brandy, and champagne; although almost any other drink combination would be satisfactory. At some receptions, cocktails with canapés, hors d'oeuvres, and other tangy foods may be served. These may be picked up by guests at a buffet or they may be passed among guests. When passed, the service is called *flying service,* from the Russians who originated the service and called the trays of foods carried around *flying platters*.

It is usual to estimate that each guest will eat two to eight pieces of food at such gatherings. The type of function, its length of time, and the variety of food offered will affect the amount of food required. Bowls of dips and platters of crisp foods, such as pickles, olives, or crudités that are easy to replenish may assist in giving flexibility. Some operations plan a runout time and, toward the end of service, leave only a few foods remaining. The number of drinks served may also vary. Men usually consume, on the average, two to four alcoholic drinks per hour of service, and women one to three. Again, the occasion decides but on the average the number of drinks consumed is falling. Light wines, beer, and nonalcoholic drinks should be included to meet the needs of guests.

Since the preparation of a large quantity of fancy canapés, sandwiches, and hors d'oeuvres can require much labor, ways should be found to reduce preparation time by purchasing items such as tiny prefilled puff shells or other items that can be filled and made to look as if they were prepared on the premises. A tray of canapés or fancy sandwiches need not have every item on it highly decorated. If a few fancy ones are properly spaced among plain ones that take little labor to produce, the effect is still appealing.

Guests in formal attire generally prefer "neat" food that will not drip, stain, or cause similar distress.

## Buffet Menus

Uniqueness in buffet menu items should be sought, in addition to familiar food that is either popular, such as salads, or comfortable, such as macaroni and cheese.

The effectiveness of a buffet depends almost wholly on the originality and presentation of the food. If these are not emphasized, the purpose is lost. Foods that look tired or that are excessively garnished, off-color, or messy will not satisfy guests.

The number of foods to put on a buffet can differ with the occasion and the meal. For a simple meal, only a few items may be used along with rolls and butter, beverage, and dessert. A dessert buffet may include only a dessert and beverage. A slightly more elaborate luncheon buffet may have four to six cold foods, including appetizers and salads, several hot entrees with a vegetable and perhaps a potato dish, a selection of bread or rolls and butter, and perhaps several desserts. There may be 20 or more cold foods on an elaborate buffet, 8 or more hot entrees, a number of different hot vegetable dishes, potatoes, a variety of hot breads and rolls, and various cakes, puddings, pies, and other desserts.

The number of items and their presentation will dictate the degree of elaborateness. Certainly an overabundance of items is a mistake and can lead to waste. The purpose of a buffet is to give guests the feeling that they can help themselves as much as they desire, but it should not lead to waste.

Some buffets require particular foods. A *smorgasbord,* the Swedish buffet, must have pickled herring or other fish, rye bread, and *mysost* or *gjetost* cheese, in addition to many cold foods; hot foods, such as *lefse* and Swedish meat balls; and a dessert, such as a pancake in which lingonberries are rolled. Cold foods are eaten before the hot ones. Dessert is served last with coffee. A true smorgasbord usually includes *aquavit* or *schnapps* (Swedish liquers) served in tiny glasses.

A Russian buffet should have caviar served from the table center, either from a beautiful glass bowl or from a bowl made of ice. Dark rye bread and butter must accompany it. The other foods should be typical of a buffet. A Russian buffet also includes small glasses of vodka.

The use of buffets for breakfast has been mentioned, and additional material will be found in the discussion on service.

### TAPAS AND TASTING MENUS

The Spanish custom of *La Tapa* is eaten between main meals so that the wait until the next repast is not unbearable. Some sources declare that this ritual began when King Alfonso the 10th, the Wise, was ill and had to eat small portions of food with wine during his recuperation. Upon recovery, he decreed that something to eat must accompany the alcohol if wine was to be served at any inn. Another theory states that tapas developed so that workers in the field and elsewhere could take a small amount of food to allow them to continue working until the big noon meal, traditionally followed by a siesta. Modern American office counterparts often have a hoard of snacks at their workspace. Once Tabernas became established, the decree held. Glasses of wine were served covered with a slice of smoked ham or cheese. It was said this both prevented foreign objects from falling into the wine and helped absorb the alcohol drunk on an otherwise empty stomach. Tapas most often served include a variety of olives, dry nuts, and cold cuts, as well as many other recipes for appetizers, some quite elegant. Tapas can and do form an entire lunch or dinner if the quantity or variety is enough to satisfy the appetite.

In China, tea became popular during the Chung dynasty. Tea leaf quality and customs around the tea drinking included a variety of accompaniments, which included dim sum. The quantity was kept small, served in tiers of bamboo steamers or on small plates so that many different varieties could be sampled. This custom has continued throughout history.

American menus have adapted these traditions to either include traditional tapas or dim sum menus (see **Exhibit 3.18**) or by creating unique "tasting" portions so that guests can sample a variety of foods. These menus might be used for afternoons, evenings, and late nights, or as a supplement to regular sized entrees.

## Menus for Patrons' Special Needs

The menu planner must focus closely on the patrons to be served. Special menus may have to be developed to meet the special tastes of specific groups. Restaurants and institutions that serve the elderly or the very young may want to consider how to meet the desires and needs of these groups, either with the regular menu or by developing a new one.

**Exhibit 3.18**
Tapas Menu: Café Ba-Ba-Reeba!

# CAFE BA·BA·REEBA!®
# COLD TAPAS

Tapas are the famous small dishes of Spain made from seafood, meats, cheeses, and vegetables. Create an entire meal selecting a variety of tapas dishes, or simply choose a selection of tapas as an appetizer and enjoy one of our paellas, calderos, brochetas or a cowboy ribeye as your main course.

**daily tapas combination plate**
*tapas combinadas del dia* ........ MARKET

| PLT. | | TAPAS |
|---|---|---|
| | **SEAFOOD** | |
| 01. | roast tomato filled with tuna<br>*tomate relleno de atún* | 3.95 |
| 03. | white anchovies, avocado & peppers<br>*boquerones con aguacate* | 4.50 |
| 05. | **citrus cured salmon, cucumber bread**<br>*salmón frío* | 6.95 |
| 07. | seafood salad, shrimp, scallops & squid<br>*ensalada de mariscos* | 7.25 |
| 09. | salmon and scallop ceviche, orange mojo<br>*ceviche de salmon y vieras* | 6.95 |
| | **CHICKEN & PORK** | |
| 11. | chicken salad with curry<br>*ensalada de pollo al curri* | 4.95 |
| 13. | **serrano ham & tomato bread**<br>*jamón serrano* | 6.95 |
| 15. | cured pork loin, manchego cheese<br>*cerdo embuchado* | 7.95 |
| 17. | serrano ham, salchichon, chorizo & manchego cheese<br>*plato de la casa* | 7.50 |
| | **VEGETABLES & CHEESE** | |
| 19. | tomato bread<br>*pan con tomate* | 1.50 |
| 21. | house marinated olives<br>*aceitunas aliñadas* | 4.50 |
| 23. | potato & onion omelette<br>*tortilla española* | 4.50 |
| 25. | marinated artichokes & fennel<br>*alcachofas al hinojo* | 4.95 |
| 27. | **garlic potato salad**<br>*patatas alioli* | 4.50 |
| 29. | roast eggplant salad with goat cheese<br>*berenjena asada con queso de cabra* | 4.50 |
| 31. | spanish artisan cheeses<br>*plato de tres quesos* | 7.95 |
| 33. | marinated manchego cheese<br>*queso manchego al romero* | 6.95 |
| | **SALAD & GAZPACHO** | |
| 35. | **spanish caesar salad, manchego cheese, crispy serrano ham**<br>*ensalada cesar* | 4.50 |
| 37. | endive, blue cheese, walnut & membrillo salad<br>*ensalada de endivias* | 5.50 |
| 39. | baby spinach salad, chorizo bits, piquillo peppers and egg<br>*ensalada de espinacas* | 4.50 |
| 41. | frisee salad, apples, marcona almonds and cheese<br>*ensalada de escarola* | 4.50 |
| 43. | traditional gazpacho soup<br>*gazpacho tradicional* | 2.95 |

# HOT TAPAS

| PLT. | | TAPAS |
|---|---|---|
| | **SEAFOOD** | |
| 02. | grilled squid in olive oil & garlic<br>*calamares a la plancha* | 5.95 |
| 04. | seared octopus & potato<br>*pulpo a la plancha* | 6.25 |
| 06. | seared salmon, vegetable, orange olive oil<br>*salmon a la plancha con vegetales, aciete de naranja* | 6.50 |
| 08. | fried calamari & tomato salsa<br>*calamares a la romana* | 6.95 |
| 10. | **seared sea scallops, raisins, couscous & pine nuts**<br>*vieras salteadas* | 8.50 |
| 12. | shrimp with garlic, olive oil & red pepper flakes<br>*gambas al ajillo* | 6.95 |
| 14. | jalapeno shrimp<br>*gambas con jalapeño* | 6.95 |
| 16. | lobster manolo<br>*langosta manolo* | 12.95 |
| 18. | escargots, garlic croutons, sherry alioli<br>*caracoles con alioli* | 6.95 |
| 20. | daily fish<br>*pescado del dia* | AQ |
| | **BEEF & LAMB** | |
| 22. | **beef tenderloin & blue cheese**<br>*solomillo con queso de cabrales* | 9.95 |
| 24. | **rioja short ribs, manchego mashed potatoess**<br>*costillitas a la riojana* | 7.95 |
| 26. | skirt steak, confit potatoes & onions<br>*faldilla de res con confitura de patatas* | 7.95 |
| 28. | beef empanada<br>*empanada de carne* | 5.25 |
| 30. | meatballs with sherry tomato sauce<br>*albondigas al jerez* | 5.50 |
| 32. | skirt steak, shallot herb butter, green onions<br>*faldilla de res a la mantequilla con hierbas* | 7.95 |
| 34. | beef skewer, horseradish cream, red onions<br>*pincho de solomillo* | 6.50 |
| 36. | braised lamb & couscous<br>*salteado de corderō* | 7.95 |
| 38. | seared lamb chops, piquillo peppers<br>*chuletas de cordero ala plancha* | 8.95 |
| | **CHICKEN & PORK** | |
| 40. | chicken empanada<br>*empanada de pollo* | 5.25 |
| 42. | chicken & ham croquettas<br>*croquetas de pollo y jamón* | 5.25 |
| 44. | **skewered chicken & chorizo sausage**<br>*pincho de pollo y chorizo* | 5.95 |
| 46. | chicken breast, spinach, raisins, pine nuts & blue cheese<br>*pollo con espinacas y queso* | 6.50 |
| 48. | roast dates with bacon & apple vinaigrette<br>*dátiles con tocino* | 5.50 |
| 50. | spicy pork kabobs with confit potatoes<br>*pincho moruno* | 5.95 |
| 52. | seared spanish sausages<br>*chorizo y morcilla de la casa* | 5.95 |
| | **VEGETABLES & CHEESE** | |
| 54. | **spicy potatoes with tomato alioli**<br>*patatas bravas* | 5.25 |
| 56. | goat cheese baked in tomato sauce<br>*queso de cabra al horno* | 6.95 |
| 58. | mushrooms stuffed with spinach & manchego cheese<br>*champiñones rellenos* | 5.75 |
| 60. | **fried green peppers & coarse salt**<br>*pimientos fritos* | 4.95 |
| 62. | mushroom empanada, porcini alioli<br>*empanada de setas* | 4.95 |
| 63. | daily soup<br>*sopa del dia* | 4.50 |

*Courtesy of Lettuce Entertain You Enterprises Inc.*

## Children's and Teenagers' Menus

Menus for young people are generally limited because youngsters usually lack food experience and tend to stick to a small number of food selections that they know and like.

Small children tend to have smaller appetites. While the food selections for children should vary, the portions should be small. Teenagers often have hearty appetites and usually are hard to fill up. Therefore, the teenage menu should be filling as well as appealing.

A highly visible, entertaining menu appeals to children and makes them feel they are getting special attention. Colorful menus, perhaps with reproductions of animals or cartoon characters (you may need permission for the latter), will appeal to young children. **Exhibit 3.19** shows an appealing children's menu. A well-planned children's menu can entertain and keep children busy working at some puzzle, drawing, or reading a cartoon, which helps keep them quiet so their families and other guests can enjoy their meals.

Offering children's food on the regular menu is often not as successful as having a special menu. A special menu is advantageous to the operation as well as to the child. That is because with a regular menu, parents may be tempted to share their own food with children and not order anything special for them. If a moderately priced child's menu is available, it leads to the child selecting his or her own food, and the total size of the guest check is increased.

The child's menu should limit selections to favorites, because confusion results if too many items are offered. The wording should be simple and straightforward. Following through on a theme is a good way to introduce interest. A menu that gives the child something to do that can be taken home—or an operation that gives a child a gift—is a big help in winning young customers. Some operations give a child something that is one of a group of items, such as a cartoon glass. Then the child wants to return to complete the set.

The selections offered on a children's or teenagers' menu should also be set for the geographic location and/or ethnic base. Some Mexican foods that are successful on a Texas menu may not be successful on a Wisconsin menu. A menu offering Cuban and South American items that is popular in Miami, Florida, might not be popular in Portland, Oregon. It is a case of suiting a menu to patrons, something covered more thoroughly in Chapter 4.

## Senior Citizens' Menus

Many operations can sell successfully to senior citizens, especially by offering discounts during slow periods and using the menu as a marketing tool. Senior citizens generally eat less than younger adults, so food portions for menu items should be smaller. In some areas seniors may be on fixed incomes so, for the most part, prices should be moderate. Some seniors cannot chew well, so soft foods should be included on the menu. Many seniors have received physicians' advice on diet. Thus, a menu that offers items that are low-fat, low-calorie, and low-sodium will likely satisfy most senior dietary needs. Some seniors are less tolerant of highly seasoned foods.

It is helpful to know something about the background of seniors who patronize an operation. For example, if they are of Germanic stock from a rural area in Wisconsin, they will probably enjoy things like chicken, pork, liver, sausage, sauerkraut, and other similar items. Knowing customers is the first and most important step to satisfying them.

Exhibit 3.19 Children's Menu: Big Bowl

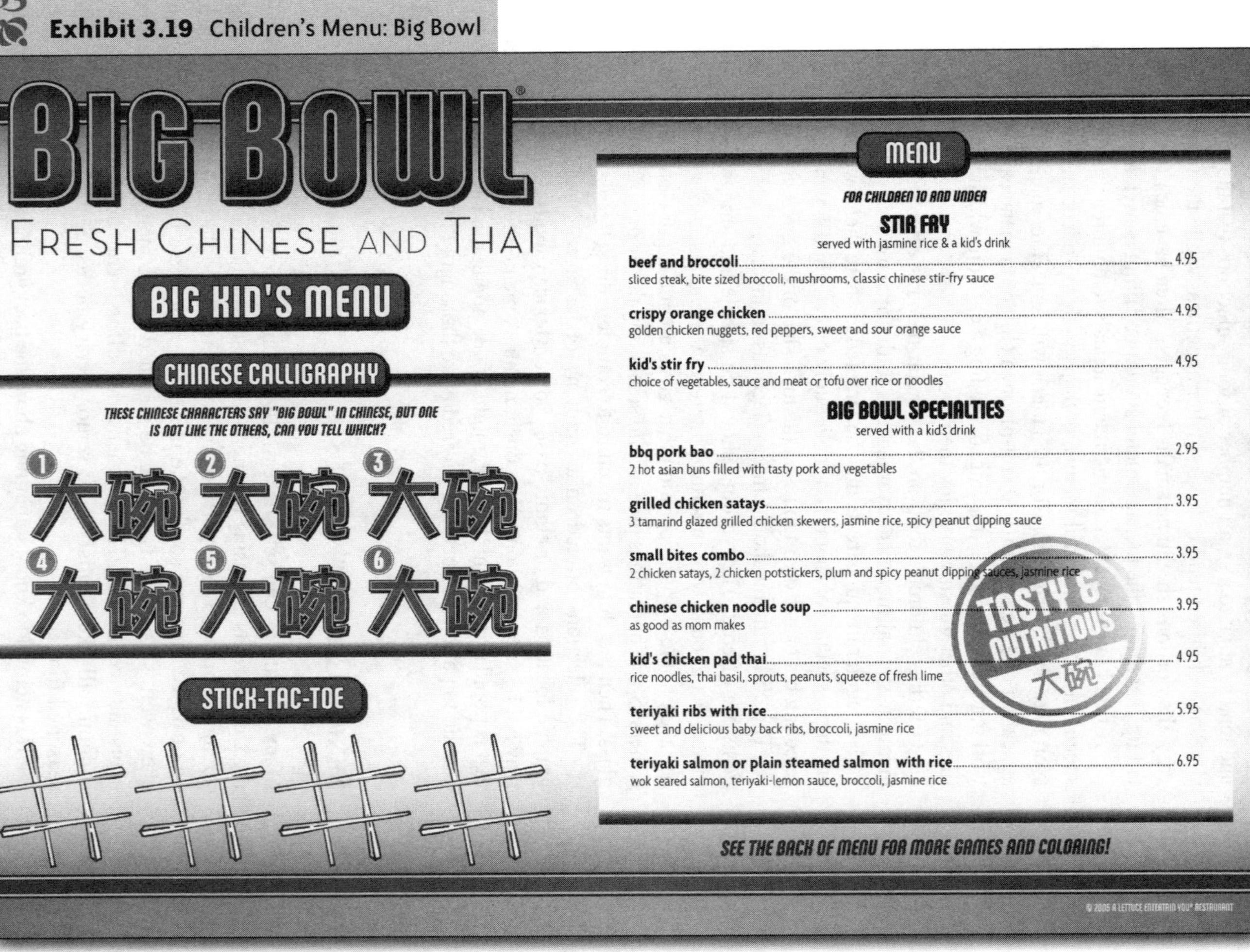


*Courtesy of Lettuce Entertain You Enterprises Inc.*

# Menus for Noncommercial and Semicommercial Establishments

Noncommercial and semicommercial operations will have menu requirements that differ from those prepared for full commercial operations. If the operation is institutional, and the patrons are a *captive market,* the menu must be evaluated, based on the fact that most patrons will eat two or more meals a day at the facility. This means that nutritional needs of patrons must be considered along with their preferences.

## Institutional Menus

Commercial operations are in business to make a profit. Institutions usually are not. They often operate from a budget that indicates the limits of expenditures, and they try to keep within these limits. *Breaking even*—bringing in enough sales to cover costs—is often the goal of an institutional operation. Most institutional foodservices work with a cost per meal allowance, or a per-day allowance. A hospital will have a cost per bed per day. Hospital budgets for operating the entire department usually take about 10 to 12 percent of the hospital's expenditures.

Many institutional menus do not have to merchandise or sell to the extent that commercial operations do, since they have a built-in clientele. Many patrons in institutions do not see the menu, since it is written for the foodservice staff only. Some menus will not have a selection. Others will offer slight variations in the basic menu, such as a choice of dessert or beverage. Institutional operations that have to compete with commercial establishments may have to do some merchandising. In any event, dull menus are not a necessary characteristic of institutional operations. A new development in hospital menus is room service that can be ordered by patients according to their dietary restrictions.

Nutritional considerations about the food served are important; however, these will depend on whether patrons must get all their meals from the operation or eat there only occasionally. If they dine there less often and have other options, the nutritional consideration becomes less important, although still a factor. Certain patrons will be on special diets, and this will have to be considered.

Whether patrons eat in the institutional foodservice all the time or just sporadically, it is important to have good variety in the foods and avoid too much repetition. Long-term cycle menus are best for institutional operations. To break monotony there should be some meals that feature seasonal or occasion menus themes, such as "Night in Venice," "Fiesta Days," "Old-Fashioned Picnic," and so forth. Such menu innovations build interest, and patrons look ahead to such events. It is also important to occasionally break the meal plan and offer items different from the typical groups of foods served.

Most institutional menus are built around the three-meal-a-day plan. Breakfast is a fruit or juice, cereal and milk, a main dish such as eggs and/or meat, hot cakes, Danish pastry, toast or bread with margarine or butter, jelly or jam, and beverage. Lunch may include an entree, vegetable, salad, bread and butter, beverage, and dessert, or be smaller, including only a soup, beverage, and dessert. The institutional dinner may have a first course, such as an appetizer or soup, followed by a main dish with vegetable and potato or other starchy food, a salad, bread and butter or margarine, a dessert, and beverage. There is usually a 10-hour span between breakfast and dinner and a 14-hour fast between dinner and breakfast, unless a snack is served in the evening.

Some institutions find that a four-meal or five-meal plan is more suitable for patron needs. In the four-meal plan, a light continental breakfast is served at 7:00 A.M. A substantial brunch follows it at 10:30 A.M. At 3:30 P.M., the main meal is served, followed by a light supper in the evening.

The five-meal plan is similar, with a continental breakfast at 7:00 A.M., a brunch at 10:00 A.M., a light snack about 12:30 or 1:00 P.M. with the main meal following between 3:00 and 4:00 P.M. A light snack is served between 6:30 and 8:00 P.M.

The four-meal or five-meal plans may reduce the labor required in the kitchen since the two big meals of the day are close enough together for one shift to prepare them. The other meals are light enough to be prepared by skeleton crews.

Some hospitals also find that the four-meal or five-meal plans permit patients to be gone for early morning examinations without losing out on a full meal. Even though the patient has had to miss breakfast, he or she can return to the room when one of the heavier meals of the day is being served.

Not all attempts to change to a four-meal or five-meal plan have been successful. The failures have usually been ones of planning. It is essential that the plan be thoroughly discussed with staff members who may be affected. The plan must have the complete support of the staff and solid backing from the administration. Communication of the plan and a complete discussion of the advantages and disadvantages should occur well in advance of implementation.

The change must consider every factor. For instance, a large state mental hospital changed to a five-meal plan and found that the patients felt they were not getting a breakfast because cereal was not a part of the first meal served. Adding either cold or hot cereals to the continental breakfast resulted in eliminating most complaints. In implementing these four-meal or five-meal plans in health institutions, the amount of sugar, flour, and fat should be watched. Sweet rolls or other breakfast pastries and evening snacks such as cookies, cakes, or other rich products cause undesirable weight gain and result in a failure on the part of patients to eat adequate amounts of other essential nutrients.

Hospitals are increasingly dispensing with paper orders for food and entering orders on handheld point-of-sale systems. Foodservice personnel input data from guest rooms and are available to answer guest questions.

### *Planning Institutional Menus*

A number of software programs have been developed that specialize in planning institutional menus. For example, a number of products for school systems integrate menu calendars, production records, costing and inventory with brand product databases, including ingredient searches, and replace functions for recipe modifications. Calendar and cycle functions, shopping lists for inventory ordering or forecasting, and printing of recipes for specified quantities are all available options. Production formats such as the *USDA Healthy School Meals Manual* and the ability to add HACCP notes are features that save much time and money over the old-fashioned paper and pen methodology.

Most planners start with the main dishes for a meal, beginning first with dinners, then lunches, and then breakfasts. Balance and variety must be sought between days and also between the meals in a day. The frequency of the types of meats to use should be established, and a program can be set up for this. For instance, in a week, beef, pork or cured pork, poultry, fish or shellfish, and a casserole dish should each be served only once. These may be varied with an occasional selection of variety meat, veal, lamb, or sausage, and eggs, cheese, or other nonmeat dishes. Similarly, a program may be set up to indicate the frequency desired for various vegetables and other foods.

After the main dishes are selected, the vegetables and potato, or other starch items, are added, followed by salads and dressings. Again, these must be balanced against the various foods used in a day and from day to day. Breads for each meal, desserts, and beverages can follow, in that order. After adding these major items, the planner may select the breakfast fruits and cereals.

After this, the menu should be checked to see that balance has been maintained, nutritional needs met, cost restraints not exceeded, and other factors, such as balanced use of equipment and skill of labor, considered. If modified diets based on this general menu are required, they should be planned by someone trained in nutrition. Menu planning software with nutritional modifications is available.

Different institutions will need different menus to suit varying operational requirements and meet patron demands. Regulatory agencies and other outside influences may also affect menu planning.

## Health Facilities

A great number of menus will be required to meet the needs of various kinds of health facilities, such as hospitals, convalescent centers, nursing homes, and retirement homes, in which there may be available nursing and medical care.

Hospitals must provide food for staff, nurses, doctors, visitors, and catered events, as well as patient meals. Different menus will be required. Some of these menus will be typical of those in other foodservices. Generally, only patient menus will differ.

A hospital's dietary staff will first prepare a general or *house* menu. This includes what foods will be served at a particular meal on a specific day. The planner must keep in mind that it is from this menu that the various modified diets the hospital must serve will be prepared. The dietitian goes through this general menu selecting items, changing them, or substituting others to meet dietary requirements. It takes an experienced, skilled person to modify a general diet. For instance, if a hospital put the items in **Exhibit 3.20** on the general menu, the dietitian would modify the menu for the various patient diets.

In a diabetic diet the cranberry jelly, potatoes and dressing, and coconut cream pie might not be allowed. The sorbet and baked apple sweetened with an artificial sweetener probably would be allowed. Some carbohydrates would be allowed on this menu, because even the most severe diabetic must still have some carbohydrates. The amount of calories allowed in the other choices on the diabetic menu would also be evaluated.

For a low-fiber diet, the dietitian would eliminate the bran muffin and would scrutinize the fiber content in the vegetables and salad, substituting something else if an item is considered too high in fiber. The baked potato would be served without the skin.

A diet of soft foods would probably require the cook to process the turkey or pork chop by chopping them up. The salmon might still be served whole because it flakes easily. The broccoli might be finely chopped, and the tomatoes might not appear in a salad but as a cooked vegetable.

A low-fat diet would substitute plain lemon juice for the lemon butter. The butter on the broccoli and that added to the squash would be omitted. The butter would be omitted from the menu

**Exhibit 3.20** Typical Hospital Menu

| | | |
|---|---|---|
| | Roast Turkey, Cranberry Jelly | |
| | Pork Chop | |
| | Poached Salmon Fillet, Lemon Butter | |
| | Baked Ziti | |
| Mashed Potato | Baked Potato | Cornbread Dressing |
| Buttered Broccoli | Cabbage Stir Fry | Pureed Butternut Squash |
| | Sliced Tomato Salad, Ranch Dressing | |
| | Bran Muffin French Roll | |
| | Butter | |
| Sorbet | Coconut Cream Pie | Baked or Fresh Apple |

and a low-fat dressing would be substituted for the regular ranch dressing. The sorbet or the apple would probably be allowed.

Menu planning for other health facilities may follow that of a typical hospital, with a general menu being written and modified diets taken from it. However, if only a few in the institution need special diets, the general menu would be written and special foods provided for those few. These foods could come largely from the general diet along with some specialized items.

Retirement homes may provide three meals a day, but some do not. Retirees may live in small apartments, which have facilities for preparing meals. Usually two meals a day are prepared there by the retirees. Thus, the foodservice facility may serve only one meal; this is often a heavy meal at noon or dinner. Patrons often need assistance during the meal. When full nursing care is given and the patron is confined to his or her rooms, a meal service might resemble that of a hospital or even hotel room service. In many retirement homes there will be a small coffee shop that serves breakfast, lunch, dinner, and snacks. Some operations may cater special events for retirees and their guests. Theme meals such as those mentioned earlier are also very popular with retirees and help to give a feeling of change and variety.

## Business and Industrial Feeding Operations

A factory or other in-plant foodservice or one in an office building may have to have several kinds of dining services. There may be an executive dining room where top executives can meet at mealtime, bring guests, and have meetings over a meal. A staff dining service and sometimes a coffee shop are also used. Meals are usually paid for by the employees, but at a marked-down rate because of employer subsidization. Some serve the meal free, particularly at remote locations such as oil rigs or aboard ships. Metropolitan Life Insurance was one of the leaders in providing meals under dietary supervision to employees. If a cycle menu is used, it must cover a fairly long period because patrons probably eat there regularly. Worker dining units are often cafeterias. Snacks, sandwiches, salads, beverages, entrees, vegetables, potatoes, breads, and desserts are served. Some foodservice operations may offer breakfast and lunch, and a few executive dining rooms may even serve dinner for late evenings.

Normally, a menu board in the cafeteria announces offerings and prices. A set meal only may be served, and no menu is prepared for patrons. The full-service dining rooms usually have printed or written menus.

## College and University Foodservices

Many college residence halls have cafeterias. Self-bussing is usual. Some residence halls have foods priced individually or together in a table d'hôte menu, and the students pay for the foods chosen. Other students may be on a board plan where the meals for a period are paid for at a set rate. There is usually no credit for missed meals. Some have systems that combine the two features, with the student purchasing a meal card and using it occasionally. Biometric technology allows fingerprint scanning. The menu must reflect such conditions.

Students are usually young and active and often want substantial meals. They will ignore foods that are not popular, and there may be a limited number of foods that they will eat. Some students still try to live on hamburgers, pizza, sodas, and French fries. Most residence hall units bow to student demands and serve what they want, hoping that in some way the student will get the nutrition needed. In other cases, students will demand natural or healthful foods, and these will have to be on the menu. Student wants vary, and it is important to try to meet nutritional needs and student desires for quality, while providing variety as well.

Many colleges and universities operate student unions. A variety of foodservice operations can exist here, including snack shops, quick-service operations, seated-service dining rooms, and even banquet and catering facilities. The menus must satisfy the students' culinary desires and, usually, meet a limited budget. However, these now may even be a part of a fixed residence hall menu.

Many campuses also have a faculty club or dining center. Usually the club's big meal is lunch, but some offer dinners. A few may have bars. Prices must be modest, and often the college or university subsidizes the operation by providing space and equipment and even some occupancy costs. These clubs are usually operated like other clubs, and promotional programs will be a part of the merchandising system to bring in members. A faculty dining center differs from a club in that it does not require membership but is open to all faculty members. Campus catering is another element of college and university foodservices. It may range from a formal dinner for the chancellor to a sports team banquet.

## Elementary and Secondary School Foodservices

Elementary schools usually are on the federal school lunch program. If so, a meal following the general meal pattern will be served. (Some schools may be only on the federal milk program.) Many school systems are now computerized and feed a large number of children. Others serve children from diverse racial, religious, and ethnic backgrounds, and their menus should reflect this patronage.

In secondary schools, students either pay for foods as they get them or use prepaid cards to pay for meals. Because of the preferences of teenagers for special foods, secondary schools may often serve popular as well as nutritious foods. For a variety of reasons, many secondary schools try to prevent students from leaving the grounds to seek meals elsewhere, such as nearby quick-service outlets. Some have "closed campuses," and the students are forbidden to leave during lunch.

At some schools, such as military academies or religion-supported schools, students live on campus. These are usually for older children and teenagers. The menus for these children should be planned with strong consideration given to their nutritional needs, since they will eat most of their meals in the school foodservice. The foods should also be those young people like. The service may be cafeteria- or even family-style, where foods are brought to the table in dishes and passed around. Some provide table service.

Many elementary, secondary, and private schools will have faculty dining rooms. For the most part, the foods will be somewhat similar to those served to students but modified for adult tastes. These foods may also be supplemented by other offerings, or there may be a completely different menu from that for the students.

### Miscellaneous Institutional Foodservices

There are many kinds of institutions, such as orphanages, prisons, associations, and religious and charitable groups, that provide food for people. Often these units function on a very limited budget. There may be no payment by the patrons. Either the government or the operation may provide the funds. In general, the provisions presented previously for institutional menu planning will apply. On a visit to Alcatraz, the infamous island prison for "tough" or escape-prone inmates, the final menu included spaghetti and baked potatoes. Tour guides state that it was believed portly prisoners would be less able to swim for shore.

## The Final Steps in Menu Planning

Once the various marketing, operational, and strategic decisions have been made, the pricing of the menu may begin. This is covered in more detail in Chapter 6. In addition, operators must continually research the popularity of menu items. Customer feedback is essential. The menu that was ideal for the establishment in the spring may be unpopular in the fall. A chef or kitchen manager must be attuned to trends and the seasonality of products. Menu analysis, which will be discussed in Chapter 8 is the key to heading off such marketing disasters. Financial considerations, covered in Chapter 12, must also be considered.

**Summary**

The task of planning and writing a menu is a daunting one. The most successful operations seem to have a menu that is *right;* that is, a menu that fits the operation. That does not happen by accident.

The menu should be thought of as the single most important document that defines the purpose, strategy, market, service, and theme of the operation. It should help sell items to patrons, and offer choices that will please a variety of tastes.

Before a menu can be planned, managers must develop a *mission statement* for the operation. This statement sums up the ultimate purpose of the operation, such as to provide high-quality, mid-priced food to a family market, or to provide three daily wholesome and nutritious meals to hospital patients. Next, managers set the objectives the operation is to attain. These can be in the form of profitability objectives, growth objectives, and market share objectives, among others. These objectives are used to develop *strategic plans* for the ongoing growth of the business.

A long-range financial plan is necessary before the menu can be developed. These long-range financial objectives are used to create budgets that must be followed by owners, managers, and supervisors.

One of the first decisions the menu planner (or menu planning team) must make is determining the number of menus an operation will use. This decision is based on the mission and scope of the operation.

Next, the type or types of menus to be used is decided. Menus fall in the following categories:

- *À la carte* menus, which offer foods separately at separate prices
- *Table d'hôte* menus, which group several items together at a single price
- *Du jour* menus, which change daily
- *Limited* menus, which offer only a few selected items
- *Cycle* menus, which rotate after a set amount of time
- *California* menus, which offer items from all meal periods at all times of the day

A single menu could include whatever derivation or combination of the above that best meets the mission of the operation.

Menu planners must take into account the meal plan to be followed—one that will fit the needs of most patrons. Each meal—breakfast, brunch, lunch, dinner, and evening meals—and each type of operation has a typical meal plan that leads to a fitting menu organization for the operation. Each course of the meal plan, such as entrees, is normally divided on the written menu into subcategories, such as fish and seafood, meat, poultry, pasta, and meatless entrees.

It is essential that patrons be given a variety of choices within each category and subcategory on a menu. Menu planners should vary flavors, textures, and cooking preparations as much as possible to please guests. This variety will also dictate the number of items included on the menu.

*Breakfast menus* are often associated with low prices and low profits unless care is taken to offer unique items to guests. Planners should strike a balance between items that patrons expect to find and those that will spark interest. *Brunch menus* traditionally combine items from both the breakfast and lunch menus. *A lunch menu* should offer guests items that are light enough, priced moderately enough, and prepared quickly enough for people who have to get back to work. Take-out and handheld foods have been popular. Leisurely items should also be available for people out on special occasions. *Afternoon menus* are intended to attract people with free afternoons—retired persons, shoppers, entrepreneurs, students, homemakers, and tourists—and typically offer sandwiches, fountain items, fruit plates, and pastries and desserts.

*Dinner menus* normally require the most elaborate dishes, organization, and variety. Of all menus, dinner menus should reflect the atmosphere, decor, theme, and patronage of the operation. *Formal dinner menus* follow the traditional courses included in classic French menus. They are appropriate for very formal occasions and may follow some—not necessarily all—classic French customs. *Evening menus* cater primarily to theater goers and others who are out for the evening. They typically offer such items as grilled sandwiches, fountain items, and desserts. *Special occasion, party,* and *reception menus* must fit the occasion and specific clientele of the event. Every effort should be made to make the meal a very unique experience for guests. *Tea menus* can range from simply tea and a light snack to an elaborate combination of teas, sandwiches, and pastries.

*Buffet menus* rely as much on presentation and service as they do on the food served, and the food itself should be unique. A traditional Swedish *smorgasbord* must have pickled herring, rye bread, Swedish cheeses, meat balls, a traditional dessert, such as lingonberry pancakes, and small glasses of aquavit or schnapps. A traditional Russian buffet includes small glasses of vodka, caviar, and dark rye bread. *Tapas,* or tasting menus, meet the needs of patrons desiring to eat lightly or wishing to try a variety of foods.

Special considerations must be taken into account when planning menus for specific groups of people. Children's menus should include a manageable number of simple items that will appeal to children. The menu itself should be fun for young diners. Menus for teenagers should include filling items that will appeal to teenage tastes. Many older guests will appreciate menu items that reflect dietary concerns, such as those low in fat, cholesterol, salt, and sugar. Softer textures and mild flavors might also be appropriate.

Menus for *institutional operations*—hospitals, nursing homes, schools, universities—require some special menu planning considerations. While commercial operations are in business to earn a profit, institutional foodservices often have as their goal to *break even,* or earn enough in sales to cover all costs. (If a not-for-profit organization does earn more than it spends, that money must go back in the business rather than being paid out to stockholders.) Since many institutional operations have a captive clientele, they are more driven to offer nutritious and varied menu choices.

*Business and industrial foodservices* must plan menus that will induce workers to stay on the premises rather than taking their business outside the office or factory. The types of food served should reflect the tastes of the patrons. College and university foodservices must please young people and their tastes. Elementary and secondary school foodservices, whether or not they follow the federal government's school lunch program guidelines, are obligated to offer children nutritious meals at costs their families can afford.

## Questions

1. Collect five menus from different types of operations. Which type of menu is each one? Are items on each menu organized well? Are there too many items? Too few items?

2. For each menu type below, discuss the types of operations in which they are appropriate, the tone they set, and the markets to which they normally appeal.

   À la carte    Limited
   Table d'hôte    Cycle
   Du jour    California

   Select two to three of these menu types and create a sample menu that combines elements of each. Describe how that might work.

3. What typical items do consumers expect to find on a breakfast menu? On a brunch menu? On a lunch menu? What specialty items have you seen on these menus that balanced expected items?
4. What are some of the important factors to consider when developing a children's menu? A menu for teenagers? A menu for senior citizens?
5. How is planning and developing an institutional menu different from developing a commercial menu? How is it similar? What are the *specific* considerations involved in developing menus for business and industry operations? For colleges and universities? For elementary and secondary schools?

## MINI-CASES

**3.1** Select an illustrative menu from the chapter or assigned menu and answer the following questions, explaining how you arrived at your conclusions:

a. What is the concept of this operation?

b. Who (what groups of people) do you believe to be the targeted market(s) for this menu?

c. Discuss whether the menu communicates effectively in terms of visual appeal, description of offerings, and its use as a sales tool.

d. Discuss whether the prices match your perceived needs of the targeted market, and of the operation.

e. Is there a sufficient range of offerings to provide customers with choice and the operation with profitability? Discuss your answer.

f. What changes would you, a restaurant management consultant, recommend to the owner? If no changes are needed, discuss what elements make this menu work for you.

**3.2** Design a menu for a surprise party for your instructor. What demographic do you believe that your instructor represents? What budget could your class afford to throw this party? Where would you have the party? What meal period would be most appropriate for this bash? What specific choices would you offer that would appeal to your class, and, most importantly, the guest of honor?

**3.3** Design a late-night menu for 11 P.M. to 3 A.M. for your community. Who would the market be at these late, graveyard-shift hours? What menu choices, decor, type of service, and price range would appeal to the persons you describe? You will be competing with others for development dollars, so be prepared to defend your concept.

How would your concept and menu differ if the hours of operation were 6 A.M. to 2 P.M.? Why? Could one operation be open from 11 P.M. until 2 P.M. the next afternoon? Would one menu be suitable for both day and graveyard markets?

**3.4** This activity is for small groups in or out of class period: Design a menu for children.

a. Decide on and define the restaurant concept, segment, and market from which the children's menu will be derived (e.g., quick service, open 10 A.M. to 2 A.M. with drive-through, and counter service for middle- to low-income families and youth).

2. Include a list of offerings, including entrees, beverages, sides, and desserts.

3. Include prices.

4. Keep in mind both sets of customers: kids (who have some say in where the family eats out and need to be kept relatively happy and quiet) and their parents (who will be paying, who want healthy kids that are not crazed with sugar intake, and who have the ultimate say in where the family dines).

5. Include what the menu will look like and any activity or extras that will be included (crayons, massive bibs, cardboard hats, treasure chest o' cheap plastic toys, etc.).

6. Be creative. Have FUN!

# 4

# CONSIDERATIONS AND LIMITS IN MENU PLANNING

## Outline

Introduction
Physical Factors
Labor Considerations
Patron and Artistic Considerations
- Guest Expectations
- Variety and Psychological Factors
- Appearance, Temperature, Texture, and Consistency
- Time and Seasonal Considerations

Rating Food Preferences
Patron Expectations
- Truth-in-Menu Standards
- Health Concerns
- Menu Pricing

Summary
Questions
Mini-Cases

## Objectives

After reading this chapter, you should be able to:

1. Identify cost constraints in menu planning and explain what considerations relate to cost.
2. List labor constraints in menu planning and explain what considerations relate to labor.
3. Identify food-purchasing constraints in menu planning and explain what considerations relate to availability.
4. Identify patron expectations and preferences and explain what considerations relate to variety, psychology, and health concerns in menu planning.
5. Explain how truth-in-menu standards relate to menu planning and patron expectations.

# Introduction

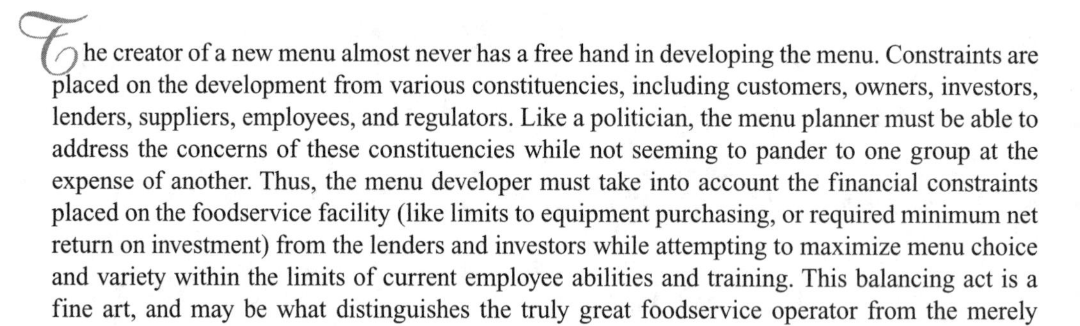

The creator of a new menu almost never has a free hand in developing the menu. Constraints are placed on the development from various constituencies, including customers, owners, investors, lenders, suppliers, employees, and regulators. Like a politician, the menu planner must be able to address the concerns of these constituencies while not seeming to pander to one group at the expense of another. Thus, the menu developer must take into account the financial constraints placed on the foodservice facility (like limits to equipment purchasing, or required minimum net return on investment) from the lenders and investors while attempting to maximize menu choice and variety within the limits of current employee abilities and training. This balancing act is a fine art, and may be what distinguishes the truly great foodservice operator from the merely competent.

Let us look at the various forces that tug at the creator of a new menu. These forces, when in balance, pull the foodservice operation down the road to success. When one or more factors are neglected or slighted, the disharmony created is as annoying as one singer in a chorus who sings flat. A menu must be in tune with all of the constraining factors.

What is the perfect balance? **Exhibit 4.1** shows the forces that affect the menu and that, when in harmony, create exceptional menus.

# Physical Factors

The menu must be written to fit the physical operation; that is, the facilities must be capable of supporting production of menu items of the right quality and quantity, and the menu must not be written so as to allow some of the facility to be overburdened while other parts are underused. A menu featuring a dinner of baked ham and escalloped potatoes, candied squash, baked tomatoes, cornbread, and chocolate cake may sound appealing, but it places too much demand on the ovens and underutilizes other equipment. This sample menu also places too much demand on too few personnel; perhaps the fry cook may wish to take this night off. Such failure to take into account physical factors may result in service breakdowns (such as making patrons wait while the over-busy oven has room enough to cook the meal.)

Clearly, some thought must be given to what equipment is on hand (or could be reasonably purchased). Menus must be planned to suit equipment capacity. A 20-gallon steam kettle is capable of producing only 16 gallons of soup. Similarly, a 24-by-36-inch griddle can produce a maximum of 80 orders of pancakes (at three cakes per order) in an hour. Planning for production of 100 orders per hour plus bacon, ham, hash browns, or eggs is another example of a failure to consider equipment limitations.

The time required to process foods through the equipment must also be considered. Using the soup example from before, to plan for more soup than the capacity of the steam kettle (say, having a menu with only one soup item on it and more than 16 gallons of soup served per meal period) means the soup must be cooked in batches. If you are cooking two batches of soup, why

**Exhibit 4.1**
Considerations in Designing a Menu

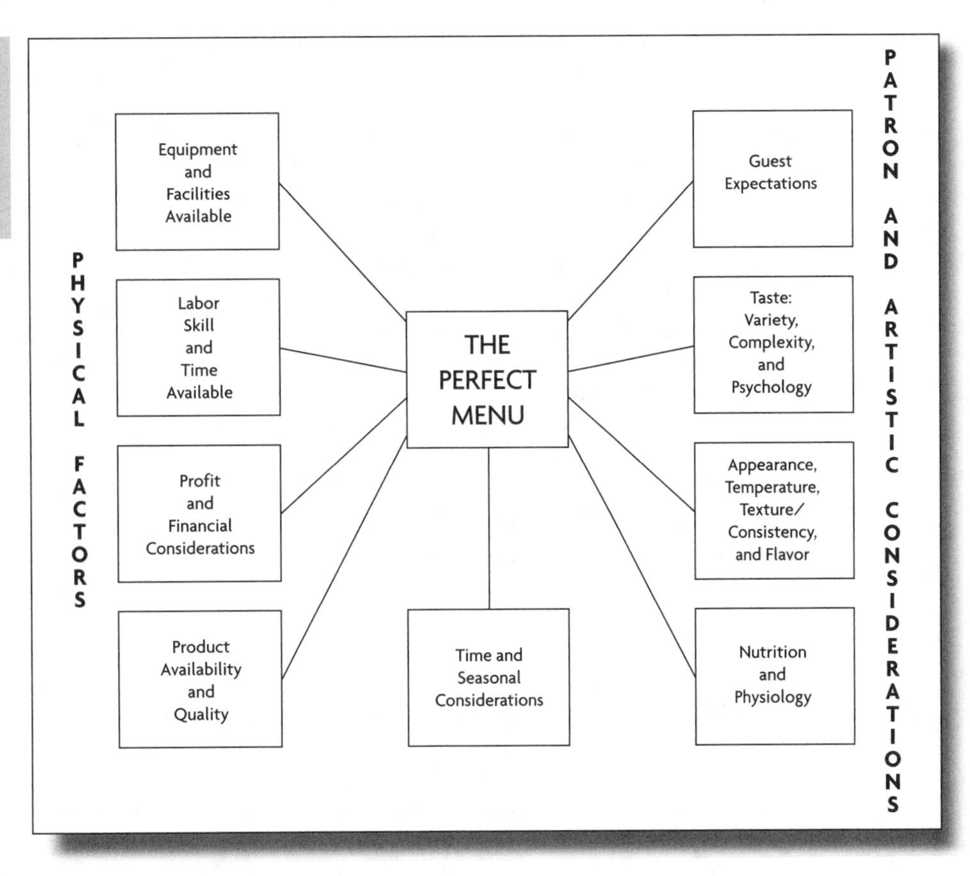


not have two different soups on the menu? Similarly, one must consider whether there is even enough time to cook more than one batch of soup before the meal is to begin. Workbenches, mixers, ovens, sinks, cooking ranges, and other equipment can handle a limited amount of work in a given time. (Not to mention the cooks, who can process food only so fast.) If a menu planner asks for more than the kitchen is able to produce, the whole system might collapse because employees cannot meet quotas if they have to wait for equipment to become available.

Storage capacity also must be considered when planning a menu. Refrigerated, frozen, and dry storage areas must accommodate foods both before and after preparation. A menu that overburdens storage abilities is disastrous for sanitation and cost reasons.

The menu planner can often alleviate equipment overload by getting input from production personnel. Planning or procedural changes regarding equipment use can often overcome perceived limitations. For instance, a seeming lack of griddle space may be overcome by cooking hashed brown potatoes in an oven and finishing smaller portions quickly on the griddle. A stew could be cooked in a roasting pan in the oven just as easily as on top of the stove. A change in the menu can avoid overload problems by using alternative cooking methods (a double bonus, as this also increases menu variety). This is a perfect example of why the menu planner must be familiar with cooking methods and preparations.

Naturally, equipment overload can often be reduced by buying preprepared, or *convenience,* foods. Many planners find that they can considerably extend the production capacity of the kitchen by purchasing this kind of food. Convenience foods have the advantage of overcoming deficiencies in labor skills as well. But nothing is free, and convenience foods may add considerably to the operation's food cost and might not offer the appropriate quality.

Of course, the menu planner's equipment considerations do not end in the kitchen. Service equipment is also a consideration. A dining area can support service for only a given number of people; and many believe that the quality of service begins to deteriorate whenever the room nears capacity. The kind of service offered matters as well. Full continental service, complete with flambé dishes prepared tableside, requires more square feet per guest than would a cafeteria-style operation. A service with numerous courses or one that requires long periods of time between ordering and serving can greatly affect the capabilities of a given space to accommodate a given number of people. Too many menu selections can also slow down service, both because of the time it takes patrons to make up their minds, and the time it takes to serve them. Even the type of staffing—experienced servers versus part-timers, how they are organized (teams versus individual servers), and what levels of expertise they bring to the job—all have an effect. These considerations must be taken into account by the menu planner. Most importantly, the menu planner must maintain a certain continuity. The service area's decor must match the menu's theme.

## Labor Considerations

Labor constitutes one of the single largest expenses in most foodservice operations. Just as the equipment, furnishings, and fixtures are assets, the smart modern manager will think of labor as a valuable asset. Just as we don't abuse the equipment by demanding more than the capacity of the griddle, so, too, we must be careful not to exceed the capabilities of our people. Unlike equipment, however, people have the capacity to learn. A good training program can teach new skills (and, hence, new capacities) to an employee. The menu planner must know what skills people possess in order to (1) create a menu that requires labor skills currently possessed by the people and (2) not create a menu that underuses employees' skills.

Training of labor can overcome the first problem, at some cost in terms of money and time. It takes time to train a person in new skills, and that time must be allocated from an already busy schedule. Also, training costs money. Even with on-the-job training the operator must realize that the person being trained and the person who is doing the training will work at a much slower pace during the training period than they may be expected to do once the trainee has been trained. By conducting a *skills inventory*—an assessment of the skills employees currently possess—the menu planner can create a menu that minimizes the amount of training and retraining required, and hence reduce the cost of implementing a new menu and the time it will take before the staff is operating as efficiently as possible under the new menu.

Underusing employees, conversely, will most likely lead to boredom. Studies have consistently shown that bored employees are less productive and more prone to absenteeism and turnover than are employees whose jobs provide challenges and are interesting. Similarly, many operations note that on slow nights they often have more problems and employees seem less coordinated than on busy evenings. The essential question is, why pay a person who has more skills than you need? In simple economic terms, it makes sense for the menu planner to write a menu that uses as many skills as possible.

The menu planner must also consider the most effective use of employees' time. If equipment is unavailable or inefficient, valuable labor is wasted. The menu planner must seek to avoid bottlenecks in production when writing the menu.

Before a food item can be placed on a menu, its availability must be known. It makes little sense to have a dish listed on the menu that is unavailable. This only causes aggravation on the part of the patron and the server. Menus planned to last a long time must not include too many seasonal items. One way of getting around this is to list a product group with the words *in season.* For instance, breakfast menus usually offer some kind of fresh fruit, so the listing could be "melon, in season" or "seasonal melon." In the United States, many food items are available year-round. But even these items often have very different prices and quality at different times of year. Cost and, often, profitability are important considerations.

Often, seasonal availability or price fluctuations can be overcome by volume purchasing. For instance, beef is often cheaper in early winter as the ranchers thin their herds. Buying large quantities and freezing them for later use will save money and ensure a constant supply of these items throughout the year. However, the costs of freezer storage, interest on the investment, and other elements must also be covered.

One subject whose importance cannot be overstated is that of food quality. Nearly every food item has a season in which its flavor, color, and texture are at their peak. Menu planners working in operations whose patrons are expecting a high level of quality must be especially sensitive to seasonal and quality constraints.

A menu should maximize revenue and minimize costs. Even in a nonprofit operation like a school or hospital, a menu planner must weigh cost and budgetary restraints when creating the perfect menu.

# Patron and Artistic Considerations

The menu must fulfill the patrons' physiological, social, and psychological needs—especially when it comes to food. To plan a menu that adequately does this requires a considerable amount of knowledge and ability.

## Guest Expectations

Often what a guest expects from a menu is unknown, even to the guest. Marketing experts use a variety of techniques to measure guest expectations. They may survey guests (or potential guests), either by telephone or through some written or computerized survey form. They may interview selected guests, either individually or in focus groups. However it is done, the goal is to encourage patrons to reveal their innermost thoughts and desires. This process is inexact, though it can help foodservice managers spot trends and have at least an idea of what patrons expect.

There is a very important difference between customers' *needs* and their *wants.* A need can be either a *physiological need,* such as the need for liquid when one is dehydrated, or a *perceived need,* or *want,* such as a "need" to own a new television set. Perceived needs are quite strong, and many industries successfully cater to these desires. The liquor industry, for instance, has experienced steady overall sales in recent years, yet premium wine sales continue to grow. Perhaps some of this strength is due to a perceived need on the part of some wine drinkers to project a certain image. Wants are often expressed through impulse sales. A patron may be quite full after a large meal, but a dessert cart display may spark a desire within the patron to order something that surely cannot fulfill a physiological need. The menu should try to satisfy both customer needs and customer wants.

In an attempt to satisfy their guests, foodservice operations have invoked ad campaigns saying everything from "You deserve a break today" to "No rules, just right." These slogans appeal to people's vague needs and expectations for satisfaction. When these needs and expectations cannot be satisfied or when they are in conflict—such as when a guest *wants* a hearty breakfast of eggs, bacon, toast smothered in butter, and milk, but also *needs* to reduce cholesterol, the menu planner must make the best effort possible toward meeting patrons needs, both physical and perceived—even if they can't both be fully satisfied.

The restaurant industry is under fire for contributing to obesity, to the point that industry lobbyists have sought to have legislation passed to prohibit lawsuits charging that restaurants have caused obesity-related health problems of patrons. Quick-service restaurants spokespersons bemoan that their sales figures for healthy choices are often not strong enough to justify their continuation as menu choices. The key seems to be finding healthy items that are satisfying to patrons and do not seem to be choices of deprivation. Anecdotal evidence suggests that low-fat items may be substituted for higher-fat items successfully if the patrons are not aware of the change, but the same patrons will reject a menu item offered as a "low-fat" alternative.

## Variety and Psychological Factors

One theory of human behavior, called *hedonism,* states that people try to maximize pleasure and minimize pain. Certainly maximizing pleasure as it relates to food means maximizing sensory input. This can be done by varying taste. Just as film makers have found that with each new thriller they must create more and more elaborate special effects, so too do menu planners find that they must have greater variety in their menus if they are to spark patrons' interest. There is a physiological reason for this: The brain reacts to new situations with much greater interest than to similar or routine situations. Just as we spend much time thinking about how to dance the latest step and almost no time thinking about how we place our feet when we walk down the street, we spend more time savoring and thinking about new taste combinations than we do when we are eating something common.

At the same time, too much variety, like too much ice cream, can cause dissatisfaction. Patrons must feel that there are items on the menu with which they are familiar, especially those patrons who are slow to experiment. However, even these common items can be prepared exceptionally well or in an unusual way. The sizzle platter (a metal plate that is heated in a broiler and causes food to sizzle when served) offers little added taste to fairly common menu items, but the unusual presentation has been used for years to add variety to menus.

There are other ways that food can meet people's psychological needs. Hunger creates anxiety and restlessness that eating quiets. This property of food is often called its *satiety value.* Some people overeat when they are frustrated or disturbed. The satiety or peaceful feeling they get from food tends to quiet and soothe them.

College and university students often show distinct ties between frustration and food. Arriving at school in the fall, most students' spirits are high and they enjoy their new experiences. At this time, the food is considered good. (This is the time for the foodservice department to save money on its budget.) However, as the newness wears off, students miss home and loved ones more. The rigors of classes, assignments, term papers, and examinations begin to create problems and frustrations, and students begin to take out their frustrations on the food they are getting. Food riots can even result. It is no coincidence that troubles on campus rise with such pressures. Students don't realize that the food they once liked now is cause for great dissatisfaction. (This is now the time to put the money previously saved on the budget into better food.)

The ties between food and psychological needs are even more apparent in prisons or reform institutions. For example, one prison riot was traced to the fact that the foodservice operation ran out of fried chicken for its weekend special and had to substitute another entree.

Many otherwise well-balanced adults have the same reactions. A business executive at a club may find the food totally unacceptable, while others at the table enjoy it. The problem may be that frustrations at the office did more to ruin the person's appreciation of the food than the chef, the waiter, or anything else the club did.

Many times institutional food of much higher quality than individuals get at home is criticized for its quality because diners miss the home atmosphere, family, or friends. People learn to eat amid familiar surroundings. There is comfort and security in eating at home where food is prepared "just so" and served on favorite or familiar dishes. This security and comfort is deeply ingrained. These feelings have been built up over a long period of time. Then, when an individual must eat elsewhere, away from his or her loved ones, there is insecurity, loneliness, and a desire for familiar things associated with food, a source of need and comfort. So it is not always the food that is lacking in quality but environmental factors associated with it. An individual who is hospitalized and ill has little appetite anyway; under the stress of illness and the loss of familiar surroundings, it is not surprising that the hospital's food is often disliked.

Fact and fancy are frequently interwoven in our notions about food. Thus, some individuals eat certain foods because they feel they are beneficial. For instance, many feel they must eat the foods they associate with home in order to stay healthy (e.g., chicken soup when ill). Others believe that raw oysters stimulate the sex drive, while saltpeter (i.e., potassium nitrate) depresses it.

Food can also have deep religious significance. Unleavened Passover bread (matzo) and kosher foods have deep religious meaning for Jews. Many Seventh Day Adventists and Roman Catholics will not eat meat at certain times. The fact that food and drink may have deep religious significance for some people should be respected.

## Appearance, Temperature, Texture, and Consistency

People use senses other than their taste buds to evaluate food. It is often said that people "eat with their eyes," so visual factors may take precedence in creating appealing food. Certainly we often see a dish well before we can smell it. Smelling the aroma actually stimulates our appetite. Once we taste a food, its temperature, texture, and consistency help determine whether we like it. Since the menu planner's goal is to provide the patron with maximum pleasure, making the food appealing is a key part of creating a menu.

Appearance includes things like color, form, and texture (in its visual sense). Colors like red,

red-orange, butter-yellow, pink, tan, light or clear green, white, or light brown are said to enhance the appeal of food, while purple, yellow-green, mustard yellow, gray, olive, and orange-yellow do not. Colors naturally associated with foods are appealing while colors associated with spoilage or unnaturalness are not. But the chef as artist does not stop here. Contrasts in acceptable colors should be sought. A myriad of bright color in a fruit cup or salad can heighten interest. Freshness in color (what an artist calls *vibrancy*) is important, while unnatural combinations or too vivid colors should be avoided.

The form taken by the food is also important. There should be an interesting assortment of forms or shapes on a plate. Variations on a plate, like a filet of baked snapper with a small heap of diced mango, red pepper, and pineapple salsa, sliced caramelized plantain with a sheen of butter, and a mound of white rice garnished with a sprig of fresh cilantro will achieve a difference in form and color that is agreeable. But too many mounds, cubes, balls, or similar shapes can cause loss of appeal, just as a picture that is too busy may distract from its central theme.

It is also important to have foods of different heights. Although this might not be possible to do on one plate (or might cause the plate to look busy), it can be achieved by having different heights on different foods. Thus, a salad can be in a deep bowl, or a dessert can be in a parfait glass in order to give contrast to the relatively flat entree plate. Tall garnishes (provided they do not overpower the dish) can also be used to give height to a plate.

It is important that at least some of the foods served retain their natural shapes. Too much pureed or chopped food can resemble baby food. This does not mean that planners should avoid interesting or novel presentations, but they should be sure to not get carried away.

Texture, such as that of kiwi fruit or fork-split English muffins, provides pleasant visual experience. A glistening or shimmering food is often more appealing than a dull one. Butter or other oil coatings, aspics and jellies, and egg liaisons all provide a shimmer.

The creative use of garnishing provides important visual activity, and having different garnishes appropriately matched to different food items is one sign that the menu planner has really put thought into creating the perfect menu.

The *flavor* of food is distinguished primarily by its aroma or odor. While our taste buds can distinguish only four tastes—sweet, sour, bitter, and salt—our noses can distinguish among different aromas with incredible accuracy. The great chef Escoffier was said to have made omelets by stirring them with a fork on which a clove of garlic was impaled. The taste of garlic was so minor, yet so compelling, that he single-handedly introduced this "forbidden vegetable" into haute cuisine.

Which flavors are pleasing or disagreeable differ with individuals. Naturally, we are more pleased with flavors with which we are familiar. The intensity of the flavor is also subject to individual preference. Intensity or *sharpness* may be an acquired taste. One may dislike sharp cheeses until one becomes used to the flavor, then one may cherish the flavor. There is a saying about olives that one must eat seven of them before one can appreciate their distinct flavor.

Our senses are sharpest around the ages of 20 to 25, and after this they slowly decline. Actually, there is a spot devoid of taste in the middle of the tongue, which grows in size as we age, perhaps from the ingestion of hot or cold foods that kill off the nerve endings. Thus, as people age, they may seek out more intensely flavored foods. Because the sweet- and salt-sensing nerves deteriorate first, older people lose their ability to taste these first.

An interesting phenomenon associated with flavor and taste is that some people cannot identify different food items by taste alone; they must see the food while tasting it. This may be why food displays engender such a favorable response from patrons. Certainly the display of foods at a buffet or in a salad bar is apt to stimulate the appetites of many patrons.

Another aspect of taste is the texture and consistency of food. Cooking food breaks down cell walls within the food, softening the texture. Unless carefully monitored, this can result in mushy, overcooked food. *Texture* is the resistance food gives to the crushing action of the jaw. Texture contrasts bring variety to the menu. A crisp, salty cracker contrasts nicely with soups; chewy turkey meat is complemented by soft, mashed potatoes.

*Consistency* refers to the surface texture of food. Okra can often have a slimy feel or consistency. Mashed potatoes made without enough liquid have a grainy, pasty consistency that is unappealing. Years ago we drastically overcooked vegetables. For instance, in the early 1900s, most recipes called for cooking asparagus for 40 minutes in water and baking soda. Most people today would find this most unappealing.

Finally, taste is a function of temperature. Serving cold foods cold and hot foods hot not only helps ensure sanitation and food quality, but also can give a pleasing temperature contrast to a meal. Mixing hot and cold foods in a meal provides further contrast and variety. A small cup of cold cucumber salad is an excellent contrast to a hot Indian curry. By serving hot foods hot and cold foods cold, we are not simply providing variety, we are showing the patron that we really care.

### Time and Seasonal Considerations

A menu planner is very much constrained by the time of day that the menu will be for. A menu with traditional breakfast items on it will find little appeal at dinnertime, except as a novelty. Others menu items are identified with certain times of the year. Others are associated with certain holidays—for example, turkey and its fixings with Thanksgiving and ham with Easter. These items can be put on menus long before and after the specific date of the holiday. Some items are associated with the four seasons. A winter menu may have a very different selection of salads, for example, than might a summer menu. Also, we would expect fewer salad offerings in winter but more soup offerings, especially up north. In sunny Florida and California, we tend not to have this constraint. In fact, tourists tend to enjoy the surprise of seeing summer salads available in the dead of winter.

## Rating Food Preferences

Patrons' food preferences must be considered when planning menus. People are influenced in their preferences by food habits acquired over a long period. Studies have been made of what foods individuals most prefer. These may be helpful to menu planners, but what people say they prefer is not always what they select from menus. The military asked what its personnel said they preferred and, then, in a further study, found that there were significant differences between what people said and what they actually selected. (See **Exhibit 4.2.**)

Many operations use taste panels made up of their own personnel to taste various dishes—which might be placed on menus—to see if the item has enough appeal to warrant featuring it. Often, score sheets are used, with the tasters rating factors such as flavor, consistency, temperature, color, and others with a scale of 0 to 9. The individual scores are totaled to see how well each dish did. If there seems to be a problem, individual scores can be checked as well as the

## Exhibit 4.2 Food Preference Questionnaire

**Food Preference Questionnaire**

Now I am going to ask you to rate the food you just ate. For each food, will you tell me if you liked it extremely, liked it very much, liked it moderately, liked it slightly, neither liked nor disliked it, disliked it slightly, disliked it moderately, disliked it very much, or disliked it extremely. This card has a list of these ratings. (Interviewer circle number.)

| | | | | | | | | | |
|---|---|---|---|---|---|---|---|---|---|
| a. What main dish? ______________ | 1 | 2 | 3 | 4 | 5 | 6 | 7 | 8 | 9 |
| b. Any other main dish? ______________ | 1 | 2 | 3 | 4 | 5 | 6 | 7 | 8 | 9 |
| c. Vegetables? ______________ | 1 | 2 | 3 | 4 | 5 | 6 | 7 | 8 | 9 |
| d. Drinks? ______________ | 1 | 2 | 3 | 4 | 5 | 6 | 7 | 8 | 9 |
| e. Breads or cereals? ______________ | 1 | 2 | 3 | 4 | 5 | 6 | 7 | 8 | 9 |
| f. Potatoes or starches? ______________ | 1 | 2 | 3 | 4 | 5 | 6 | 7 | 8 | 9 |
| g. Salads? ______________ | 1 | 2 | 3 | 4 | 5 | 6 | 7 | 8 | 9 |
| h. Soup? ______________ | 1 | 2 | 3 | 4 | 5 | 6 | 7 | 8 | 9 |
| i. Desserts? ______________ | 1 | 2 | 3 | 4 | 5 | 6 | 7 | 8 | 9 |

For breakfast, ask only for main dishes, beverages, breads and cereals, and fruits.

Overall, how would you rate the meal you just ate, using the same scale? (Circle)

1 2 3 4 5 6 7 8 9

How did this meal compare with other Army meals you have had?

☐ Much better? ☐ About the same? ☐ Much worse?

☐ A little better? ☐ A little worse?

Respondent's name ______________ Number ______________

Interviewer ______________

*Source: United States Army*

scores for specific factors to see why a dish might not score well. Often scores are interpreted as follows:

| | |
|---|---|
| 9 | Like extremely |
| 8 | Like very much |
| 7 | Like moderately |
| 6 | Like slightly |
| 5 | Neither like nor dislike |
| 4 | Dislike slightly |
| 3 | Dislike moderately |
| 2 | Dislike very much |
| 1 | Dislike extremely |
| 0 | Discard |

Not all individuals have a good sense of taste, so panel members should be selected carefully. Some have poor taste perception, lacking the ability to distinguish subtle differences of one or more of the four basic tastes: sweet, sour, bitter, and salt. They might have poor flavor memories so they cannot carry over flavors to evaluate differences between two or more samples. Smokers often have taste problems. Alcohol dulls the taste, as can a cold, fatigue, and stress.

A young person might not be as good a judge as an older one, because flavor identification requires experience. Older people often have more experience, but may lose good taste perception.

Taste panels are best conducted at 11:30 A.M. and 4:30 P.M., when people are apt to be most hungry. The room in which the taste panel works should be quiet and comfortable. Each judge should have a separate place in which to taste and score, unless a panel discussion is desired as the tasting goes on. A score sheet is shown in **Exhibit 4.3.**

Where only flavor, consistency, texture, and temperature are being evaluated, judges may be blindfolded and asked to judge a food without being able to see it. Often the elimination of appearance can make quite a difference in what a judge thinks of a food. The saying, "We eat with our eyes" is often proven in such a test, since a food found not too appealing when tasted and seen may be quite acceptable when one does not see it. However, the opposite result can also be true.

Sometimes guests are given complimentary samples to see how well they like new items. Sampling is usually not as formal as a taste panel, and guests may simply be asked if they like it or not.

# Patron Expectations

When most patrons look at a menu they conjure up specific images of what they expect. They should not be disappointed. If the menu item is accompanied with a short description, this can help in indicating what it is. Service personnel can also do much to indicate this.

The purpose of a menu is to communicate to the patron what is offered and, in commercial operations, the price. Menu terms should be simple, clear, and graphic, presenting an exact description of what the patron is to get. Patron expectations and what is served should coincide. The patron may misinterpret unclear terms. Or, the menu writer may mean one thing, but the

**Exhibit 4.3** Scoring Coconut Flan

**Item: Coconut Flan**

**Menu No. 512**

**Date** ________

**Scorer LHK**

| **Characteristic to Score** | **Score (0 to 9)** | **Comments** |
|---|---|---|
| Flavor | 7 | Little eggy; may have been baked at too high a temperature; either lower temperature or bake in pan of water. Coconut flavor present without overpowering. |
| Color | 8 | Good; also good sheen. |
| Texture | 6 | Slightly watery and some openness showing some syneresis. Coconut soft and chewy. |
| Form | 8 | Good; soft, yet solid. Holds shape. |
| Temperature | 6 | Served chilled, or try slightly warm. |
| Consistency | 6 | Slightly tough perhaps because of too high a baking temperature or too much heat. |
| **Total Score** | 41 | |

**General Comments:**

Cut down on baking temp. and bake in another pan of water; try reducing eggs & replacing with yolks to tighten up more and retard breakdown. Serve warm

patron may interpret it differently, especially if there are regional meanings applied to cooking terms or certain food flavors. There should be no ambiguity or confusion. Nor should descriptions be overstated.

If foreign words are used, the writer must be sure they are explained and understood by patrons. An explanation may be given in English below a term. *Keep the language simple* is a cardinal rule in menu planning.

Restaurant critic John Rosson has said that readers should be able to read menus quickly and understand what the terms mean.[1] He discourages the use of ambiguous terms, such as "meats dressed with the chef's special sauce" or, for a pie, "(made from) a secret family recipe, handed down from generation to generation," saying that they seem phony and may even be insulting. Rosson advises the use of descriptions that clarify without being overdone. Terms that mean little or nothing to the average diner should not be used. Examples are items called "The Navajo Trail" and "The Ocean Blue." The first might involve beef and the other fish, but they could just as easily be a Mexican platter and a plate of tempura seaweed. The item's name should leave little doubt as to what will be served. It should be noted that occasionally, successful restaurateurs, such as the late, great Michael Hurst and Rich Melman, have included "teaser" menu names designed to appeal to the adventurous and to evoke questions: Notably, "Don't Ask, Just Order It" menu items have created feelings of fun and spontaneity.

When a menu says the steaks come from prime beef, this means the beef grade is prime and not a lower grade. A *bisque* is a cream soup flavored with shellfish, yet one company produces a canned soup product called "tomato bisque," which is essentially false. Every menu planner should be sure of the terms used and what they mean when they are put on the menu. Too much license is frequently taken in the use of terms when writing menus. Selecting words to describe a food and give it glamour must be done carefully. To try to make a curry sound more exotic by calling it "Curry of Chicken, Bombay," without using the typical ingredients that a curry from Bombay, India, would have, is not encouraged. The right curry seasoning, accompaniments (especially the small, dried fish called Bombay duck), and other factors should be a part of the menu item.

Often menus present the name of the item in large letters and then give a description in smaller letters, as in **Exhibit 4.4.** Further clarification of the dish might also come from the server, who might explain that the rice is served as a mound in the middle of a large bowl containing the gumbo. (The method of item presentation will be discussed further in Chapter 7.)

## Truth-in-Menu Standards

In some cities and states, *truth-in-menu,* or *accuracy-in-menu* laws govern menu descriptions. These laws are the results of governmental interest in menu irregularities, such as indicating one item on a menu and actually serving another. These irregularities are perceived as *misrepresentations,* which may benefit the operator at the expense of the consumer. Some of these laws are state interpretations of Food and Drug Administration standards. They act as guides to menu planners when it is time to write the menu.

**Exhibit 4.4** Menu Description

**Shrimp Gumbo with Rice**

A thick soup made of shrimp, tomatoes, okra, and other vegetables delicately flavored with filé, a seasoning containing ground sassafras leaves, served over a mound of fluffy rice.

The federal government's Pure Food, Drug, and Cosmetic Act of 1938 forbids the use of any pictorial or language description on a container that misrepresents the contents. No grade or other term can be used that is not representative of the product. For example, if a label says, "Georgia's choicest peaches" and the peaches are not graded at least choice, the product is considered mislabeled because of the similarity of the terms *choicest* and *choice.* If any nutritional claim or nutrient value is given, the claimed value must be a part of the food. (For more on truth-in-menu guidelines, see Appendix A.)

The federal government also has developed *standards of identity* that define what a food must be if a specific name is used for it. For example, food labeled as egg noodles *must* contain at least 3.25 percent dry egg solids. If the word *cheese* is used, the product must contain 51 percent or more of milk fat, based on the dry weight of the product. The term *cheese food* must be used for food that cannot meet this regulation. Standards of identity exist for all foods except those considered common, such as sugar. Many states have adopted laws that clarify federal standards.

On January 1, 2006, the U.S. Food and Drug Administration came out with mandatory labels, which will list the amount of heart-unhealthy trans fat and the presence of eight major potential allergens. New York University nutrition professor Marion Nestle says the prospect of having to list these fats has forced food producers to stop using partially hydrogenated oils, the primary source. At this writing, restaurants are not required to list the trans fats in their foods. But it may be that eventually every use of partially hydrogenated oil will be replaced. The labeling change is "based on scientific evidence that trans fat intake is associated with increased risk of cardiovascular disease," says Barbara Schneeman, director of the FDA's Office of Nutritional Products, Labeling, and Dietary Supplements In December of 2006, the New York City Board of Health voted to amend the city health code to restrict the use of artificial trans fat in any foodservice establishment or by any mobile food unit commissary. Several states have also proposed partial or total statewide bans on trans fats. The National Restaurant Association is concerned about the hardships imposed on operators with the phase out of artificial trans fat in New York City to be completed by July of 2008. Concerns center on the quick time-span, the availability of replacement fats, the education of operators and the taste and quality of the foods made with the replacement fats. This is a complex issue that will require good communication among the affected parties to achieve the common goal of moving away from the use of artificial trans fat.

## Health Concerns

Medical experts have long warned that weighing too much increases the risk of diabetes, heart disease, and cancer. The U.S. Department of Agriculture released another version of the food pyramid in April 2005. It includes a figure walking up the steps, representing the need to couple exercise with a healthy diet. Further, a Website *www.MyPyramid.gov* allows people to customize the chart based on age, gender, and physical activity level (see **Exhibit 4.5**). The American Heart Association came out with dietary guidelines in 2000 seen in **Exhibit 4.6.** They continue with advice on how to eat at home, giving tips on planning meals, shopping, and preparing food, as it is deemed simply too tempting and, by implication, dangerous, to eat out. Extensive guidelines are offered on how to negotiate the danger inherit in eating at a restaurant, with specific advice on various cuisines and types of restaurants (see **Exhibit 4.7**).

Clearly, Americans know that they *need* to eat healthy; they just have problems with *wanting* to.

**Exhibit 4.5** MyPyramid

| GRAINS<br>Make half your grains whole | VEGETABLES<br>Vary your veggies | FRUITS<br>Focus on fruits | MILK<br>Get your calcium-rich foods | MEAT & BEANS<br>Go lean with protein |
|---|---|---|---|---|
| Eat at least 3 oz. of whole-grain cereals, breads, crackers, rice, or pasta every day<br><br>1 oz. is about 1 slice of bread, about 1 cup of breakfast cereal, or 1/2 cup of cooked rice, cereal, or pasta | Eat more dark-green veggies like broccoli, spinach, and other dark leafy greens<br><br>Eat more orange vegetables like carrots and sweetpotatoes<br><br>Eat more dry beans and peas like pinto beans, kidney beans, and lentils | Eat a variety of fruit<br><br>Choose fresh, frozen, canned, or dried fruit<br><br>Go easy on fruit juices | Go low-fat or fat-free when you choose milk, yogurt, and other milk products<br><br>If you don't or can't consume milk, choose lactose-free products or other calcium sources such as fortified foods and beverages | Choose low-fat or lean meats and poultry<br><br>Bake it, broil it, or grill it<br><br>Vary your protein routine — choose more fish, beans, peas, nuts, and seeds |

For a 2,000-calorie diet, you need the amounts below from each food group. To find the amounts that are right for you, go to MyPyramid.gov.

| Eat 6 oz. every day | Eat 2 1/2 cups every day | Eat 2 cups every day | Get 3 cups every day; for kids aged 2 to 8, it's 2 | Eat 5 1/2 oz. every day |
|---|---|---|---|---|

**Find your balance between food and physical activity**

- Be sure to stay within your daily calorie needs.
- Be physically active for at least 30 minutes most days of the week.
- About 60 minutes a day of physical activity may be needed to prevent weight gain.
- For sustaining weight loss, at least 60 to 90 minutes a day of physical activity may be required.
- Children and teenagers should be physically active for 60 minutes every day, or most days.

**Know the limits on fats, sugars, and salt (sodium)**

- Make most of your fat sources from fish, nuts, and vegetable oils.
- Limit solid fats like butter, stick margarine, shortening, and lard, as well as foods that contain these.
- Check the Nutrition Facts label to keep saturated fats, *trans* fats, and sodium low.
- Choose food and beverages low in added sugars. Added sugars contribute calories with few, if any, nutrients.

U.S. Department of Agriculture
Center for Nutrition Policy and Promotion
April 2005
CNPP-15

*Courtesy of U.S. Department of Agriculture*

## Exhibit 4.6 Dietary Guidelines

Healthy food habits can help you reduce three of the major risk factors for heart attack—high blood cholesterol, high blood pressure and excess body weight. They'll also help reduce your risk of stroke, because heart disease and high blood pressure are major risk factors for stroke. The American Heart Association Eating Plan for Healthy Americans is based on these new dietary guidelines, released in October 2000:

- Eat a variety of fruits and vegetables. Choose 5 or more servings per day.
- Eat a variety of grain products, including whole grains. Choose 6 or more servings per day.
- Include fat-free and low-fat milk products, fish, legumes (beans), skinless poultry and lean meats.
- Choose fats and oils with 2 grams or less saturated fat per tablespoon, such as liquid and tub margarines, canola oil, and olive oil.
- Balance the number of calories you eat with the number you use each day. (To find that number, multiply the number of pounds you weigh now by 15 calories. This represents the average number of calories used in one day if you're moderately active. If you get very little exercise, multiply your weight by 13 instead of 15. Less-active people burn fewer calories.)
- Maintain a level of physical activity that keeps you fit and matches the number of calories you eat. Walk or do other activities for at least 30 minutes on most days. To lose weight, do enough activity to use up more calories than you eat every day.
- Limit your intake of foods high in calories or low in nutrition, including foods like soft drinks and candy that have a lot of sugars.
- Limit foods high in saturated fat, trans fat and/or cholesterol, such as full-fat milk products, fatty meats, tropical oils, partially hydrogenated vegetable oils and egg yolks. Instead choose foods low in saturated fat, trans fat and cholesterol from the first four points above.
- Eat less than 6 grams of salt (sodium chloride) per day (2,400 milligrams of sodium).
- Have no more than one alcoholic drink per day if you're a woman and no more than two if you're a man. "One drink" means it has no more than ½ ounce of pure alcohol. Examples of one drink are 12 oz. of beer, 4 oz. of wine, 1½ oz. of 80-proof spirits, or 1 oz. of 100-proof spirits.

Following this eating plan will help you achieve and maintain a healthy eating pattern. The benefits of that include a healthy body weight, a desirable blood cholesterol level and a normal blood pressure. Every meal doesn't have to meet all the guidelines. It's important to apply the guidelines to your overall eating pattern over at least several days. These guidelines may do more than improve your heart health. They may reduce your risk for other chronic health problems, including type 2 diabetes, osteoporosis (bone loss), and some forms of cancer.

**Exhibit 4.7** Healthy Options for Dining Out

| Healthy Options for Eating Out |
|---|
| ◆ If you're familiar with the menu, decide ahead of time what to order and avoid temptation. It helps to be the first to order. |
| ◆ Removing butter from the table and drinking two full glasses of water assist in healthy eating |
| ◆ Reach for Melba toast or whole grain rolls instead of fat laden muffins or croissants—just say "no" to butter. |
| ◆ If you're unfamiliar, ask questions about preparation and ingredients and request low fat substitutes when available. |
| ◆ Order dressings and sauces on the side and use them sparingly, if at all |
| ◆ Order vegetables and side dishes, ask to have butter and sauces left off |
| ◆ Be choosy at salad bars. Say "no" to cheese, marinated salads, pasta salads, chopped eggs, croutons and fruit salads with whipped cream. Say "yes" to fresh greens, raw vegetables, garbanzo beans and low fat dressing or a squeeze of lemon |
| ◆ Baked, boiled or roasted potatoes with salsa or pepper and chives, yogurt or low fat cottage cheese are much healthier than with butter or sour cream. Avoid fried potatoes! |
| ◆ Order seafood chicken or lean meat. Remove all visible fat or skin. Have your entrée broiled, baked, grilled steamed or poached, never fried. |
| ◆ When grabbing a bite at Quick Service Restaurants, have burgers topped with pickle, onion, catsup mustard, tomato or lettuce rather than cheese, special sauces or bacon. Avoid fried fish sandwiches! |
| ◆ Grilled chicken sandwiches, fajitas etc. are preferable to standard burgers or tacos, pretzels trump chips and bagels beat Danish. Have juice or low/no-fat milk instead of a shake. |

*Adapted from* www.americanheart.org *©2006, American Heart Association, Inc.*

## Menu Pricing

Menus must often be priced to meet patrons' expectations. Some patrons are very price conscious, and a menu must meet their expectations of perceived value. Most patrons want to get the most for their money. The price paid should represent an adequate value in patrons' minds, and they will not be happy if they do not feel the value is there. However, menus must also be priced to cover costs in institutional operations and ensure profits for commercial ones. Since pricing is such an important subject, it will be discussed in depth in Chapter 6.

## Summary

The perfect menu is one that weighs the various constraining factors to menu design and ideally balances them. Many of the factors involved in menu planning seem contradictory. Like a good politician, the menu planner must produce a solution that keeps everyone happy. Patron needs and desires must be balanced against the needs of the owners to make money. Equipment constraints must be balanced against the need for a variety of choices. The skills and abilities of the staff must be balanced against the desire by the patron for new and interesting choices. An ethical balance of indulgent choices and healthful ones should be offered to assist their patrons to maintain their health.

## Questions

1. What are some personnel considerations that will affect the success of a menu?
2. Why might a patron who orders a meal of broiled chicken, mashed potatoes and gravy, and a fruit cup in a restaurant not like the same meal when in a hospital?
3. Why is it important to know whether a menu can help satisfy the physiological, psychological, and social needs of patrons?
4. Look at an actual menu's items and descriptions. Are explanations adequate in predicting what will be served? How do items served together complement each other?
5. Name five important considerations discussed in this chapter that anyone planning a menu should give special emphasis. Why did you select these and not others?

## MINI-CASES

4.1 You have been convicted of a capital crime and have exhausted all appeals. You live in a state where the death penalty is legal, and that is what you have been sentenced to.

You are scheduled to die tomorrow. Tonight, as is the custom, the state has offered to allow you to order up your "last meal."

You may order whatever you wish; however, you are to stick to food and alcoholic (and/or nonalcoholic) beverages only. (No cake with a gun baked in it, etc.) The focus is on what you would want to eat if you were never going to be able to eat again.

What would your meal consist of? (Feel free to order whatever you want, regardless of whether it is more food than is necessary for one dinner).

4.2 You (and your assigned partner, if applicable) have been stranded on a desert island. Fortunately, your island is equipped with a gas stove with burners, pots and pans, and a buried gas tank. You have a cistern to catch fresh water. There is no electricity or refrigeration. The island is rather sandy.

Once a month, a plane will drop a large crate into the ocean, next to the island, with whatever foods or beverages you request. You may not request any items that are not food or beverages (no magazines, refrigerators, electrical supplies). (No, the plane will not rescue you.)

You will be living on the island for the next two years. What supplies will you request for your monthly drop? Whatever supplies you decide on, the list will be the same every month until you are rescued.

# 5

# COST CONTROLS IN MENU PLANNING

## Outline



## Objectives

After reading this chapter, you should be able to:

1. Perform the following basic cost calculations: food costing, portion costing, recipe costing, and labor costing.
2. Describe what is meant by *prime costs* and explain how these cost factors affect menu planning.
3. Relate various control techniques to their associated costs and describe how these methods function.
4. Compare and contrast cost control factors unique to commercial and nonprofit institutions.

# Introduction

Of all the limitations placed on menu planners, costs are probably the most challenging. Normally, menu planners are concerned mostly with food and labor costs. But other costs are also important, such as the cost of energy, equipment, nonfood supplies, and storage. Whenever possible, the menu should reflect the overall costs of running an operation, whether the operation is institutional or commercial.

The menu planner has to know the operational costs to stay within a price budget or cost allowance. The costs of running a commercial operation must be more than covered in the budget if a profit is to be made. Cost is also the basis for pricing many menus. If costs are not known, selling prices are impossible to reasonably determine. The planner of the institutional menu who pays no attention to costs will not be employed long. This is because institutional or noncommercial establishments must operate within a budget, and costs are tightly controlled, both tangible and intangible.

# Obtaining Operational Costs

For costs to be controlled, they must be known. Managers must identify costs, determine their magnitude and frequency, and then, if they are too high, take steps to reduce them. It is important to recognize that locating such information is not, in itself, a control. It only indicates that control is needed. Such information about costs indicates a symptom, not a cure. Reports such as budgets, daily food cost reports, precosting sheets, financial statements, and payroll figures are tools used by foodservice managers to analyze data and root out any food-cost problems.

Once a menu is planned, priced, and analyzed, it can be implemented, and a series of operational functions can be set in motion. These functions are purchasing, receiving, storage, issuing, preparation/production, service, and cleanup. During each of these steps, controls must be established to assure that only planned costs occur and that the menu price or meal charge, if any, gives the desirable return. Today computers are used to do most of this.

In the *purchasing* function, the proper kind of food and its cost, quantity, and quality must be determined and the right supplier found. If the wrong kind, quantity, or quality of food is bought at the wrong cost, the menu will not have the desired end result.

Many aspects of the purchasing function can be computerized for greater efficiency. The operations chef and purveyors are helpful in developing product specifications to fit the standardized recipes used. When these recipes are input into available software systems that are linked with production scheduling and front-of-house point-of-sales systems, utilized items are subtracted from inventory. Management is signaled when it is time to order more recipe components. Similarly, recipe component costs can be monitored so that management is alerted when food and supply item prices have climbed above a range deemed acceptable. Substitutions can be suggested that are acceptable.

Efficiency during the *receiving* is essential to see that exactly what was ordered is obtained.

This also starts a series of accounting and inventory steps that are important to cost control. Handheld devices that can read barcodes so that inventory can be captured electronically can be purchased. Data from the handheld devices can be uploaded into the primary computer system, eliminating the paper and pencil method.

Controls must be placed on *storage* procedures so there will be no loss as a result of spoilage, contamination, or theft.

Product *issuing* must be exact to control costs of foods moving into preparation or production.

*Preparation* and *production* are usually controlled by standardized recipes and proper application of procedures. Careless acts such as overprocessing potatoes or overcooking a roast results in waste and loss of product quality.

Two food cost figures are often used. One is the *total* or *overall cost of goods sold,* or *total food cost.* The other is a food cost for individual menu items or for a group of menu items. Both have different uses, so they will be discussed separately.

## Total (Overall) Cost of Goods Sold

Total or overall food cost is usually stated as both a dollar figure and as a percent of total sales. The dollar value is most often calculated from beginning and ending inventory and total food purchases, as shown. The percent can be calculated from this total figure, if the total sales for the same period are known.

The total food cost can also be obtained if a *daily food cost report* is maintained. The cost of goods sold and food cost percentage are kept from the start of the period, accumulated up to the last day of the period. The *daily food cost* is obtained for this report by totaling the value of issues from storerooms and adding direct deliveries.

| | |
|---|---|
| Beginning inventory | $ 62,016.25 |
| + Purchases plus transportation, delivery, and other charges | + 113,183.22 |
| Total food available for period | $ 175,199.47 |
| – Ending inventory | – 59,866.21 |
| Total food cost | $ 115,333.26 |

To find the total *food cost percentage*, divide the total food cost by total sales.

$$\frac{\text{cost of good sold (food cost)}}{\text{total sales}} = \text{food cost percentage}$$

If total sales were $350,000, then the food cost percentage would be 32.95 percent ($115,333.26 ÷ $350,000 = 0.32952, or 32.95%).

Comparative data are also usually given, such as food cost a month ago, a year ago, and so on. **Exhibit 5.1** shows part of a daily food cost report.

If the labor cost is included, the word *labor* is added to the title and similar figures showing labor cost are presented. The report is usually compiled by the food and beverage or accounting departments. This is done with one of a variety of available software programs, saving considerable time and creating enhanced accuracy if the information put into the system is correct. If poor data are used, the results are not useful. Copies go to management of the operation, management in the foodservice department responsible for food cost (and labor cost, if compiled), and accounting, if compiled by the food and beverage department.

An even more complex daily food cost report can be prepared by breaking down food cost into different categories, such as dairy, fish, meats, and bakery. This helps pinpoint food cost production problems by showing which category is out of line. A total overall figure is also given.

Suppose issues total $12,212.34 and direct deliveries (direct from receiving dock to the production area for use) are $900.87. Then the total food cost can be calculated as follows:

Storeroom issues + Direct deliveries = Total food cost
$12,212.34 + $900.87 = $13,113.21

If sales for that day are $45,840, the food cost percentage is 28.6 percent ($13,113.21 ÷ $45,840 = 0.2860).

Sometimes similar dollar and percentage figures are obtained in advance of actual operation by *precosting,* or estimating future costs. If management knows that daily sales average $35,000 and it makes the estimate for 30 days, then total sales will be $1,050,000. A menu is planned to run a 30 percent food cost, where projected total food cost is $315,000. Precosting figures helps managers in overall planning. For instance, suppose a large chain is planning to run a special on prime rib, but wants to estimate what the impact on food cost will be if the ribs jump in cost by $0.05 a pound. The result can be seen by plugging data into the program.

Another way of precosting is called *averaging.* It is only helpful in giving a figure on food cost when a limited menu is used and the menu is separated into very similar or like items in each group of items. To average, one should have a good idea of what the sales mix is on the sale of various items sold. One must also know the food cost percentage for each member of a group of menu items. **Exhibit 5.2** shows how to find the overall food cost for one day's sales. Note that once the percent of dollar sales for each category and its percent food cost are known, the calculation is merely one of multiplication and addition.

Often in *budgeting,* we estimate future total food cost. This may be done based on previous figures, or a percentage of a specific allocation of funds. As an example of the first, a hospital knows that its food cost runs $6.25 per patient-day, and based on an estimate of the number of

**Exhibit 5.1** A Daily Food Cost Report with Comparative Data

| | | | Total Food Cost | | Sales | | Food Cost Percent | |
|---|---|---|---|---|---|---|---|---|
| Date | Direct Purchases | Storeroom Issues | Today | To Date | Today | To Date | Today | To Date |
| 5/1 | $159.15 | $545.68 | $704.83 | | $2,113.23 | | 33.4 | |
| 5/2 | 206.22 | 611.17 | 817.39 | $1,522.22 | 2,844.11 | $4,957.34 | 28.7 | 30.7 |
| 5/3 | 185.34 | 589.13 | 774.47 | 2,296.69 | 2,788.10 | 7,745.44 | 29.6 | |

**Food Cost Percent**

| | Last Month | | Last Year | |
|---|---|---|---|---|
| | Daily | Accumulated | Daily | Accumulated |
| 4/1 | 31.2 | 30.8 | 32.3 | 31.8 |
| 4/2 | 32.1 | 31.0 | 31.9 | 31.7 |
| 4/3 | 33.0 | 32.6 | 32.1 | 31.8 |

**Exhibit 5.2** Averaging Food Costs

| Item | Sales | Percent of Dollar Sales | Percent of Food Cost |
|---|---|---|---|
| Sandwiches | $875 | 40.8 | 39.0 |
| Smoothees | 410 | 19.1 | 32.0 |
| Daily side | 550 | 25.6 | 27.0 |
| Beverages | 312 | 14.5 | 30.0 |
| Total | $2,147 | 100.0 | |

Then, the following calculation can be made to get average or overall food cost:

| | | |
|---|---|---|
| Sandwiches | $0.408 \times 0.39 = 0.159$ | |
| Smoothees | $0.191 \times 0.32 = 0.061$ | |
| Daily side | $0.256 \times 0.27 = 0.069$ | |
| Beverages | $0.145 \times 0.3 = 0.044$ | |
| | | 0.333, or 33.3% Overall food cost |

There are four steps in making this calculation:

1. Predict total sales and the percent contribution each item or group makes to sales.
2. Calculate the food cost percentage for each item or group.
3. Multiply each item's or group's percent food cost by its percentage of sales.
4. Total these results for an average food cost percentage.

patients expected during an operating period, makes the calculation $6.25 × 350,000 estimated patient-days = $2,187,500 total food cost budgeted. For the second, an example might be where a prison allocates in its budget $4,000,000 to the foodservice department for operation for one month. Historically, the department knows its food cost is usually 55 percent, labor 28 percent, and other expenses 17 percent. The total food cost allocation is therefore $2,200,000 (0.55 × $4,000,000).

The simplest way of compiling total food cost is the *spindle method.* Here a foodservice takes all purchase invoices, tacks them onto a spindle, and adds them up later, the total becoming the food cost. While only an approximation, because it does not consider foods in inventory, it gives management a ball-park idea of the total food cost, especially if kept over a substantial period of time. **Exhibit 5.3** shows how such a simplified method works out, giving both dollar values and percentage values based on sales. With computerized systems increasingly taking over accounting report functions, the spindle method is fading into history.

## Individual or Group Item Food Cost

The cost of food items may be obtained from (1) the purchase price of foods purchased per portion or per item; (2) calculations based on the number of portions derived from a single purchase unit; (3) recipe costing; and (4) yield tests.

**Exhibit 5.3** Approximate Daily Food Costs and Percentages

| | Sales | | Food Purchases | | Food Cost Percent | |
|---|---|---|---|---|---|---|
| **Date** | **Today** | **To Date** | **Today** | **To Date** | **Today** | **To Date** |
| Oct. 1 | $882.75 | | $254.16 | | 29 | |
| Oct. 2 | 901.30 | $1,784.05 | 312.86 | $567.02 | 35 | 32 |
| Oct. 3 | 792.45 | 2,576.50 | 301.20 | 868.22 | 38 | 34 |
| Oct. 4 | 922.20 | 3,498.70 | 295.18 | 1,163.40 | 32 | 33 |
| Oct. 5 | 1,006.10 | 4,504.80 | 452.13 | 1,615.53 | 45 | 36 |
| Oct. 6 | 831.35 | 5,336.15 | 286.17 | 1,901.70 | 34 | 36 |

### *Calculating Portion Cost*

It is relatively easy to find the cost of a food portion if it is purchased on a portion basis. For example, if five 10-ounce sirloin steaks are purchased for $22.50, it is easy to determine that the portion cost is $4.50. This simple calculation is often used.

### *Calculating Portions Obtained*

When a unit that contains a specific number of portions is purchased, the portion cost is obtained by dividing the cost of the unit by the number of portions. This is simple, except when there are food losses to calculate and additions in weight because of added ingredients, absorption, or other factors. In such cases, the percent of loss or decrease must be considered if an accurate food portion cost is to be obtained.

For instance, if a case containing six #10 cans of artichoke bottoms costs $47.40, the cost per can is $7.90 ($47.40 ÷ 6 cans = $7.90). If there are 35 bottoms per can, the portion cost is $7.90 ÷ 35 artichoke bottoms, or $0.23 per bottom.

Allowing for cooking loss in cost calculations can complicate the job. If preblanched frozen potato strips cost $0.30 a pound and they lose a third of their weight in frying, the cost per pound of finished potatoes is $0.45 ($0.30 ÷ 67% yield (0.67) = $0.45). The cost of the frying oil absorbed must also be considered; most frozen preblanched potatoes absorb about 6 percent of their weight in oil. Thus, 1 pound of finished potatoes has about an ounce of new frying oil in it (16 oz × 0.06 = 0.96 oz).[1] If the cost per pound of frying oil is $1.00, 1 ounce costs $0.06 ($1.00 ÷ 16 oz = $0.062). Thus, for 17 ounces of finished product—16 ounces of potatoes and 1 ounce of fat—there is a cost of $0.45 ÷ 0.062, or $0.51. An ounce then costs $0.51 ÷ 17 oz, or $0.03 per ounce. If a 2-ounce portion is served, the cost per portion will be $0.06.

### *Recipe Costing*

If used correctly, a standardized recipe will give a known quantity and known quality of food. It should name all ingredients and the amount of each that is needed. It is then easy to calculate the cost of each ingredient and add up these costs to get a total food cost for the recipe. Since a standardized recipe gives the *yield* or resulting number of portions, all one has to do is divide the yield into the total cost to get a *yield cost* or *portion cost.* **Exhibit 5.4** shows a standardized recipe costed out using computer software. In some cases, 2 percent of the total ingredient cost is added to the ingredient cost to pay for the cost of seasonings, frying fat, and so forth. Other operations add up to a 10 percent for seasonings, frying fat, garnish, or crackers. **Exhibit 5.5** shows a standardized recipe obtained from a purveyor's online purchasing Website. **Exhibit 5.6** illustrates a subrecipe costed out using software. **Exhibits 5.7** and **5.8** are also examples of software-generated recipes, which are then costed out with information programmed into the software. **Exhibit 5.9** further analyzes the nutritional content of the costed recipe, illustrating the variety of uses for recipe software.

It is usual to round off ingredient costs to the nearest cent. Portion costs are usually carried out to a tenth of a cent. Many operations have a practice of pricing out all recipes and then checking

**Exhibit 5.4** Standard Recipe Costed Out with Software

**Menu Item: Dave's Pork Surprise**

| | |
|---|---|
| | Recipe Number: 15 |
| Special Notes: | Recipe Yield: 24 |
| Boston butt net | Portion Size: 5 oz. |
| All ingredients weighed as EP | Portion Cost: ______ |

| **Ingredients** | | | **Ingredient Cost** | | |
|---|---|---|---|---|---|
| **Item** | **Amount** | **Unit** | **Unit Cost** | **Unit** | **Total Cost** |
| Boston butt | 10 | lb. | $5.90 | lb. | $59.00 |
| Jones Spicy Sauce | 4 | oz. | 8.00 | lb. | $32.00 |
| Onion | 8 | oz. | 1.20 | lb. | $9.60 |
| Water | ¼ | C | N/A | N/A | N/A |
| Salt | 2 | T | 0.40 | lb. | $0.80 |
| Pepper | 1 | t | 12.00 | lb. | $12.00 |
| Garlic | 1 | clove | 0.60 | clove | $0.60 |
| Pineapple juice | ½ | C | 3.78 | gal. | $1.89 |
| Total | | | | | **$115.89** |

| | | | |
|---|---|---|---|
| Total Recipe Cost: | $115.89 | | |
| Portion Cost: | $4.83 | Date Cd: | 4/13 |
| Previous Portion Cost: | N/A | Previous Date Costed: | N/A |
| Selling Price: | $9.75 | | |
| Food Cost Percentage (by portion): | 49.5% | | |
| Food Cost Percentage Goal: | 31.5% | | |

Food and Beverage Cost Control, *3rd edition; Miller, Jack E., Lea R. Dopson, David K. Hayes; Copyright©2005; Reprinted with permission of John Wiley & Sons, Inc.*

**Exhibit 5.5** Standardized Recipe Using Purveyor Online Ordering Features

**Stuffed Pork Chop with Black Bean Stew and Rice**
**Tomatillo Salsa and Smoked Tomato Sauce**

*Recipe created by the Culinary Institute of America*

| Ingredients | Measure |
|---|---|
| **Filling** | |
| Oil | 5 oz. |
| Onions, minced | 4 oz. |
| Chorizo sausage, diced | 4 oz. |
| Chili peppers, minced | 3 oz. |
| Cayenne pepper | ¼ tsp. |
| Ground black pepper | ¼ tsp. |
| Ground Coriander | ¼ tsp. |
| Eggs | 1 ea. |
| Bread Crumbs | 4 oz. |
| Jalapeño Jack cheese | 3 oz. |
| Pork Chops, 6 oz ea. | 10 ea. |
| Garlic, chopped | ½ oz. |
| Salt | to taste |
| **Smoked Tomato Sauce** | |
| Wood Chips | 2 cups |
| Tomatoes, canned plum | 1.5# |
| Oil | 4 oz. |
| Onions, minced | 4 oz. |
| Garlic, chopped | ½ oz. |
| Tomato puree | 8 oz. |
| Bay leaf | 1 ea. |
| Salt | to taste |
| Ground black pepper | to taste |
| **Black Bean Stew** | |
| *Black beans (pre-soaked) | 8 oz. |
| Water | 1½ qt. |
| Ham hock (smoked) | 1 ea. |
| Oil | 3 oz. |
| Garlic, chopped | ¼ oz. |
| Onion, minced | 2 oz. |
| Chorizo | 2 oz. |
| Salt | to taste |
| Pepper | to taste |
| **Cilantro Rice** | |
| Oil | 1 oz. |
| Onions, minced | 1 oz. |
| Rice, long grain | 2 cups |
| Salt | to taste |
| Pepper | to taste |
| Chicken stock or water | 4 cups |
| Cilantro, chopped | 2 Tbsp. |
| **Tomatillo Salsa** | |
| Tomatillos | 8 oz. |
| Oil | 2 Tbsp. |
| Green chilies | 3 oz. |
| Tomatoes, skinless, seedless | 4 oz. |
| Onions, minced | 2 oz. |
| Limes, juiced | 1 oz. |
| Cilantro, chopped | 1 Tbsp. |
| Salt | to taste |
| Pepper | to taste |
| **At Service** | |

**Procedure**

1. Heat 3-oz oil in a pan.
2. Add onions, peppers and Chorizo, and sauté until tender.
3. Add cayenne, black pepper, and coriander and continue to sauté for 2 to 3 minutes.
4. Pour mixture into a large mixing bowl and allow to cool.
5. Stir in breadcrumbs, egg, and cheese.
6. Make a slit into the side of the pork, chops to create a pocket. Fill the pocket with 2 oz. of filling and press firmly on top of the chop to even it out.
7. Drizzle chops with remaining oil, and season with garlic, salt and pepper. Hold refrigerated for service.
8. Sprinkle wood chips in the bottom of a heavy roasting pan lined with foil.
9. Place tomatoes on a wire rack over wood chips. Place roasting pan on a high burner until chips begin to smolder. Cover pan loosely with foil, and allow to smoke for 10 minutes. Remove tomatoes and hold.
10. Heat oil in a saucepan. Add onions and garlic and sweat for 5 minutes.
11. Add tomatoes, tomato puree, and bay leaf. Bring to a boil and allow to simmer for 20 minutes.
12. Puree and strain the sauce and season to taste with salt and pepper. Hold warm for service.
13. Place beans, water, and ham hock in a saucepan and bring to a boil. Allow to simmer until the beans are tender.
14. Heat oil in a medium size pot. Add Chorizo, onions, and garlic and sauté until tender.
15. Add beans and season with salt and pepper.
16. Remove ½ of the beans and puree them. Fold pureed beans back into mixture. Hold warm for service.
17. Heat oil in a medium size pot. Add onions and sweat until tender.
18. Add rice and toast.
19. Season with salt and pepper.
20. Add hot stock or water. Bring to a boil. Cover and place in a 350°F oven for 20 minutes. Remove from oven and rest for 5 minutes. Uncover and fluff with a fork.
21. Stir in chopped cilantro and season to taste. Hold warm for service.
22. Lightly oil tomatillos and chilies. Place on a wire rack over an open flame. Allow tomatillos and chilies to char on the outside, turning them as necessary for even cooking.
23. Place charred tomatillos and chilies in a bowl and mash with a spoon. Add remaining ingredients and mix well. Season to taste with salt and pepper. Chill and hold for service.
24. Grill chops to order to in internal temperature of 165°F.

**Yield:** 10 servings (one 6 oz. chop, 3 oz. sauce, 3 oz. rice, 3 oz black bean stew, 2 oz. salsa)

**Preparation time:** 2 hours

*Canned beans may be substituted.

**Exhibit 5.6** Subrecipe Costed Out Using Costing Software

| Tomatillo Salsa | | |
|---|---|---|
| **Quantity** | **Unit** | **Ingredient** |
| 8 | ounce | Tomatillos |
| 2 | tablespoon | Oil, vegetable |
| 3 | ounce | Chile pods, California and New Mexico |
| 4 | fluid ounce | Tomatoes, Pld., Sd., Ch. |
| 2 | ounce | Onions, each large |
| 1 | ounce | Lime juice, yield of 1 each |
| 1 | tablespoon | Cilantro, fresh |
| **Recipe Cost:** 1.13 | | |

**Exhibit 5.7** Standard Recipe

**Potato Dumplings**

**Ingredients:**

2.5 lb Boiled potatoes, peeled, cold
12 oz Flour
2 tsp Salt
2 each Eggs
4 oz Butter
4 oz Dry bread crumbs

**Procedure:**

1. Grate the potatoes into a mixing bowl.
2. Add the flour and salt and mix lightly until just combined.
3. Add the eggs and mix well to form a stiff dough. Work in more flour if necessary.
4. Divide the dough into 20 equal portions. Roll each piece into a ball. Refrigerate 1 hour. Dumplings may be made ahead up to point.
5. Heat the butter in a sauté pan and add the bread crumbs. Sauté for a few minutes, until the crumbs are toasted and brown. Set aside.
6. Place the dumplings into a pot of boiling salted water. Stir so that they rise to the top and don't stick to the bottom of the pan. Simmer 10 minutes.
7. Remove with a slotted spoon and place in a single layer in a hotel pan (or onto serving plates).
8. Top with the toasted buttered bread crumbs. Serve 2 pieces per order. (Dumplings may also be served with melted butter or with pan gravy.)

**Portion Size:** 5 oz
**Yield:** 50.0 oz

**Exhibit 5.8** Standard Recipe Costed Out Using Software

| Potato Dumplings | | |
|---|---|---|
| **Quantity** | **Unit** | **Ingredient** |
| 2.5 | pound | Potato, peeled, shredded |
| 12 | ounce | Flour, all-purpose |
| 2 | teaspoon | Salt, regular |
| 2 | each | Eggs, whole, shelled, pooled |
| 4 | ounce | Butter |

**Recipe Cost:** 2.79

Professional Cooking, *6th edition CD-ROM; Gisslen, Wayne; Copyright©2007; Reprinted with permission of John Wiley & Sons, Inc.*

these every six months to see whether they are accurate. Some simplify a recalculation every six months by comparing it to the *national price index*. If the index rises or drops a specific percentage, the total recipe cost is raised or lowered by this percentage and the portion or yield cost is recalculated. Even using this method, it is advisable to do a complete revision every year.

It is often necessary to combine portion costs. While a steak may cost $4.10 if it is served with a salad, french fries, roll and butter, and beverage, these must be added to the steak cost to get a cost for the entire menu item. Often, the exact cost of all added food is calculated and added to give the total cost. Some operators simply calculate the average cost of extras and add this figure to similar items.

When patrons help themselves at a salad bar, there may be some question as to how to cost out a meal. The usual procedure is to keep an account of the cost of foods in the salad bar and keep another count of patrons served from it. Dividing the number of patrons into the total cost of the salad bar will give an average cost per patron. This figure is then added to the basic entree cost.

**Exhibit 5.9** Recipe Software Nutritional Analysis of Potato Dumpling Recipe

**Nutrition Facts**

Serving Size: 5 oz
Servings: 10

**Amount per 100g**

| | | | |
|---|---|---|---|
| **Calories** | 202.60kcal | **Protein** | 4.44g |
| **Total Fat** | 6.25g | **Vitamin A** | 174.34IU |
| **Saturated Fat** | 3.69g | **Vitamin B-6** | 0.21mg |
| **Polyunsaturated Fat** | 0.47g | **Vitamin B 12** | 0.03mcg |
| **Monounsaturated Fat** | 1.55g | **Vitamin C** | 12.79mg |
| **Cholesterol** | 14.89mg | **Vitamin E** | 0.19mg |
| **Sodium** | 234.68mg | **Calcium** | 25.35mg |
| **Potassium** | 310.80mg | **Magnesium** | 22.62mg |
| **Total Carbohydrate** | 32.17g | **Iron** | 1.80mg |
| **Dietary Fiber** | 2.30g | | |

Professional Cooking, *6th edition CD-Rom; Gisslen, Wayne; Copyright©2007; Reprinted with permission of John Wiley & Sons, Inc.*

### *Conducting a Yield Test*

A *yield test* is made to see how much edible food is obtained from raw, unprocessed items, plus cost per portion. Often those making the test must know the amount of product obtained in some common measures or containers. Some of these are given in Appendix B, along with the amount in some common measures frequently used in yield testing.

Making a yield test may be simple or complex. A simple example is where a menu planner wants to know the food cost of a 2-ounce portion of cranberry sauce, if the cranberries are purchased fresh. Five pounds of cranberries costing \$4.70, five pounds of sugar at \$1.25, and 1 quart of water are used; thus, the cost is \$5.95. The yield test shows that the yield is 11 lb, 12 oz, or 188 oz, or 94 portions. The cost of a 2-ounce portion is therefore \$0.063 (\$5.95 + 94 = \$0.063). (Since only the food cost is desired, the cost of labor, heat energy, etc., is not added in.)

A more complex yield test occurs often in cutting out meat portions, where fat, bone, trim, and other items are obtained in addition to the desired portions. A cutting shrink also occurs. **Exhibit 5.10** shows the results of a cutting test made on a No. 180 boneless strip loin to get boneless strip steaks. Note that the byproducts, including the suet, have a value that must be deducted from the original cost. In this test, the cost of the labor is included so the buyer can get an idea of the cost per 10-ounce steak to compare with already-cut 10-ounce steaks that might be purchased instead.

In some cases, evaporation and trim contribute to weight loss. It is therefore wise to not always take the original purchase price as the cost of the item in a yield test, but to take the cost of the item as it goes into the test.

Getting a product down to its *edible portion* (EP) state from its *as purchased* (AP) state in yield tests does not always produce the ultimate *as served* (AS) cost.

Once yields are known, menu planners may not need to repeat them again and again. However, some products can produce very different yields when they are used from time to time. For instance, very young, tender spinach cooks down to a much smaller quantity than older spinach with coarser leaves and texture, so it is wise to make spot checks.

**Exhibit 5.10** Precooking Yield Test on Boneless Strip

| **Boneless Strip, No. 180, 12 lb, \$3.80/lb** | | | **\$45.60** |
|---|---|---|---|
| Yield: | | | |
| Trimmed meat | 16 oz | \$2.60/lb | \$2.60 |
| Suet | 12 oz | 0.08/lb | 0.06 |
| Cutting loss | 4 oz | | |
| | 32 oz (2 lb) | | 2.66 |
| Strip steaks: 10 lb, or 16 10-oz steaks | | | \$42.94 |
| Cost of labor @ \$0.30/lb | | | 3.00 |
| Cost of steaks | | | 45.94 |
| Cost per lb | | | 4.59 |
| Cost per 10-oz steak | | | 2.87 |

*virtualrestaurant.com*

A No. 180 boneless strip is trimmed of 12 ounces of fat before roasting. The total cost upon use is \$45.60 − \$0.06, or \$45.54. A 25 percent shrink occurs during roasting, and there is a further 5 percent carving loss when the meat is portioned for service. The yield cost of a 6-ounce AS portion would be:

12 lb − 12 oz fat = 11 lb 4 oz or 180 oz × (100% − 25% shrink) = 135 oz

135 oz × (100% − 5% cutting loss) = 128.25 oz = 21.4 6-oz portions

\$45.54 cost of meat = \$2.13 per portion

21.4 portions

# Obtaining Labor Costs

As mentioned previously, labor cost is a major factor in menu planning for both commercial and noncommercial operations. Finding the right quantity and quality of employees at the right cost is a challenge to all foodservice managers.

The menu planner must also be aware of labor cost constraints. Labor costs can be high if the menu requires elaborate preparations. Also, menu items that are complicated to prepare require skilled labor, which is not always readily available. Scheduling is another factor not easily controlled by the menu. Rather, it is controlled by forecasting how many patrons can be expected for various meal times.

The labor factor can prove less of a constraint if good hiring procedures are used and thorough training programs are implemented. Employees that feel valued are more likely to feel loyalty to an operation. This may take the form of reduction in waste and turnover and an increase in motivation to perform efficiently.The menu planner has considerable control over how much labor must be used in food preparation and, to a degree, how much may be used in service. If some foods on the menu require complex preparation and production labor, the cost goes up. Simple preparation will save on labor. It is also possible to use purchasing to reduce labor needs. Purchasing frozen instead of fresh leaf spinach can save on washing the latter, and buying shelled rather than unshelled walnuts also saves labor. Keep in mind, though, that quality must meet patron expectations.

Convenience foods can be used to reduce labor needed to produce menu items. In a number of cases, the products are as good as or even better than products made from scratch. The foods also standardize quality and often standardize portions. However, the food cost of the convenience product is often higher than that of the similar products made on the premises, and menu planners must evaluate this extra cost to see whether the convenience product is within cost restraints. Besides labor saved by the convenience product, menu planners also need to add in the cost of energy and perhaps other costs that enter into the preparation of items.

Managers must always keep an eye on service requirements. Certain menu items require much more labor than others. Some items require more tableware and so increase not only service time but also the time needed to wash and put away extra dishes.

Labor cost is made up of wages, salaries, payroll taxes, employee meals, and other costs and benefits. Normally, labor cost for a commercial operation is calculated as a dollar cost and then stated as a percentage of sales.

Often other factors, such as the labor cost per meal, per individual item, per patron served, or the average labor cost per employee may be required to accurately determine total labor costs. Many institutions that do not work on percentages use a dollar labor cost per meal or per patron.

Different kinds of labor costs will be determined. *Direct labor cost* is the cost directly involved in producing, serving, or otherwise handling a menu item. We often calculate the direct

If labor is $665 and sales are $2,500, the labor cost is 26.6 percent of sales ($665 ÷ $2,500 = 0.266 × 100 = 26.6 percent).

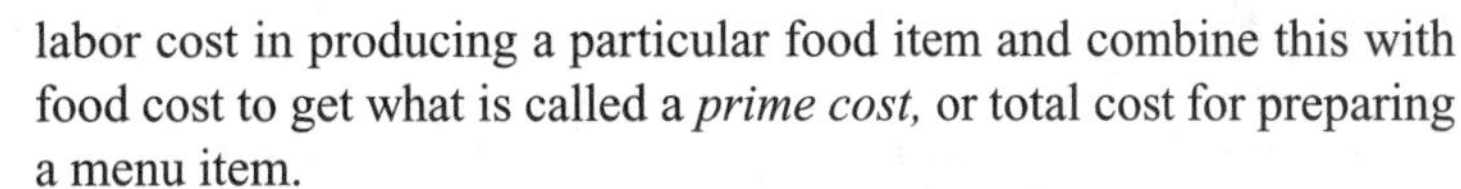

labor cost in producing a particular food item and combine this with food cost to get what is called a *prime cost,* or total cost for preparing a menu item.

Labor cost can vary in foodservices. Some clubs have a labor cost that is more than 50 percent of sales because they must have considerable labor on hand, regardless of whether members are there or not. A take-out unit, however, may have a labor cost of 15 percent. Deciding what is proper must, therefore, be left up to each manager.

In some cases, such as in hospitals, instead of using a labor cost figure stated in dollar values, a figure representing items produced per hour of labor is used.

If a small hospital in one week used 5,664 hours of labor to produce 1,416 meals, the meals produced per labor hour used would be 4 (5,664 ÷ 1,416 = 4).

However, other types of units—and even some hospitals—might want to know how many units of a particular menu item are produced per labor hour.

## Controlling Costs

If a pantry worker produces 60 sandwiches in 4 hours, the worker productivity figure in making sandwiches is 15 per labor hour used.

For any type of foodservice operation, controlling costs is vital. Generally, both commercial and noncommercial operations budget costs. They have slightly different reasons for doing so, but cost control is vital to the success of the menu. If labor and food costs consistently cannot be controlled, the menu may not be right for the facility.

### Institutional Cost Control

Nonprofit operations usually do not use a percentage of food cost for information and control as much as commercial ones do. When they do calculate a percentage of raw food expenditures against income or the budgeted allowance, the food cost frequently runs 50 percent or more of total costs. This can happen when some institutions are subsidized and costs other than those for food and labor may be minimized.

Normally, institutional foodservices, such as hospitals, operate on budgeted cash allowances per meal, per bed, or per person per day for an established period. Amounts in dollars are budgeted for food, labor, and other costs. These will usually be allocated for a month, year, or some other set period, and can then be broken down into an allowed cost per day for food, labor, and other costs.

### Commercial Cost Control

Commercial foodservices usually operate on a cost allowance based on a percentage of sales. Often, a food cost allowance may be 35 percent of sales or of the selling price, and the labor cost allowance 30 percent of sales or the selling price, for a combined 65 percent. Many authorities

Suppose a yearly allowance of $3,555 per person is made for food and all other costs. This allowance is broken down as food, 50 percent; labor, 34 percent; and other costs, 16 percent. The budget will then allow $9.74 per day for food and other per-person costs divided for spending as $4.87 for food, $3.31 for labor, and $1.60 for other costs. This daily cost might be further divided, as shown in **Exhibit 5.11.**

**Exhibit 5.11** Daily Per Person Food Allowance

| | Breakfast | Lunch | Dinner | Total |
|---|---|---|---|---|
| Food (50%) | $1.08 | $1.08 | $2.80 | $4.96 |
| Labor (34%) | 0.75 | 0.75 | 1.75 | 3.25 |
| Other (16%) | 0.40 | 0.40 | 0.73 | 1.53 |
| Total (100%) | $2.23 | $2.23 | $5.28 | $9.74 |

An institution might divide the $9.74 equally between the three meals of the day, for an allowance of $3.25 per meal. This might be broken down per meal to $1.65 for food, $1.01 for labor, and $0.59 for other costs. The annual allowance would be $3,555, with $1,759 allocated for food, $1,176 for labor, and $416 for other costs per person.

agree that exceeding 65 percent may spell trouble for an operation. Some even recommend that the combined percentage allowance be lower because of the consistent rise in the percentage of other costs, such as for supplies or rent. **Exhibit 5.12** shows a profit-and-loss statement for a commercial operation. While food and labor costs fall within the norm, other costs do not, and the profit is low (1.4 percent). Other costs, some beyond the control of the menu planner, are eating up the profit.

Precosting has become one way in which the menu planner can check to see how a menu might do. The menu can then be changed if the precosting information does not give the desired answer. Good precosting requires a good forecast of the sales for menu items—this forecast is often called the *sales mix*—and then a precise indication of costs for these menu items. Some operations are able to use precosting to make quite accurate forecasts of what will happen.

With the increased use of computers, the process of precosting has been simplified. The computer can do much of the detailed work and come up with a fairly good prediction of what will occur. There are a number of precosting programs available to foodservice operators.

## Controlling Food Costs

Many factors can cause high food costs. Improper purchasing, poor inventory control, inaccurate forecasting, waste, a lack of portion control, poor receiving procedures, a failure to follow standardized recipes, poor production procedures, a lack of good promotion and service, poor security, and improper selection of menu items can all cause undesirable food costs.

### *Forecasting*

*Forecasting* is a necessary step in menu planning. After menu items are selected, it is helpful to forecast how many of each will sell. This way the planner can forecast whether the menu will

**Exhibit 5.12** Profit-and-Loss Statement

| | | | | |
|---|---|---|---|---|
| **Sales** | | | | |
| Food | $366,412 | 73% | | |
| Bar | 138,222 | 27 | | |
| Total sales | | | $504,634 | 100.0% |
| **Cost of sales** | | | | |
| Food | $128,944 | 35% | | |
| Bar | 29,978 | 22 | | |
| Total cost of sales | | | $158,922 | 31.5% |
| Gross profit | | | $345,712 | 68.5% |
| **Operational expenses** | | | | |
| Wages | $151,402 | 30.0% | | |
| Vacation and retirement | 10,093 | 2.0 | | |
| Payroll taxes | 5,407 | 1.0 | | |
| Employee meals | 10,067 | 2.0 | | |
| Salaries | 12,617 | 2.5 | | |
| Supplies | 15,008 | 3.0 | | |
| Maintenance and repair | 7,511 | 1.5 | | |
| Laundry and uniforms | 11,608 | 2.3 | | |
| Advertising | 2,418 | 0.5 | | |
| Dishes, glassware, etc. | 7,710 | 1.5 | | |
| Heat, light, and power | 15,001 | 3.0 | | |
| Office expenses | 7,065 | 1.4 | | |
| Total operational expenses | | | $255,547 | 50.6% |
| **Occupancy expenses** | | | | |
| Capital costs (interest, mortgage) | $15,140 | 3.0 | | |
| Rent | 35,327 | 7.0 | | |
| Insurance | 2,487 | 0.5 | | |
| Depreciation | 20,187 | 4.0 | | |
| Taxes | 10,008 | 2.0 | | |
| Total occupancy expenses | | | $83,149 | 16.5% |
| Total operational and occupancy expenses | | | $338,696 | 67.1% |
| Net profits | $7,016 | 1.4% | | |

make the necessary profit for a commercial operation or satisfy the budgetary needs for an institutional foodservice.

There are rhythms in patronage in almost every type of foodservice. For example, payday may bring out a lot of workers; there is always less illness on that day, and in-plant or office foodservice operations can increase production. College dormitories often are nearly empty on weekends, with students going home or eating out. Resorts and operations on well-traveled highways can plan for an increase in business during summer, when vacations start, or when holiday travel begins.

Hospitals can expect lower counts just before a holiday because patients delay entering so they can be home on that day. A racetrack can expect a capacity crowd on the day of a big race, and a stadium will expect one on the day of a big football game. People tend to eat out more on certain holidays than on others. Mother's Day is a big day for many foodservices, while Christmas is not—many close for the latter. There is also a rise and fall in business done by days of the week, depending on location and patronage. Forecasting in these cases is somewhat simplified, but it is always somewhat uncertain. Some software forecasting programs are available for the computer. An example is shown in **Exhibit 5.13.**

A calculation can be made based on the *probability* of increased or decreased patronage. For instance, suppose a club is running a six-day bridge tournament. The club has presold 350 tickets, and the maximum number of patrons the operation can serve per day is 600. The manager estimates the number served per day between 350 and 600. The manager then estimates the probability of each number actually being sold. (These probabilities must add up to 100 percent.) Each estimate of people served is multiplied by its probability to arrive at an average factor. These factors are averaged out to the number most likely to be served each day, in this case 470. These numbers are plotted on a *probability chart* as shown in **Exhibit 5.14.**

If a manager can get some kind of firm base or estimate, it is easier to make a good forecast. In the example, there was a base of 350 tickets to use. An airline has a passenger count based on ticket sales; a banquet manager will have a reservations count as well as experience on what to expect as far as no-shows and walk-ins.

Weather can affect a probability forecast. For foodservice operations in busy shopping areas or near large office buildings, a clear, warm, bright day can bring people out. If it is cold, rainy, or snowing, the place may be almost empty.

*Promotions* can also act to swell volume. A drive-in may offer a bargain price for several days on a product, and the manager should know that the operation will have to gear up to take care of it. A hotel may have a special promotion in its dining areas, offering an early-bird price between 4:30 and 6:30 P.M. The hotel has to be prepared to handle the increased business such a promotion may bring in. The problem is, of course, to assess *how well* the promotion will work. To hold a promotion, have it work—in terms of increased patronage—and then not be able to handle the business properly can be as effective in *turning away* business as a well-handled promotion can be in creating it.

The menu planner should realize that good forecasting is necessary to indicate how much food should be produced. Poor forecasting can cause over- or underproduction. Overproduction results in waste and loss; underproduction means that some patrons may be denied something they want.

### Portion Control

Controlling the size of portions can control food costs. Management is responsible for establishing portion size for the menu item and seeing that this portion is served.

**Exhibit 5.13** Weekly Sales Projection Worksheet: Produced Utilizing Software

Restaurant Name

| | | Sales | | | | |
|---|---|---|---|---|---|---|
| | | **Customer Count** | **Check Average** | **By Meal Period** | **Daily Total** | **% of Week** |
| Monday | Breakfast | 0 | $— | $— | | |
| | Lunch | 50 | $8.00 | $400 | | |
| | Dinner | 70 | $22.00 | $1,540 | $1,940 | 10.70% |
| Tuesday | Breakfast | 0 | $— | $— | | |
| | Lunch | 50 | $8.00 | $400 | | |
| | Dinner | 75 | $22.00 | $1,650 | $2,050 | 11.30% |
| Wednesday | Breakfast | 0 | $— | $— | | |
| | Lunch | 55 | $8.00 | $440 | | |
| | Dinner | 80 | $22.00 | $1,760 | $2,200 | 12.10% |
| Thursday | Breakfast | 0 | $— | $— | | |
| | Lunch | 55 | $8.00 | $440 | | |
| | Dinner | 85 | $24.00 | $2,040 | $2,480 | 13.60% |
| Friday | Breakfast | 0 | $— | $— | | |
| | Lunch | 65 | $8.00 | $520 | | |
| | Dinner | 105 | $26.00 | $2,730 | $3,250 | 17.80% |
| Saturday | Breakfast | 0 | $— | $— | | |
| | Lunch | 55 | $10.00 | $550 | | |
| | Dinner | 120 | $26.00 | $3,120 | $3,670 | 20.10% |
| Sunday | Breakfast | 0 | $— | $— | | |
| | Lunch | 90 | $10.00 | $900 | | |
| | Dinner | 75 | $23.00 | $1,725 | $2,625 | 14.40% |
| | | | | WEEKLY TOTAL | $18,215 | 100% |

*Adapted from Restaurant Research Group, Lexington, MA;* www.rrgconsulting.com *for* virtualrestaurant.com

There are standard portions for many different foods. For instance, a standard portion of vegetables is usually 3 ounces, or about one-half cup. An oven-baked potato, or a portion of mashed potatoes, is usually 5 ounces. A standard portion of sauce is usually 2 ounces. Meat portions will vary according to the type of operation and meat. Normally, menu planners calculate three portions to the pound as purchased (AP) of a roast that has some bone in it. This gives about three 3-ounce portions. If stuffing is served with the meat, the meat portion may be only 2 ounces, with

**Exhibit 5.14** Probability Chart

| | Maximum Number Served | Probability | Average Factor |
|---|---|---|---|
| Day 1 | 600 | 10% | 60.0 |
| Day 2 | 550 | 15% | 82.5 |
| Day 3 | 500 | 20% | 100.0 |
| Day 4 | 450 | 25% | 112.5 |
| Day 5 | 400 | 20% | 80.0 |
| Day 6 | 350 | 10% | 35.0 |
| Totals | 2,850 | 100% | 470.0 |

2 or 3 ounces of stuffing. Ground meat items usually give four portions to the pound AP and an as-served (AS) weight of about 3 ounces per portion.

A table of standard-size portions is usually found in many of the standard recipes used by foodservices.[2] Foodservice operations should also compile their own lists and see that employees follow them. Portion sizes can be indicated on production schedules, recipes, or service charts.

There are many ways to achieve the right size portions. Items can be dished into standard cups or dishes, or a portion scale can be used. Portioning tools, such as scoops and ladles, can also be used to control portion size. The size of the scoop indicates how many portions the scoop gives *per quart* when it is full and level. Thus, a No. 12 scoop 2⅔ oz will give 12 scoops from a quart, if full and level. When speed is important in dishing, employees may not take the time to see that a scoop is full and level. For this reason, it is better to use a smaller scoop, let it be rounded, and obtain faster portioning. Actually, if rounded just slightly, the portion in a No. 12 scoop is about 3 ounces. A slightly rounded basting spoon also gives about a 3-ounce portion and is good for dishing vegetables. Ladle sizes are usually numbered by the ounces they contain when full and level. Standard measures can also be used. Sometimes it is useful to work with volume, not weight measure, in portioning.

Individuals who work in food production should be trained to watch portioning and see that the right number of portions is prepared. Some operations have stainless-steel measuring rods that, when inserted into a steam-jacketed kettle, indicate the number of gallons in the kettle. Thus, if an operation plans to have 400 portions of soup, each portion to be 8 ounces, the rod should show at a certain mark that there are 25 gallons of soup in the kettle. Since evaporation can occur in cooking, this measure should be made just before service. Some operations install mechanical measuring devices that measure and shut off automatically.

Pan weights should also be carefully worked out for proper portions of specific items. Suppose chicken paella is to be dished into 12-by-20-inch steam table pans and baked. Each portion is to be 6 ounces, and 32 portions are to come from each pan. Panning instructions should indicate that 1½ gallons of paella mixture (six quart-ladles full) are to be put into each pan, to be divided in four columns and eight rows.

Some panning instructions in recipes may be stated in weights. If this is so, then it is wise to stamp metal pans with their weight so that employees know the net weight of the pan when filled. Sometimes an average weight is taken for a certain type of a pan and used as a preproduction weight. Thus, a 17-by-25-inch roasting pan may be stamped "4½," indicating that it weighs 4½ pounds. If 20 pounds of au gratin potatoes are to be baked in such a pan, giving 40 8-ounce portions, the pan after filling should weigh 24½ pounds.

As much as possible, foods should be marked or scored to indicate portions. Thus, if food in a 12-by-12-inch pan is to be cut into four columns and six rows, markings should be put on the food before or after baking. For instance, a piecrust can be marked lightly before baking to indicate cutting into four columns and six rows. There are many food markers available for use in cutting pies and cakes.

Hidden factors can disrupt portioning. Although a No. 12 scoop gives 12 scoops per quart of solid material, such as potatoes, it will not give this in ice cream or sorbet because these pack down in the scooping and the volume is reduced. Instead of 12 scoops per quart, one is more apt to get 7 when serving ice cream or sorbet.

Items can lose weight and volume in cooking or baking. One may think there are 3 gallons of sauce in a kettle, because that was the original volume, but it might be only 2½ gallons because of the evaporation loss. There is usually a 4 to 16 percent baking loss in baking bread or cakes. Thus, if a baker expects to have a loaf of bread weighing 16 ounces, the pan weight of the dough must be 18 ounces.

### *Selecting Items to Meet Cost Needs*

In planning a menu, one may wish to have a certain item on the menu but find that it cannot be included because it is too high in price. Often, an adequate, similar substitution can be made. The substitution should be appropriate to the operation and maintain patron value perception. For example, instead of sirloin steak, a cafeteria menu planner might offer a Swiss steak. (For a steakhouse, however, such a substitution will probably not be possible.) It is also possible at times to reduce other costs associated with the item. A less expensive garnish or sauce may bring about the desired change. If the item is combined with other foods, as on a table d'hôte menu, lower-cost foods to serve with the item can be selected.

Many menu planners work with lists of suggested menu items that have been grouped according to cost. If one needs to have an item in a particular cost range, one can find the group in such a list and select one of the items. Sometimes it is possible to call a purveyor and see if something is on the market that meets not only the need, but also the price restriction. Thus, one may wish to put lemon sole on the menu, but the price may be too high. An inquiry may find that petrale sole or sand dabs would be an adequate substitute.

Usually, item selections needed to meet costs require balancing one cost against the other to get the best possible trade-off. For example, one may have an item on the menu that costs a bit more than desired and balance it with a lower-cost item. One must be sure in doing this that the two items balance out in popularity with guests as well. If the selections are swayed to direct sales toward the higher-food-cost item and the lower food cost item is neglected, the menu planner is in trouble.

## Controlling Labor Costs

Labor costs are usually controlled by allocating a specific amount of labor per unit produced, per number of meals per day, per hour, per covers served, or per dollar sales. Commercial operations try to allocate labor costs as a certain percent of sales. Forecasts of labor needs are often made based on expected business. From this a labor budget can be made. *Scheduling* is a crucial step in controlling the amount of labor used. Some operations try to improve labor productivity so they can hire fewer people and still get the same results. This requires good selection, training, and motivation of employees.

The amount of labor used is usually obtained from a time clock. This may be verified by a department head or supervisor and then sent to payroll. At the end of each pay period, the information can be gathered and sent to payroll.

Computers can be used to record employees' actual work time. The computer can also be used to identify the code number of the employee checking in, the code number of the job, the pay for that job, and the deductions, so it accumulates employee hours and can come up with final payroll calculations. With a printer the computer can also prepare payroll checks.

### *Allocating Costs*

Some operations allocate payroll dollars or hours on the basis of dollar sales. **Exhibit 5.15** shows the dollar sales for a week in an operation's dining and bar areas. One hour of labor is allocated for every $50 in sales, and a payroll dollar allowance is made at $7.00 per hour.

Sometimes an allocation may be made on the number of positions for every dollar amount of sales. For instance, for every $75,000 in sales per year, one position is allowed. (A job position may mean more than one employee.)

Some units allow one worker per a specific number of check covers. For instance, there may be one server for every 20 to 25 breakfasts or lunches served and one for every 15 to 20 check covers at dinner. Donald Greenaway, a former executive vice president of the National Restaurant Association and a hospitality instructor, advised that a host or hostess should be able to take care of 200 people at a meal, and a cook should be able to prepare 100 meals during an eight-hour shift.

Formulas have been developed from which staffing requirements can be established. As previously noted, one of the most common standards used today for noncommercial operations is the number of meals produced per labor hour used. (See **Exhibit 5.16.**) If an operation produced 15,000 meals in a month and wanted to produce four meals for every hour of labor used, 3,750 hours of labor would be allocated.

The quantity of labor used between service and back-of-the-house areas varies for different types of operations. A cafeteria or buffet operation often uses less labor for service than a full-service restaurant or club. Some full-service restaurants have found that an allocation of seven hours for back-of-the-house compared to ten hours for service is adequate, but no standard can be set precisely; every operation should work out its own formula.

Some operations develop labor budgets that allocate a given number of hours for various departments per period. Thus, a small foodservice could allow 739 hours for a month as follows; chef and manager, 162; first and second cooks, 244; assistant cooks or helpers, 275; consulting dietitian, 16; kitchen or lunchroom workers, including dishwashers and storeroom personnel, 26; and accounting, 16. Some may allocate labor on the basis of a percentage for various types of workers. For example, in an institutional cafeteria, management may get 10 percent of the total hours; service and cashiers, 20 percent; clerical and accounting, 3 percent; food production, 44 percent; janitorial and cleanup, 12 percent; maintenance, 3 percent; and miscellaneous, 6 percent. If a labor budget is set up, it should be extremely flexible and allow for shifting labor between departments.

## Exhibit 5.15 Labor Allocations by Dollar Sales

| Unit and Meal | Sun 9/10 | Mon 9/11 | Tues 9/12 | Wed 9/13 | Thurs 9/14 | Fri 9/15 | Sat 9/16 | Weekly Total |
|---|---|---|---|---|---|---|---|---|
| **Dining Room** | | | | | | | | |
| Breakfast | $300.00 | $370.00 | $380.00 | $390.00 | $390.00 | $430.00 | $540.00 | $2,820.00 |
| Lunch | 240.00 | 360.00 | 350.00 | 420.00 | 420.00 | 420.00 | 470.00 | 2,710.00 |
| Dinner | 440.00 | 600.00 | 770.00 | 770.00 | 730.00 | 790.00 | 800.00 | 4,740. |
| Total | $980.00 | $1330.00 | $1400.00 | $1540.00 | $1540.00 | $1650.00 | $1830.00 | $10,270.00 |
| Labor Hours | 20.00 | 26.00 | 28.00 | 31.00 | 31.00 | 33.00 | 37.00 | 207.00 |
| Payroll | $140.00 | $189.00 | $196.00 | $217.00 | $217.00 | $231.00 | $259.00 | $1449.00 |
| **Catering** | | | | | | | | |
| Breakfast | $320.00 | | $40.00 | $80.00 | $80.00 | | $60.00 | $480.00 |
| Lunch | | $260.00 | 80.00 | 80.00 | 120.00 | $80.00 | 200.00 | 820.00 |
| Dinner | 40.00 | 100.00 | 50.00 | 80.00 | 240.00 | 820.00 | 520.00 | 1,850.00 |
| Total | $360.00 | $360.00 | $170.00 | $240.00 | $440.00 | $900.00 | $740.00 | $3,250.00 |
| Labor Hours | 7.00 | 7.00 | 3.00 | 5.00 | 9.00 | 18.00 | 16.00 | 65.00 |
| Payroll | $49.00 | $49.00 | $21.00 | $35.00 | $63.00 | $126.00 | $112.00 | $455.00 |
| **Bar** | | | | | | | | |
| Total Sales | $240.00 | $500.00 | $650.00 | $540.00 | $600.00 | $750.00 | $800.00 | $4,080.00 |
| Labor Hours | 5.00 | 10.00 | 13.00 | 11.00 | 12.00 | 15.00 | 16.00 | 82.00 |
| Payroll | $35.00 | $70.00 | $91.00 | $77.00 | $84.00 | $105.00 | $112.00 | $474.00 |
| Total Sales | $1,580.00 | $2,190.00 | $2,220.00 | $2,320.00 | $2,580.00 | $3,300.00 | $3,410.00 | $9,600.00 |
| Total Labor Hours | 32.00 | 44.00 | 44.00 | 47.00 | 52.00 | 66.00 | 69.00 | 354.00 |
| Total Payroll | $224.00 | $308.00 | $308.00 | $329.00 | $364.00 | $462.00 | $483.00 | $2,478.00 |

Formulas used to allocate labor can give broad estimates of requirements. A number of years ago, John F. Johnson developed a formula for staffing cafeterias that indicated that the number of employees on the staff (Y) equaled 2.99 plus 0.82 times the number of thousands of meals served per month (X). In this formula, $Y = 2.99 + 0.82X$. Thus, a cafeteria serving 90,000 meals in a month would be allocated $2.99 + (0.82 \times 90) = 76.79$ employees. Some hospitals find that a better formula for them is $Y = 2.99 + 0.9X$. This allows for the additional labor used to handle special diets and meal deliveries: Professor Broton of Cornell, using Johnson's technique, also developed formulas for hotels, clubs, schools, and hospitals, as follows:

| | |
|---|---|
| Hotels | $Y = 2.34 + 2.2X$ |
| Clubs | $Y = 2.34 + 2.2X$ |
| Schools | $Y = 6.44 + 0.92X$ |
| Hospitals | $Y = 4.01 + 1.08X$ |

**Exhibit 5.16** Average Number of Meals Produced per Labor Hour

| Type Operation | Meals Produced per Hour |
|---|---|
| Hotels and clubs | 1.25–1.75 |
| Restaurants | 1.50–3.00 |
| Cafeterias | 3.50–8.50 |
| School lunches | 11–13 |
| College dormitories | 11.5 |
| Hospitals | 3–6 |
| Nursing homes | 5 |
| Large state hospitals | 11.6 |

Broton indicated his formulas were for full-time employees working a combined average of 206 hours per month.

In allocating labor, it is necessary to differentiate between the number of employees on the payroll and the number of positions allowed. There will be more employees on the payroll than allowed positions because a single position frequently must be filled by more than one person in the seven days of the week, and because of the various shifts. Say, for instance, that an operation requires 56 labor hours to cover one shift position. This means that 1.4 employees must be on the payroll to cover it (56 ÷ 40 hours per week = 1.4). However, this does not provide for days off, holidays, or vacations, so it is usual to have 1.5 to 1.6 workers on the payroll to cover a seven-day position. Many employees work about 232 days a year, or about 64 percent of 365 days. Some operations allow 1.1 workers on the payroll for a five-day (40-hour) position to cover sick leave, absenteeism, and vacations. Where a considerable amount of prepared foods are used, these formulas do not hold. They overestimate how much labor would be required.

### *Scheduling Labor*

In too many foodservices, scheduling and forecasting the required amount of labor are poorly done. Few make attempts to ascertain whether the labor scheduled is needed or whether more or less might be required. Fewer make evaluations on whether it is effectively used. This never translates to effective cost control.

Schedules for workers are used for different purposes. One type shows when workers are to be on the job and is usually called a *work schedule.* A second type shows days off or when vacations are to be taken, and a third shows the production tasks for workers on the job. It is possible to combine the first and last types.

A work schedule should identify each employee and indicate time at work for a specific period. This is done differently, depending on the type of schedule and the particular operation. A production work schedule frequently shows the following:

- Period covered
- Work to be done
- Who does the work
- Amount to produce
- Recipe to use
- Portion size
- Meal or completion time required
- Comments
- Slack-time assignments

Operations should use whichever schedules serve their needs; several are shown here that illustrate this diversity. **Exhibit 5.17** shows a production schedule for an institutional operation; it could be used, with some modification, for a commercial unit. Sometimes a schedule is set up that does not indicate hours to be worked, but work to be done. Thus, management does not set the time that the worker can put in, but sets the work or goals to be accomplished. **Exhibit 5.18** shows such a work schedule by assignment. A typical schedule for work in a commercial dining room—with workers assigned by station—is presented in **Exhibit 5.19. Exhibit 5.20** is a time-line schedule developed for an institutional foodservice with a number of dining services. A count of patrons in an operation at a certain time and an estimate of the labor needed can result in what is termed a *bar schedule graph.* This shows when certain employees should be on the job.

**Exhibit 5.17** Production Schedule

Date ____________________

| Meal and Item | Worker | Amount | No. Portions | Portion Size | Comment |
|---|---|---|---|---|---|
| **Breakfast** | | | | | |
| Carrot juice | F | 12 46-oz | 87 | 6 oz | Do Tues PM |
| Pomegranate grapefruit juice | F | 1 No. 10 | 12 | 6 oz | Do Tues PM |
| Dates | F | 1 case | 16 | 3 dates | Do Tues PM |
| Oatmeal (R-1) | B | 4 gal | 60 | 1 c cooked | |
| Dry cereals | F | (Assorted) | | 1 pkg | Send to floors |
| French toast | B | | 84 | 3 half slices | |
| Syrup | G | | 84 | 1 packet | |
| Butter | G | | 84 | 1 pat | |
| Coffee (G-2) | B | 4 gal | 60 | 1½ c | Send in pitchers |
| Milk | G | 138½ pt. | | ½ pt. | |
| Tea | B | 2 gal hot $H_2O$ | 30 | 1½ c $H_2O$ with 1 tea bag | |

**Other assignments**

**Morning**

Clean refrigerators 1 and 3 and storeroom.

Pre-prep:
- Wash lettuce and separate leaves for dinner salads.
- Peel 40 lb AP potatoes for dinner.
- Make marinara sauce for pasta for Thursday (2 gal) (C-11 recipe).
- Pick over dried beans for Thursday lunch soup.
- Take turkey meat from bones stored in refrigerator 2.
- Check salad dressings and make up those necessary.

Lunch

**Exhibit 5.18** Work Schedule by Assignment

Date ______________________

| Cook | Prep Cook | Pantry and Baker |
|---|---|---|
| Make coffee and hot cereal | Pick turkey meat from bones | Put ice into bins |
| Prepare french toast (B-8) | Pare potatoes | Cut butter and margarine |
| Prepare marinara sauce | Wash lettuce and fix leaves | Help dish breakfast |
| Mash potatoes | Help dish breakfast | Make and dish up lunch |
| Cook green beans | See cafeteria counter is properly filled | |
| Help dish lunch | Prepare 1½ qt diced pepper | Make 3 pans cornbread (S-10) |
| Help clean refrigerators | Chop 1 c pimiento | Make 2½ gal pancake batter |
| Help clean storeroom | Clean up cafeteria counter | Batter for Thurs. breakfast (S-8) |
| Do pre-prep as instructed for next day | Make salad for night | Make lemon sorbet (R-15) |
| Bread calamari | Help clean refrigerators and storeroom | Dish sorbet |
| Get vegetables ready for dinner | Help set up lunch counter | Set up salads |
| Pan baked potatoes and get ready for baking | Do preparation as instructed | Help dish lunch and supervise lunch counter |
| | Help dish up lunch | Do preparation for apple pie for Thurs. |
| | Clean lunch counter | Clean own unit and do own pots and pans |
| | Wash pots and pans as required | Make 10 dozen garlic knot rolls for dinner (S-17) |
| | Clean cook's shelves | Cut 30 lemon wedges for tea |

By taking a look at such a bar graph and looking at the patron count, one can quickly note whether there is a sufficient number of workers when most of the work must be done.

It is extremely important that employees be properly informed of days off and when they can expect to take vacations. Employees have the right to know at least a week ahead which days they can be off to arrange personal time. Nevertheless, days off should be guided as much as possible by business activity and operational needs. Workers must be there when the work is there. However, management should attempt to keep scheduling as flexible as possible so that workers can, in times of personal need, take time off.

Planning vacations should be done with care. There are certain times when labor requirements may be less, and this is the time to schedule vacations. It may be possible, with proper planning, to take care of much vacation time without having to hire additional labor. Planning ahead can do much to please workers and keep the operation running smoothly.

There are state and federal regulations that control the number of hours that teenagers and, at times, children can work. The type of work performed may also be controlled.

The number of hours worked will be controlled, and all hours worked over a total time must be paid in overtime. There are different overtime rates. In scheduling employees, it is important to avoid overtime payment, especially at high penalty rates. Workers should not be allowed to be

**Exhibit 5.19** Commercial Operation Schedule

**GOLD ROOM SCHEDULE**

| Morning Shift 7 to 3:30 | Station | Afternoon Shift 3:00 to 11:30 |
|---|---|---|
| Maria Sanchez | 1 | Scott Noell |
| Ben Hallman | 2 | Gloria Rutter |
| Sachi Metra | 3 | Caimile Roberts |
| Sarah Smith | 4 | Julia Dilan |
| Natalie Merck | 5 | Naavah Samuels |
| Roger Widner | 6 | Jim Dishaw |
| E. Turner | 7 | Helen Peller |
| Lenge Killian | 8 | Mary Sage |
| Pablo Cruz | 9 | closed |
| Razi Razak | 10 | closed |
| Sam Baker | 1, 2 & 3 | Marilyn Lowers |
| John Hepner | 4, 5 & 6 | Richard Kosse |
| Gary Dentor | 7, 8 & 9 | Stan Stanford (7 & 8 only) |
| Sandra Holman 11–3:30 | 11 | closed |
| Valerie Mason 11–3:30 | 12 | closed |
| T. Crane 11–3:30 | 14 | closed |
| Randall Hill 11–3:30 | 11, 12 & 14 | closed |
| | | |
| | | |
| | | |
| | | |
| | | |
| | | |

Date Feb. 20 Signed George DiNatale

## Exhibit 5.20 Scheduling Worksheet

**MY RESTAURANT**
**Blind Staffing Worksheet**

**Prepared by: Steve Jones** **Date: June 22, 2002**

**Day of Week** **Thursday**

| | Owner | Manager | Chef | Kitchen | Kitchen | Kitchen | Kitchen | Waiter | Waiter | Waiter | Waiter | Waiter | Bar | Host | Busser | Total (Hrs) |
|---|---|---|---|---|---|---|---|---|---|---|---|---|---|---|---|---|
| 6–7 AM | | | | | | | | | | | | | | | | 0 |
| 7–8 | | 1 | | | | | | | | | | | | | 1 | |
| 8–9 | | 1 | | | | 1 | | | | | | | | | | 2 |
| 9–10 | | 1 | | | | 1 | | | | | | | | | | 2 |
| 10–11 | | 1 | | | | 1 | | | | | | | | | | 2 |
| 11–12 | | 1 | | | | 1 | | | | | | | | | | 2 |
| 12–1 PM | | 1 | | 1 | | 1 | | | | | | | | | | 3 |
| 1–2 | | 1 | | 1 | | 1 | | | | | | | | | | 3 |
| 2–3 | | 1 | 1 | 1 | | 1 | | | | | | | | | | 4 |
| 3–4 | 1 | | 1 | 1 | 1 | 1 | | 0.5 | | | | | | | | 5.5 |
| 4–5 | 1 | | 1 | 1 | 1 | | 1 | 1 | 0.5 | | | | | | | 6.5 |
| 5–6 | 1 | | 1 | 1 | 1 | | 1 | 1 | 1 | 1 | | | 1 | 1 | | 10 |
| 6–7 | 1 | | 1 | 1 | 1 | | 1 | 1 | 1 | 1 | | | 1 | 1 | 1 | 11 |
| 7–8 | 1 | | 1 | 1 | 1 | | 1 | 1 | 1 | 1 | | | 1 | 1 | 1 | 11 |
| 8–9 | 1 | | 1 | | 1 | | 1 | 1 | 1 | 1 | | | 1 | 1 | 1 | 10 |
| 9–10 | 1 | | 1 | | 1 | | 1 | 1 | 1 | 1 | 1 | | 1 | 1 | 1 | 11 |
| 10–11 | 1 | | 1 | | 1 | | 1 | 1 | 1 | 1 | 1 | | 1 | 1 | 1 | 11 |
| 11–12 | | | | | 1 | | 1 | | | | 1 | | 1 | | 1 | 5 |
| 12–1 AM | | | | | | | | | | | | | 1 | | | 1 |
| TOTAL | 8 | 8 | 9 | 8 | 9 | 8 | 8 | 7.6 | 6.5 | 6 | 3 | 0 | 8 | 6 | 6 | 101 |

| Category | RATE | # HRS | TOTAL $ |
|---|---|---|---|
| Owner | $18.00 | 8 | 144.00 |
| Manager | $12.00 | 8 | 96.00 |
| Chef | $15.00 | 9 | 135.00 |
| Kitchen | $9.00 | 33 | 297.00 |
| Waiter | $2.50 | 23 | 57.50 |
| Bar | $5.00 | 8 | 40.00 |
| Host | $10.00 | 6 | 60.00 |
| Busser | $7.00 | 6 | 42.00 |
| | | 101 | 871.50 |

| | |
|---|---|
| TOTAL HOURS SCHEDULED | 101.0 |
| TOTAL DAY'S WAGES | 871.50 |
| EST PAYROLL TAX & BENEFITS | 130.73 |
| GROSS WAGES | 1,002.23 |
| PROJECTED SALES for DAY | 3,100.00 |
| LABOR PERCENTAGE | 32.3% |

absent from the job and then, later, attempt to qualify for overtime because they worked special days. Good control of scheduling can do much to reduce loss from manipulation by employees.

Forecasting labor requirements is very valuable. Management can usually work with department heads to establish accurate labor needs. Computerization of forecasting and scheduling has greatly simplified management tasks.

After actual events have occurred, it is always necessary to compare forecasted and actual figures. Too much labor may have been scheduled, or not enough, or perhaps it was not used to the best advantage. There will always be a range of error; but if there is accurate historical information (for example, on business trends, advance reservations, house occupancy, bookings, local or regional events, and other factors, such as the season or internal needs), error can be reduced if the forecast is evaluated afterward.

Computer programs can be used to budget labor. Whoever is responsible for assigning labor inputs the data and copies the supervising office. Then the actual time used is shown on a revised sheet. If there is too great a difference, the person responsible for the labor assignment must explain the discrepancies.

### *Improving Productivity*

Authorities agree that high labor waste occurs in foodservices and that more emphasis should be given to improving the use of labor. Currently, programs for doing this are directed toward four areas: hiring and training; using labor-saving devices and foods requiring less labor to prepare; improving layouts; and improving forecasting and scheduling of labor.

Many operations use poor methods in selecting employees, with the result that they get employees who do not have the requisite skills, knowledge, or motivation to do an adequate job. Furthermore, after hiring inadequate employees, these operations do little to see that they become more effective employees.

Attention must be given to hiring practices that minimize problems and fill positions with those likely to do the best job. A *job specification* is written to indicate the qualifications, skills, and experience of employees hired for a specific position. This should be used in interviews and for seeking potential employees to improve the selection process. The job specification usually covers the job and some of its work requirements, skills, education, and knowledge required.

A *job description* should be written for each position to indicate the tasks of each position. This should be shared with employees during their orientation, during which employees become oriented to the company, position, supervisor, and co-workers.

Training can also do much to improve job performance. Employees should be given the opportunity to learn on the job or take classroom or correspondence courses to become better employees and be of more value to the company. Job evaluations assist management and employees in knowing how well employees are performing and what might be needed for improvement.

Most workers have never been taught how to make their jobs easier yet more productive. Efficient workers often become so because they are basically efficient by nature and learn the shortcuts quickly. However, workers can be taught how to work more efficiently. By planning work, arranging work areas to be more efficient, and reducing motions and energy in work, managers can increase their employees' productivity while reducing turnover. The field of knowledge on how to improve jobs is called *work simplification,* or *human engineering.* Food services that have worked to improve conditions frequently find that not only does productivity increase, but so does employee morale. Considerable material on this subject is available for use in foodservice.

## Other Controls

Managers should continually examine the nature of all fixed operational costs. Some costs thought to be fixed may be what are called *programmed costs* (that is, they may be only programmed into the operation), and these sometimes can be controlled or changed. For instance, the cost of electricity, water, heat for cooking, or the cost of cleaning supplies may be programmed into operation as fixed costs, but if lights are turned off more frequently, supplies are used properly, leaking faucets are fixed to reduce the use of hot water, and broilers, griddles, and steam equipment are turned off when not in use, some of these costs may be reduced. A small change may make a considerable difference in the amount required to pay for these fixed costs.

Although some organizations are seeking to cut costs by eliminating employee meals, it might serve managers well to recall that those in the hospitality business perhaps, ought to want employees to be well-fed. If not driven by conscience, the cost of not feeding employees might be higher in pilferage, as the staff members nibble throughout their shift or take food home. This behavior is not without consequence in sanitation terms. Some operations provide communal preshift meals that serve to promote team harmony and rally the troops. Others offer deep discounts on nearly all menu items, allowing employees to choose what they would like, deducting the cost from the employee check. Employee meals are not food but labor cost if the price charged employees is below prime cost.

Being willing to "sell out" versus "run out" of a special prevents waste through overprepping, or the business cost of serving a product to guests that is inferior in an attempt not to waste any product.

Employee empowerment is successfully being taken to new levels by a number of large corporations that are allowing managers to buy in and become part owners of a particular operation. This creates a more personal sense of responsibility and may assist in dropping the turnover rate as managers stay due to ownership and hourly employees see what they might aspire to.

Part timers can also be a huge asset to foodservice operations, lending flexibility, helping to alleviate overtime, and infusing enthusiasm from a fresh perspective.

## Budgeting Costs

Many foodservices prepare *budgets* to act as guides to direct operations and indicate cost allocations. The budget shows expected income and how it will be spent. Budgets should be based on *realistic expectations* and not hopes. Budgets should be flexible enough to adjust to changing conditions, but only up to a point. If changed too often, a budget becomes *subject to* operating conditions rather than *controlling* them. Some operations have budgets based on variations in income and costs.

In some cases, an institution may modify a cost allowance by setting up a *food allowance* per day or per period for a specific quantity of foods from different food groups. This allowance is based on the quantity of food from each group required to provide each individual with an adequate diet. Thus, for each person, a specific grain amount; milk or its equivalent; fish, poultry, or meat; vegetables; and fruit is allowed per day, per week, per month, or per period. The cost of each of these food groups is determined, and the total becomes the cost allowance per person. American military services operate on such a system. It is usually called a *ration allowance.* In many states, tax-supported institutions, such as hospitals and prisons, use the ration system.

Some ration allowances are based on weekly allowances developed each year by the U.S. Department of Agriculture (USDA) based on current prices. There are three levels for these allowances: low, moderate, and liberal. Most state institutions use the low-cost plan but some, such as California, add a bit more meat to give palatability and acceptability. When a ration system is used, only the cost of the food is obtained. Other costs, such as those for labor, must be added to food costs to get total operation costs.

The USDA operates the school foodservice program and recommends that a General Meal Pattern meal contain servings of meat or a meat alternate, vegetables and/or fruits, bread or a bread alternate, and milk. (See **Exhibit 5.21.**) School foodservices are actively trying to reduce fats and use sodium and sugar in moderation in the student diet, and menus have reflected this change.

Experience is helpful in establishing what a budget should be, but the budget should not be based on performance alone. Using past figures may compound past mistakes and faulty calculations. Previous budgets should be guides only, used to indicate what costs were under past conditions. If a budget is based on the past, it may be good for past conditions but not necessarily for the present.

Today many operations wipe out past budget figures at the beginning of a new period and start fresh. They first project expected conditions and make up budget figures based on them. The method of not using any past figures in budget planning is called *zero-based budgeting.* If such a budget is well researched and the information is presented with alternatives for action, management has a better chance of evaluating factors regarding menu planning and other issues requiring management decision. For instance, if budgetary projections are properly made, it may be much easier for a manager to decide whether it would be a better idea to purchase all bakery goods or to remodel the bakeshop.

## Exhibit 5.21 A Menu Planner for Healthy School Meals

| Major Features of Food-Based Menu Planning: Comparing Traditional and Enhanced | |
|---|---|
| **TRADITIONAL** | **ENHANCED** |
| **1. Nutrition Goals** | |
| Menus must meet nutrition goals when averaged over a school week and analyzed by the state agency during a state nutrition review. | Same as *Traditional* |
| **2. Nutrient Standards and Age/Grade Groups** | |
| ◆ **For school-age students (K–12)**<br>*LUNCH:*<br>There are two established age/grade groups: grades K–3 and grades 4–12. There is also an optional recommended age/grade group: grades 7–12.<br>*BREAKFAST:*<br>There is one established age/grade group: grades K–12. | ◆ **For school-age students (K–12)**<br>*LUNCH:*<br>There are two established age/grade groups: grades K–6 and grades 7–12. There is also an optional recommended age/grade group: grades K–3.<br>*BREAKFAST:*<br>There is one established age/grade group: grades K–12. There is also an optional recommended age/grade group: grades 7–12. |
| ◆ **For preschool children**<br>*LUNCH AND BREAKFAST:*<br>There are two preschool age groups: ages 1–2 years and ages 3–4 years. | ◆ **For preschool children**<br>*LUNCH AND BREAKFAST:*<br>There are two preschool age groups: ages 1–2 years and ages 3–4 years. |
| **3. Criteria for a Reimbursable Meal** | |
| Provides the required food components and food items in the correct serving sizes to meet the appropriate *Traditional* meal pattern.<br>◆ Four food components for lunch<br>◆ Five food items for lunch<br>◆ Three or four food components for breakfast<br>◆ Four food items for breakfast | Provides the required food components and food items in the correct serving sizes to meet the appropriate *Enhanced* meal pattern.<br>◆ Four food components for lunch<br>◆ Five food items for lunch<br>◆ Three or four food components for breakfast<br>◆ Four food items for breakfast |
| **4. Meal Structure for Lunch** | |
| The following are minimum requirements for school-age students by age/grade group for each of the four components: Meat/Meat Alternate; Grains/Breads; Vegetables/Fruits; and Milk. | |
| ◆**Meat/Meat Alternate:**<br>*Grades K–3:* 1½ oz.<br>*Grades 4–12:* 2 oz.<br>*Optional (recommended)*<br>*Grades 7–12:* 3 oz. | ◆**Meat/Meat Alternate:**<br>*Grades K–6:* 2 oz.<br>*Grades 7–12:* 2 oz.<br>*Optional (recommended)*<br>*Grades K–3:* 1½ oz. |

*United States Department of Agriculture (USDA)*

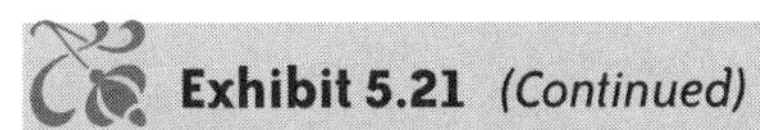

**Exhibit 5.21** *(Continued)*

| TRADITIONAL | ENHANCED |
|---|---|
| **4. Meal Structure for Lunch** *(continued)* | |
| ◆**Grains/Breads:**<br>*Grades K–3:* 8 serv. per week; minimum 1 serv. per day<br>*Grades 4–12:* 8 serv. per week; minimum 1 serv. per day<br>*Optional (recommended)*<br>*Grades 7–12:* 10 serv. per week; minimum 1 serv. per day | ◆**Grains/Breads:**<br>*Grades K–6:* 12 serv. per week; minimum 1 serv. per day<br>*Grades 7–12:* 15 serv. per week; minimum 1 serv. per day<br>*Optional (recommended)*<br>*Grades K–3:* 10 serv. per week; minimum 1 serv. per day |
| ◆**Vegetables/Fruits:**<br>At least two different fruits and/or vegetables must be offered.<br>*Grades K–3:* ½ cup per day<br>*Grades 4–12:* ¾ cup per day<br>*Optional (recommended)*<br>*Grades 7–12:* ¾ cup per day | ◆**Vegetables/Fruits:**<br>At least two different fruits and/or vegetables must be offered.<br>*Grades K–6:* ¾ cup per day plus additional ½ cup per week<br>*Grades 7–12:* 1 cup per day<br>*Optional (recommended)*<br>*Grades K–3:* ¾ cup per day |
| ◆**Milk:**<br>*For all age/grade groups:*<br>8 oz. fluid milk as a beverage | ◆**Milk:**<br>*For all age/grade groups:*<br>8 oz. fluid milk as a beverage |
| **5. Meal Structure for Breakfast** | |
| The following are minimum requirements for school-age students by age/grade group for each of the three or four components: Meat/Meat Alternate and/or Grains/Breads; Juice/Fruit/Vegetable; and Milk. | |
| ◆**Meat/Meat Alternate and/or Grains/Breads:**<br>*Grades K–12:* Two servings of Meat/Meat Alternate (1 ounce per serving) or two servings of grains/breads or one of each | ◆**Meat/Meat Alternate and/or Grains/Breads:**<br>*Grades K–12:* Two servings of Meat/Meat Alternate (1 ounce per serving) or two servings of grains/breads or one of each<br>*Optional (recommended)*<br>*Grades 7–12:* Same as grades K–12 plus one additional serving of grains/breads. |
| ◆**Juice/Fruit/Vegetable:**<br>*Grades K–12:* ½ cup | ◆**Juice/Fruit/Vegetable:**<br>*Grades K–12:* ½ cup<br>(Same for grades 7–12, Optional) |
| ◆**Milk:**<br>*Grades K–12:* 8 oz. fluid milk as a beverage or on cereal or both | ◆**Milk:**<br>*Grades K–12:* 8 oz. fluid milk as a beverage or on cereal or both<br>(Same for grades 7–12, optional) |

*United States Department of Agriculture (USDA)*

**Exhibit 5.21** *(Continued)*

| TRADITIONAL | ENHANCED |
|---|---|
| **6. Offer versus Serve for Lunch** | |
| Required for senior high schools.<br>High school students must take no fewer than three of the required five food items. They get to choose which item(s) to decline.<br>Optional for lower grades. (School food authorities decide whether to have OVS.) | Same as *Traditional* |
| **7. Offer versus Serve for Breakfast** | |
| Optional for senior high schools.<br>Optional for lower grades.<br>Students may decline one food item from any food component. | Same as *Traditional* |
| **8. Standardized Recipes** | |
| A record and copy of recipes used must be available during state nutrition review and analysis. | Same as *Traditional* |
| **9. Processed Foods** | |
| A record of products used must be on file.<br>Child Nutrition label or other documentation will assist in determining food credit of food components and serving sizes.<br>Nutrition facts labels and/or manufacturer's nutrient data sheets will be needed during state nutrition review. | Same as *Traditional* |
| **10. Production Records** | |
| Program regulations require schools to keep food production and menu records. | Same as *Traditional* |
| **11. Child Nutrition Labeling** | |
| Child Nutrition (CN) labels on products show the product's contribution toward meal pattern requirements. | Same as *Traditional* |

*United States Department of Agriculture (USDA)*

## SUMMARY

One of the most important factors in menu planning is cost. It is important after establishing a price on a menu to see that both food and labor costs are covered.

All types of operations must control costs. Commercial operations must do so to attain a healthy profit. Noncommercial operations must cover costs, usually to stay within a budget. Food costs may be affected by factors outside the food buyer's control, such as weather, labor strikes, and demand. Yet the operator can control costs by using methods such as forecasting sales, controlling portions, and deliberately selecting items to meet cost needs.

Labor costs are controlled partly by external factors, such as labor demand and minimum wage laws. However, improving productivity is an internal way the operator can control costs. Careful scheduling is crucial.

Other costs, such as those for energy and cleaning supplies, should be evaluated to see how they can be controlled.

## QUESTIONS

1. A take-out operation has the following in dollar sales, percent of dollar sales, and food cost percentage for the listed items. What is the average food cost?

| Item | $ Sales | % of Sales | % Food Cost |
|---|---|---|---|
| Pasta | $ 2,250 | 17.0 | 30 |
| Salads | 1,866 | 14.0 | 28 |
| Pizzas | 4,023 | 30.3 | 38 |
| Wings | 2,436 | 18.4 | 27 |
| Beverages | 2,673 | 20.2 | 29 |
| | $13,248 | 99.9 | |

Average food cost = ________________

2. Determine the portion cost for each of the following.
   a. 24 apples per box; one box costs $12.82; portion cost: ________________
   b. Tart à la mode:
      12 portions per tart; cost per tart $8.80; portion cost: ________________
   c. Gelato 32, 4-ounce servings per gallon; 2.5 gallons cost $100,
      portion cost: ________________

3. A leg of lamb weighs 10 pounds. It is boned, rolled, roasted, and carved. The losses are boning and rolling, 35 percent; cooking shrinkage, 30 percent; and carving loss, 5 percent. What is the amount of servable meat after carving? How many 3-ounce portions are obtained?

4. According to Johnson and Broton, if a club served 14,820 meals in a 30-day period, how many employees should be allocated to the payroll?

5. Food cost is $864.22 and sales are $2,564.22. What is the food cost percentage?

6. A Polenta recipe calls for the ingredients listed below. What is the total food cost, and the cost per portion if 28 two-ounce portions are obtained?

| | | |
|---|---|---|
| Shallots chopped | 4 tsp (10ml) | .99 |
| Olive oil | 2T | .25 |
| Water | 4 lb | .0 |
| Cornmeal | 12 oz | .76 |
| Salt + pepper | To taste | .05 |

## MINI-CASES

5.1 Find a recipe for a vegetarian burger and cost it out. Now search for a comparable commercial product and find out its cost. What other information would a manager need to make a "make or buy" decision? List the questions that would need to be answered and describe the steps that could be taken to find answers.

5.2 Would it be preferable to employ two high school students as buspersons who have no experience working in restaurants, but who are willing to work for minimum wage with no benefits, or to employ one very experienced busperson who is able to work much more efficiently but wants a higher hourly rate and may want benefits down the road? What are the pros and cons of each?

# 6

# MENU PRICING

## Outline



## Objectives

After reading this chapter, you should be able to:

1. Characterize the most common pricing techniques used in the foodservice industry.
2. Identify historical perspectives on pricing and explain why non-cost pricing methods are no longer practical.
3. Discuss the factors of pricing that influence product selection, including value perception and economic influences.
4. Explain how pricing psychology and market research play a role in menu pricing.
5. Characterize the most common pricing techniques based on costs used in the foodservice industry, including food cost percentage, pricing factor/multiplier, prime cost, and actual costs.
6. Characterize other pricing techniques in the foodservice industry, including gross profit, one-price, marginal analysis, cost-plus-profit, minimum charge, and pricing based on sales potential.
7. Identify software's role in assisting with food pricing.

# Introduction

After a menu is planned, each item on it has to be priced. There are a number of items to be considered in pricing, including costs, the market, and type of operation.

*Prices must cover costs.* This requirement is essential for both commercial operations and institutions that charge for meals but do not operate for profit. However, noncommercial operations may seek supplemental help from government or charitable contributions to balance losses. Commercial units do not have this insurance and will pay the consequences of faulty pricing. Profit averages 3 to 7 percent of sales, so there is not a wide margin for error.

The market is a major factor in pricing. Some patrons want only low prices; others seek moderate ones; some will be willing to pay higher prices. *Perception of value* will also vary from patron to patron; it is sometimes difficult to meet the desires of each group. Many operators use a "what the market will bear" approach to pricing. However, such a simplistic strategy may not meet operational needs and may drive patrons to competitors.

Different kinds of operations use different markups. A *markup* is the difference between the cost of a product and its selling price. Some operations can use a low markup and depend on high volume to give an adequate income with which to operate and make a profit. Others will use a higher markup and require a lower volume to attain a profit. Others will use a higher markup and require a lower volume to attain a profit. Menu prices must not only cover the costs of food and labor but must often include other significant cost factors such as atmosphere, rent—especially in a prime location—and advertising.

# Non-Cost Pricing: Historical Perspectives

A number of methods that are not based on cost have historically been used in pricing menus. Some, such as pricing based on tradition, competition, or what the market will bear, are in this category. Surprisingly, they were used often and, even more surprisingly, were sometimes successful. Some are difficult to use. Finding a price based on what the market will bear is an involved process and takes a lot of testing and observation. Others, such as traditional pricing or pricing against competition, are relatively simple.

## Pricing Based on Tradition

Many operations look at traditional prices, often those set by the leaders in a particular market niche. Tradition in pricing may relate not only to a specific price but also to the pricing structure and market. Many operations find they cannot get as high a markup for California wines as they can for European ones, although the California product may be of equal or better quality. Tradition has it that the domestic product is usually lower in price, and, therefore, the pricing structure must be varied to suit this tradition.

Different types of operations will also find that the pricing structure they must follow is one in which a lower markup must be taken than that of another type of foodservice. Thus, while two operations may sell exactly the same thing, one will be able to set prices on a different basis from the other.

Some operators may find that prices have been the same for so long that they become traditional with their customers. When they attempt to change these prices, they may meet with very strong sales resistance and customer dissatisfaction. Thus, they may decide not to change the price on a particular item but get a higher markup on other items that are not so restricted.

It is possible for a price to become traditional because a leader in the market charges this price. Thus, an industry leader like McDonald's sees its prices often become traditional among similar operations. With changing costs, this is an impractical and risky method of pricing.

## Competitive Pricing

One of the most common pricing methods is to base the price on what the competition charges. Although it is wise to pay attention to competitors' prices, it is unwise to base your prices *completely* on them. What the competition charges may bear no relation to your costs. Pricing in this manner wrongly assumes that prices satisfactory for the competition's customers are also satisfactory for your customers. In addition, copying competitors' prices does little to differentiate an operation.

If competitive pricing is indicated by research, a unit should study its own costs to analyze how it can price menu items to produce a more favorable response than that produced by the competition's prices. A competitor's price may be studied to reveal information about the food, labor, and other operating costs. Food and labor are standard commodities that will have a known price. A close scrutiny of these may indicate what the competition might be doing to achieve a favorable price structure. A study of competitive pricing and its effect on one's own business is also warranted. In some cases, prices may have to reflect the influence of competition. Value bundling is a competitive strategy adopted by quick-service restaurants, reluctantly but in response to customer demand.

## What the Market Will Bear

A method of pricing used by some companies is to design a product and then test market it at a given selling price. The product may be put on several markets at different prices, and customer reaction will be studied to ascertain which is the best price.

Marketing specialists state that one of the best routes to success is to develop a product the market wants and then price it so that a healthy profit is made. With proper pricing and strong demand, these specialists say there is a high chance of developing a successful market. They recommended studying the value of the product *in the minds of patrons* and then charging accordingly. Some products may have to be priced only slightly above cost to be accepted by the market; others can have a much higher margin.

Establishing a selling price based on what the market will bear is not just a trial-and-error method; it should be based on sound research. To some extent, prices can be tested to see how well patrons accept them; however, not too much experimentation per item can be done. Perhaps three or four prices can be tested on an item, and then the testing must be stopped. Otherwise, patrons might reject the item because of the instability of the price.

Pricing according to what the market will bear has become a popular method with a number of industries. It is a perfectly legitimate system. If a customer attaches a certain value to a product and is willing to pay a higher price for it, there is no reason why it should not be marketed at that price. Patrons are not interested in what it costs to produce and market an item. They are much more interested in getting something they feel represents a good value for their money. If the price is within the value they have in mind, they are happy. If they feel that a meal priced at $4 is worth $4, even though its cost to the operation is much less, there is nothing wrong with charging $4 for it. However, if a meal costs $4 to produce and patrons consider it worth only $2.50, they will not buy it, even if it represents a lower price than cost.

Success in pricing according to what the market will bear is enhanced if some attempt is made to show patrons the true value of the menu item. It is often difficult for patrons to see the value in atmosphere, service, fine tableware, linens, and foods that are somewhat out of season or brought in at an additional cost. If a way is found to get patrons to understand that these increased costs increase value, they may be willing to pay a higher price. Having a differentiated product is one way of getting patrons to see and appreciate value. The foodservice with exceptional service, decor, atmosphere, and dining experience, including fine presentation that patrons would not get at home, can make patrons feel they are getting good value for their money.

## Value Perception

*Value perception,* or what patrons think of the desirability of a product compared with its menu price, is an important factor in menu pricing, since prices largely influence patrons' thinking about the value of menu items. A high price may be associated with high quality and a low price with lower quality. Some people want to go to a place where prices are high, feeling that they will be provided with an ultimate experience. A businessperson may take potential customers to a luxury restaurant to impress them with lavish pampering and hope this will transfer into a lucrative relationship. The menu must tie in with any attempt to present high prices and luxury by using proper menu wording and item presentation.

The way patrons perceive value is often manipulated favorably by getting buyers to think there is something special about a product that competing products do not have. We call this development of special, unique characteristics in a product *differentiation.* If a product can be favorably differentiated, buyers will want it rather than a competing product, and the seller has better control of the market and pricing. Through good advertising, buyers can be persuaded that it is better in some way than competitive products.

Foodservices can differentiate products and services in various ways. Often several special items on the menu can do the trick, such as a unique chef creation, a marvelously indulgent dessert, or a unique cocktail. Location, atmosphere, and decor are also used to differentiate one operation from another. A smiling maitre'd greeting new guests with a warm welcome and calling returning patrons by name can differentiate the service from that of a competing operation. When a foodservice achieves differentiation, patrons may pass up competitors, going a long way just to eat at a particular place.

## Pricing Psychology

Studies have been made on how menu planners set, and how patrons react to, menu prices. Menu planners usually try to avoid whole numbers and try to shade numbers just below them. A price of $6.95 or even $3.99 is perceived as less than $7 and $4. Number "5" is the most-often used ending digit. Some authorities say that a price ending in .99 is more suited to quick-service menus, and 0 and 5 as ending digits suit full-service menus better. Some fine-dining establishments use reverse psychology, casually pricing items with the number indicating the dollar value of the guest cost: "six, "twenty-three," and so on.

Price length also has importance; many menu pricers hesitate to use four-digit prices if they can avoid it. They feel that a price of $9.95 is better than $10.95, that three digits appear as a much lower price to patrons than four.

Patrons tend to group prices by range and think of them as single-figure amounts. For example, prices from $0.86 to $1.39 are considered to be about $1.00; from $1.80 to $2.49, about $2.00; and from $2.50 to $3.99, about $3.00. Prices from $4.00 to $7.95 are thought of as being about $5.00. Instead of raising a price into the next full price range, the menu planner may try to hit the upper limit of the range the current price is in. That is, instead of raising a menu price from $2.25 to $2.55, the planner may raise it to $2.45. A price of $7.75 is preferable to $8.25 because the former is in the "about $5" range and the $8.25 is in the "about $10" range.

Patrons do not seem to like wide ranges in menu prices. They want prices grouped together within the price range they want to pay. If too wide a price range occurs, patrons will tend to select the lower-priced items.

Patrons do not always view price increases in the same way. A price increase from $5.95 to $6.25 is seen as a bigger jump than from $6.25 to $6.75.

Some operations find that patrons resist buying after a certain price is reached. A Montana restaurant built as a stockade and located on a high mountain pass found there was resistance to any dinner price above $10. Items priced below $10 were popular. Items over that were not. To resolve this, the table d'hote meals were dropped and all items were listed à la carte at seemingly lower prices. The highest-priced steak with a salad, potato, and roll and butter was $8; a popular Trappers Stew was priced at $6.50, and so on. If patrons wanted appetizers, soup, or desserts, they paid extra. The strategy worked. Most checks now came to more than $10 per person. The $10 wall was broken.

## Market Research

Market research should point out what kind of market exists and what consumers will pay. If a gourmet restaurant opens, it must be located where it will attract customers with money. Adequate market information can lead to greater precision in setting prices. A *base price* and *top price* can be defined and the menu planned to work within this range. Of course, the product must be perceived to be worth the price paid.

# Economic Influences

Many methods are used to price menus. Some managers calculate their costs as a percentage of sales and then calculate the selling price. Others base selling prices on factors other than costs.

Foodservice operators should be aware that laws of economics and commerce work both to the benefit and harm of businesses. The two most basic economic laws affect every operation: (1) when supply is limited, prices tend to rise, and when supply is plentiful, prices tend to drop; and (2) when demand is high, prices rise, and when demand is low, prices drop. Thus, every menu planner should try to plan menus that create high demand. Operations that can restrict supply are few, but when they can, they have a chance to charge enough to make a good profit and still hold their business. Dropping prices may create more demand, while raising them may reduce it. Some operations reduce prices and hope to increase demand and, while making a smaller profit, make it up with increased volume. Others raise prices, reducing demand but earning a greater profit. The south Florida area has a large demand for food and housing during the winter months, and many operations do well as a result. Many restaurants have waiting lines for lunch and dinner. In the summer, both tourists and residents leave, and many operations drop their prices, cut their staffs, and retrench in every way because of the lack of demand. Some even close.

A smart menu planner knows that when the supply of an item is plentiful, costs will be low, and a better profit can be made. Thus, in the late spring and early summer, lamb is plentiful, relatively inexpensive, and of good quality and, therefore, is a good menu item. Turkey is plentiful in the late fall and early winter. When the smelt run is on in the Great Lakes, local foodservices offer them as *all-you-can-eat* menu items. This promotion works well. Changing menu items based on seasonality can serve the operation's need for profit and the guests desire for freshness and variety.

If possible, menu planners should be sure that there will be an adequate supply of all menu ingredients. Sometimes, an item is on the menu and the prices of the items used for it become so costly that the item is served at a loss. Often such an item is removed from the menu. Some prices, such as those for fresh fish and seafood, vary so much that a menu planner lists only *market price,* indicating the price will depend on the operation's cost.

Competition can reduce the number of patrons coming to an operation. Some may try to increase demand by dropping prices. However, this can be dangerous, because if competitors also drop their prices, both fight for the same demand, earning less from it.

Advertising and special promotions create demand. When a big-name act appears in a Las Vegas casino, food, beverage, and room prices may rise because the demand for them is high. Many operations put on special promotions to create a higher demand during certain periods. A quick-service chicken operation may advertise a special price on a bucket of fried chicken during a slow period. Such a promotion not only increases demand but also introduces the product to people who might not otherwise become patrons.

When forces of supply and demand increase and decrease in relation to increases and decreases in prices, the market is called *elastic.* If it does not, it is *inelastic.* If patrons pay no attention to prices and purchase items regardless of price, or refuse to purchase others even when prices drop, the market is said to be *inflexible.* However, if patrons increase demand when prices drop, and decrease demand when prices rise, the market is called *flexible.* A market where neither the price nor supply and demand are changing is called *steady.*

A menu pricer should know whether a market is flexible, inflexible, or steady, and how patrons are likely to respond to price changes. Often a test can be made by changing prices briefly to test patron response. The test should be repeated several times over a period.

Thus, menus must reflect and take advantage of the influence of economic laws. A menu is one of the best contacts a foodservice can have with its patrons. It can help create demand and take advantage of the flow of supplies.

## Pricing Based on Costs

One of the most common methods of pricing is for the planner to list costs of the food for each item on the menu—using methods discussed in Chapter 4—and then mark up the final figure to obtain a selling price. For instance, if an à la carte item has a food cost of $2 and the operation wants a 30 percent food cost percentage, the base selling price would be $6.66 ($2 ÷ 0.30 = $6.66). Per pricing psychology, it might be nudged to $6.75 or $6.95. If the operator wants to maintain a 30 percent food cost in relation to sales, all foods would be marked up by dividing food cost percentage into food cost.

The disadvantage of this method is that it assumes that other costs associated with preparing menu items remain the same with every menu item. On the contrary, menu items vary considerably in the cost of labor, energy, and other factors needed to produce and serve them.

For instance, a steak is on the menu for $16.95, while a pasta dish sells for $11.95. The labor cost to make a portion of pasta is $4.25 and the labor cost to prepare the steak is $3.50. The pasta's labor cost is 42.7 percent of the selling price and the steak's is 20.6 percent.

Assuming other costs have the same ratio to food cost for each menu item is a mistake. This leads to undesirable pricing that fails to cover costs on some items and can work against the sale of others. Costs must still be a paramount factor in pricing, and all costs—not just one or two—should be considered.

Pricing experts also claim that food cost pricing does not work because many foodservice operators do not know *all* their costs. These critics point out that many hidden food costs, such as spoilage, are not determined.

### *Derived Food Cost Percentage*

The most common method is to use a *derived food cost percentage.* Simply divide the dollar cost of the food by the desired percentage (food cost ÷ food percentage). If a menu item has a total food cost of $2.73, and the operator wants a food cost percentage of 35 percent, the calculation is $2.73 ÷ 0.35 = $7.80.

### *Pricing Factor or Multiplier*

Sometimes managers may convert a desired food cost percentage into a *pricing factor* or *multiplier.* For this, one divides the desired food cost percentage into 100 percent. This formula gives a factor by which a food cost is multiplied to get a selling price. For instance, suppose a food's cost is $2.73 and the desired food cost percentage is 35 percent; 100% ÷ 35% = 2.86, and 2.86 × $2.73 = $7.81, the selling price. **Exhibit 6.1** provides the multipliers for many common foot cost percentages, but any other multiplier can be determined by dividing 100 by the desired food cost percentage. The multiplier can then be used to calculate selling price. Thus, if an operator wants

**Exhibit 6.1** Pricing Factors (Multipliers)

| Food Cost % | Factor | Food Cost % | Factor | Food Cost % | Factor |
|---|---|---|---|---|---|
| 20 | 5.00 | 30 | 3.33 | 40 | 2.50 |
| 21 | 4.76 | 31 | 3.23 | 41 | 2.44 |
| 22 | 4.55 | 32 | 3.13 | 42 | 2.38 |
| 23 | 4.35 | 33 | 3.03 | 43 | 2.32 |
| 24 | 4.17 | 34 | 2.94 | 44 | 2.27 |
| 25 | 4.00 | 35 | 2.86 | 45 | 2.22 |
| 26 | 3.85 | 36 | 2.78 | 46 | 2.17 |
| 27 | 3.70 | 37 | 2.70 | 47 | 2.12 |
| 28 | 3.57 | 38 | 2.63 | 48 | 2.08 |
| 29 | 3.45 | 39 | 2.56 | 49 | 2.04 |
| | | | | 50 | 2.00 |

a multiplier based on a combined food and labor cost of 65 percent, the calculation would be 100% ÷ 65% = 1.54. If the combined food and labor costs were $5.20, the selling price would be 1.54 × $5.20 = $8.01, probably set at $8.00 or $7.95.

### *Variable Cost Pricing*

A *variable food cost pricing* method is sometimes used for à la carte menu items. For instance, an operator may assign food cost markups for menu items, as shown in **Exhibit 6.2.** This variable pricing can also be used to arrive at table d'hôte menu prices. Using the food costs in **Exhibit 6.3** and the percentages in **Exhibit 6.2**, a table d'hôte meal can be priced out.

Labor costs also can vary with different menu items. In these cases, menu items are divided into high (H), medium (M), and low (L) labor costs. Different percentages of food costs are then assigned to these. For instance, for high-labor menu items, a food cost of 25 percent is assigned; for medium, 35 percent; and for low, 40 percent **Exhibit 6.4** indicates how the pricing of three items turns out compared with an overall price based on a 33.3 percent food-cost-only calculation. This method weighs food cost and allows the price to reflect the influence of labor as a cost.

### *Prime Cost Pricing*

A selling price can be based on a dollar *prime cost value* divided by a cost percentage or multiplied by a factor. Prime cost is raw food cost plus *direct labor,* or labor spent in preparing an item. Thus, if food cost is $1.00 and direct labor is $1.25, prime cost is $2.25. An operation that wants a 35 percent food cost and a 25 percent direct labor cost has a prime cost percentage of 60 percent.

**Exhibit 6.2** Variable Cost Markups

| | | | |
|---|---|---|---|
| Appetizers | 25% | Beverages | 40% |
| Salads | 40 | Desserts | 35 |
| Entrees | 35 | Breads and butter | 30 |
| Vegetables | 40 | Miscellaneous | 35 |

Direct labor time is usually obtained by timing work. A selling price based on prime cost is obtained by establishing a desired combined food and direct labor cost percentage. Suppose 60 percent is the desired prime cost percentage and a cook takes 1.5

**Exhibit 6.3** Table D'hôte Meal Costed Out

| | Food Cost | Allocated Menu Price |
|---|---|---|
| Appetizer | $1.35 | $5.40 |
| Salad | 1.78 | 4.45 |
| Entree | 4.84 | 13.83 |
| Vegetables | 1.12 | 2.80 |
| Roll and Butter | .47 | 1.57 |
| Dessert | 1.96 | 5.60 |
| Beverage | .57 | 1.43 |
| Total Food Cost | $12.09 | $35.08 |

hours to make a recipe that gives 30 portions. The food cost of the recipe is $126.56. The cook is paid $8.75 per hour; 1.5 × $8.75 gives a direct labor cost of $13.12, which, added to $126.56, gives a prime cost of $139.69. The prime cost per portion is then $139.69 ÷ 30 portions = $4.66. The selling price would be $4.66 ÷ 0.60 = $7.76.

### *Combined Food and Labor Costs*

Prices may sometimes be based on a combined food and labor cost percentage. For instance, if you want a 35 percent food cost and a 30 percent labor cost, the combined cost would be 65 percent. If food cost for an item is $3.94 and all labor costs are $4.60, then using these figures the selling price would be ($3.94 + $4.60) ÷ 0.65 = $13.14. For this method, it is necessary to use a dollar value for labor cost that is based neither on a percentage of the selling price nor on a percentage of sales. It has to be a labor cost that is specific for the item.

### *All or Actual Cost Pricing*

Sometimes a method called by several different names—*all cost, actual cost,* or *pay-yourself-first*—is used in operations that keep detailed and accurate cost records. Cost are divided into food, labor, and operating cost units. A dollar value and the desired food and labor costs are obtained for each item. A desired profit percentage is also established. **Exhibit 6.5** illustrates how this method is used to price a dinner choice on a menu. Selling price equals 100 percent; desired profit equals 10 percent; food cost equals 35 percent, or $3.12; labor cost equals 30 percent, or $2.44; operating cost equals 25 percent, or $1.81. The formula is food cost plus labor cost plus operating cost plus 10 percent profit equals 100% (or selling price). Using these figures, $3.12 + $2.44 + $1.81 = 100% − 10% = $7.37 ÷ 90, a selling price of $8.20 is obtained. Again, *actual cost figures must be used,* not percentages based on sales or selling price. This method is useful only with a good cost accounting system. Many smaller foodservices do not have the accounting system to do this.

**Exhibit 6.4** Selling Costs Based on Various Labor Costs

| Item | Labor Requirements | % Food Cost Assigned | $ Food Cost | Selling Price Based on Labor and Food Costs | 33.3 % Selling Price |
|---|---|---|---|---|---|
| Stew | H | 25 | $2.33 | $9.32 | $7.06 |
| Steak | M | 35 | 4.20 | 12.00 | 12.72 |
| Milk | L | 40 | .32 | .80 | .97 |

**Exhibit 6.5** All Cost Pricing

| | | |
|---|---|---|
| Food cost | 35% | $3.12 |
| Labor cost | 30 | 2.44 |
| Operating cost | 25 | 1.81 |
| Total | 90% | $7.37 |
| Profit | 10 | .74 |
| Total | 100% | $8.11 |

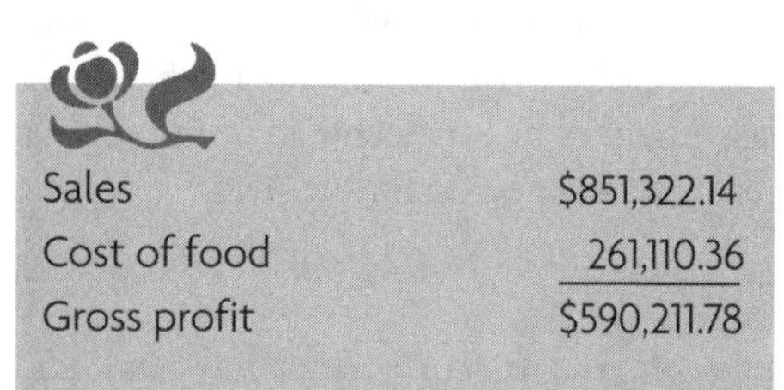

| | |
|---|---|
| Sales | $851,322.14 |
| Cost of food | 261,110.36 |
| Gross profit | $590,211.78 |

Suppose the number of guests served during this period was 108,113. The average gross profit per guest is $590,211.78 ÷ 108,113 = $5.46. The dollar cost for an item is then added to this average dollar gross profit to get a selling price. For instance, suppose the food cost for four items was $2.10, $3.13, $2.85, and $4.00. The selling price would then be calculated as in **Exhibit 6.6.**

When using a gross profit average, be sure it includes an adequate profit. If it does not, the desired profit should be added in a dollar value. Thus, if the profit were $0.27 of the $5.46 figure but management wants a per-patron profit of $0.85, about 10 percent of sales, $0.85−$0.27, or $0.58, is added to the $5.46 figure, bringing the gross profit average up.

## Gross Profit Pricing

In this method, a gross profit dollar figure is taken, usually from the profit and loss statement for a certain period. This is divided by the number of guests served during that time to get an average dollar gross profit per guest.

Gross profit pricing is a useful method because in many operations the cost of serving each patron is much the same after food costs are considered. It tends to even out prices in a group.

## The One-Price Method

In some foodservices, the overall cost of menu items is the same, such as a doughnut shop where all doughnuts and beverages cost about the same. This is the one-price method in action. The operation can charge just one or a few prices to simplify things. The small differences will usually even out. A nightclub with a cover charge can also use the one-price method because the cost of what is served is nominal when compared to other costs, such as entertainment, music, and decor.

Another kind of operation that might charge one price regardless of the item selected is one in which selling food is not a primary purpose. This operation could be a tavern that makes all sandwiches one price. Or it could be a casino, where the objective is to get people in to gamble, and if food at one price helps do that, the operation benefits.

## Marginal Analysis Pricing

Retail operations, including foodservices, may use the *marginal analysis* pricing method. This is an objective method in which the maximum profit point is calculated. The selling price chosen will be the one that establishes this maximum.

Say a quick-service operation wants to set the most favorable price for its milkshakes in order to maximize profit. It estimates that at various prices it will sell a certain number of milkshakes, as shown in **Exhibit 6.7.** Fixed costs are $80.00, and variable cost per milkshake is $0.40. Thus, 100 shakes cost 100 × $0.40 food cost plus $80.00. That is, (100 × 0.40) + $80 = $120.

From the marginal profit column we can see that the best-selling price is $1.10 with 410 sold; the next best is $1.20. **Exhibit 6.8** (on page 176) shows how this marginal analysis problem appears in graph form.

The greatest distance between the costs and sales lines is at

**Exhibit 6.6** Gross Profit Pricing

| Item | Food Cost | Average Gross Profit | Selling Price |
|---|---|---|---|
| A | $2.10 | $5.46 | $7.56 |
| B | 3.13 | 5.46 | 8.59 |
| C | 2.85 | 5.46 | 8.31 |
| D | 4.00 | 5.46 | 9.46 |

points *a* and *b,* where 410 milkshakes are sold to bring $451 in sales at a cost of $244, giving a profit of $207.

## COST-PLUS-PROFIT PRICING

In the *cost-plus-profit* pricing method, a foodservice may decide it needs a standard profit from every patron who enters. The rationale behind this is that every customer who comes through the door costs the operation money, no matter what is ordered. The operation may reason that it wants a set amount of profit from every patron. Therefore, total costs may be calculated and then a set amount is added to this. For instance, an operation that wants to make $600 a day in profit may find it has an average of 400 patrons per day. Then, when pricing, the food, labor, and operating costs are added together, plus a $1.50 profit. This covers all costs and should result in the desired profit of $600. Next, an average labor cost value and operating cost value are determined from the profit and loss statement, and both are divided by the number of patrons served in that period. If labor costs, including all benefits, are $80,511, operating costs are $50,336, and 83,001 patrons are served, then the average labor cost per patron is $0.97 and the operating cost per patron is $0.61. The selling price calculations for items A, B, C, and D are shown in **Exhibit 6.9.** In this case, the low operating and labor costs give low selling prices. This method tends to even out selling prices.

## MINIMUM CHARGE PRICING

Pricing based on a minimum price to cover costs and give a desired profit in a commercial operation is much the same as calculating the price based on costs plus profit. The rationale for this method is also much the same. That is, every customer costs a certain amount to serve, and by having a minimum charge these costs will be covered. A hospital or nursing home might use such a

**Exhibit 6.7** Marginal Analysis Projection

| Selling Price | Number Sold* | Total Sales | Total Costs | Marginal Profit |
|---|---|---|---|---|
| $1.50 | 100 | $150.00 | $120.00† | $30.00 |
| 1.40 | 190 | 266.00 | 156.00 | 100.00 |
| 1.30 | 275 | 357.50 | 190.00 | 167.50 |
| 1.20 | 340 | 408.00 | 216.00 | 192.00 |
| 1.10 | 410 | 451.00 | 244.00 | 207.00 |
| 1.00 | 450 | 450.00 | 260.00 | 190.00 |
| 0.90 | 525 | 472.50 | 290.00 | 182.50 |

*Estimated from marketing studies or by other means.
†$80 fixed cost + (100 sold × $0.40 each) = $120.

**Exhibit 6.8** Marginal Analysis Graph

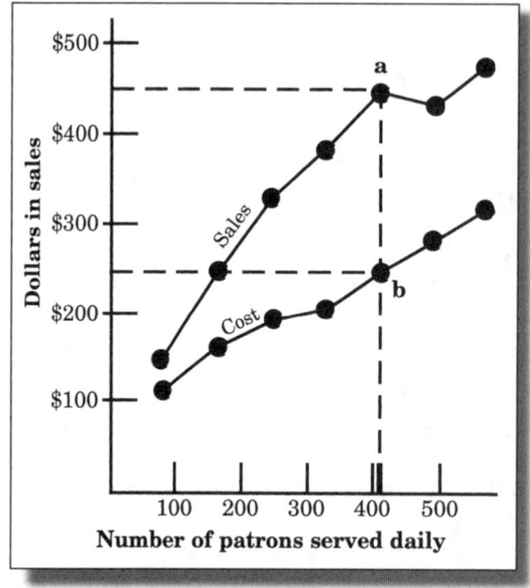


method. A private club might have a minimum for certain rooms where food and beverage service is provided. Such a policy is also common in some commercial dining rooms. Pricing may be arranged on a menu card to make it impossible to obtain service below a certain price. For instance, the Four Seasons Restaurant in New York City may not want to see a price of less than $25 for lunch. Such operations will test out various luncheon combinations. Whatever menu items produce the desired result will be priced accordingly.

Some operations may state on the menu card that there is a check minimum. If a customer does not order enough food to cover this minimum, the check will still include the minimum payment. Clubs may require that members spend a specific amount during a certain period on foods and beverages so that the foodservice department receives enough income to operate. If the member does not spend this amount during the period, the balance is added to the bill.

A *cover charge* is a set price that is added to customers' bills, regardless of what menu items are purchased. This cover charge establishes a base from which costs for space, atmosphere, entertainment, and other costs will be paid. The cover charge is used by nightclubs and other operations where entertainment or dancing may be an attraction.

## Pricing Based on Sales Potential

Some menu planners believe that pricing should reflect factors in addition to food and labor costs, including how an item is expected to sell. They divide items into high cost (HC) or low cost (LC), high risk (HR) or low risk (LR), and high volume (HV) or low volume (LV). Thus, one menu item may be labeled HC-LR-HV and another LC-HR-HV. Ratings of LC, HV, and LR are

**Exhibit 6.9** Cost-Plus-Profit Pricing

| | Food Cost | | Labor Cost Per Patron | | Operating Cost Per Patron | | Profit | Selling Price |
|---|---|---|---|---|---|---|---|---|
| A | $2.10 | + | $0.97 | + | $0.61 | + | $1.50 | $5.18 (or $5.25) |
| B | 3.13 | + | 0.97 | + | 0.61 | + | 1.50 | 6.21 (or $6.25) |
| C | 2.85 | + | 0.97 | + | 0.61 | + | 1.50 | 5.93 (or $5.95) |
| D | 4.00 | + | 0.97 | + | 0.61 | + | 1.50 | 7.08 (or $7.15) |

considered favorable, while HR, HC, and LV are considered unfavorable. Eight combinations are possible.

HR-LC-HV    HR-HC-LV
HR-LC-LV    HC-LR-LV    HV-LR-LC
HR-HC-HV    HC-HV-LR    LR-LC-LV

If a plus (+) is assigned for a favorable factor and a minus (–) for an unfavorable one, the following matrix is obtained, based on the previous combinations.

– + +    – – –
– + –    – + –    + + +
– – +    – + +    + + –

If this formula is used when pricing, the highest markup would be assigned to any item having three minuses or two minuses and a plus, a lower markup to one minus and two pluses, and the lowest markup to the one with three pluses. Perhaps the highest markup would be based on a 25 percent food cost, the next on a 30 percent food cost, and the last on a 40 percent cost. This type of pricing would ensure a profit margin for low-volume, high-cost items. **Exhibit 6.10** shows how an operation might classify items, and indicates the desired food cost markup.

## Pricing Aids

Pricing can be made easier if tables and computer printouts are available that give food costs of various menu items, with prices based on the operation's food cost percentage already tabulated. Software programs are available from some purveyors for their customers and are commercially sold to assist in performing accurate calculations. Many are bundled into packages that include inventory functions.

**Exhibit 6.10** Menu Item Markups

| Menu Item | Risk | | Cost | | Volume | | % Markup |
|---|---|---|---|---|---|---|---|
| | High | Low | High | Low | High | Low | |
| Steak | — | | — | | + | | 30 |
| Apple tart | | + | — | | | — | 30 |
| Chicken and wild rice | | + | | + | + | | 40 |
| Chocolate ice cream | | + | | + | + | | 40 |
| Lobster | — | | — | | | — | 25 |
| Vegetables | | + | | + | | — | 35 |
| Fish | — | | | + | + | | 35 |
| Tofu | | + | | + | | — | 35 |

These figures can be used to record risk, cost, and volume levels so the menu pricer can properly allocate markup.

## Evaluating Pricing Methods

Few operations use only one pricing method. Most use a combination to best meet their needs and those of their patrons. Pricing based only on competition is generally not a good method, but considering the prices of competition when establishing one's own menu prices is advisable. Using a standard markup over an accurate cost determination can give a fairly precise price basis and usually assures an adequate profit or budget performance. However, varying margins over cost based on what the market will bear should also be considered, as should pricing based on volume. Perhaps markups should be varied to promote merchandising and entice patrons with *loss leaders* (items that have a low markup but can bring in extra business). Additionally, various items may be priced to cover the high food cost of items with prices based on local competition. Thus, a quick-service operation may not make as much on its hamburgers and hot dogs as it does on milkshakes, carbonated beverages, and fries, but in the end the achieved sales mix can give a very desirable markup level.

And, finally, *pricing is not something that is done and then is over.* There is a need to evaluate prices constantly, to study how customers react to prices, and to gather data on costs. Too often no follow-up occurs, and when a revision of menus and prices is necessary, there is a lack of adequate information to do a good job. Gathering, compiling, collating, and filing pricing data are as much a part of the pricing function as the establishment of prices.

## Pricing for Nonprofit Operations

The previous discussions on menu pricing for commercial operations are applicable to noncommercial operations, except that most do not need to allow for profit in their prices. They are required to meet costs and perhaps make a slight margin above costs as a safety factor that can be accumulated to bridge times when costs are not covered. Costs may be paid for in part by the patron, with the balance paid by federal and state agencies or charitable donations.

When nonprofit operations price items to break even, accurate cost information must be available from operating records. The information must be timely, so that action can be taken promptly to bring costs into line when they vary from desired levels. Many institutional operations estimate costs for a period and are given an allowance from the total budget to cover this estimate. When costs for a future period must be estimated, information from federal agencies, economic indexes, and economic price predictions of price changes can be used.

Nonprofit operations usually establish prices on the basis of a budgeted amount per meal, per day, per week, per month, or per period. Some nonprofit budgets allow a cash allowance for food, labor, and operating expenses. Other operations may first get a total food allowance for a period, and then a cash allowance is worked out from this. This food and cash allowance system is called a *ration allowance.* (See the discussion on ration allowance budgeting in Chapter 5.)

For example, say an operation has total costs of $22,800 for 15,571 meals served during the period, with 1,240 of these meals being eaten by employees.

$22,800 √ 15,571 = $1.464 per meal
$1.464 × 0.50 = $0.732 food cost per employee meal
1,240 × $0. 732 = $907.68 total food cost for employee meals

# Pricing Employee Meals

In in-plant and company foodservices, whether employees pay for meals or the company subsidizes them, costs must be calculated so they are covered.

Employee meals are often considered a benefit and, thus, an operating expense. Therefore, when the cost of food for employee meals is included in the cost of food used, it must be deducted. The value for deducting an employee meal can be based on: (1) the actual cost of the meal, (2) experience, (3) a standard charge made in the area for meals, or (4) an arbitrary amount. The actual cost may be standard menu prices with a discount given to employees. In some areas it may be a practice to deduct a predetermined amount for employee meals, a plan followed by a number of foodservices. The arbitrary amount may be only a nominal charge but one that the foodservice thinks is adequate to cover its costs, or at least a major portion of them.

Most operations take their average food cost percentage and use this to arrive at a value for the food used for employees' meals. Thus, if the food cost is 35 percent and a total value of $800.19 is assigned to the employees' meals for a certain period, the value of the food in these meals for that period will be 0.35 × $800.19, or $280.07.

When employees eat the same meals as patrons, the cost of employees' meals may be calculated using the following steps:

1. Ascertain the number of employee meals.
2. Calculate the individual meal cost.
3. Consider food for employees' meals as 50 percent of the cost of a meal.
4. Multiply the number of employees' meals by the food cost per employee meal.

Based on this information, it would be simple to arrive at a nominal menu price. This information will also provide the data for deducting meals as a benefit. The operator will want to bear in mind the legal and ethical issues arising from deducting for meals that are not taken or provided.

# Changing Menu Prices

At times a change in a menu price is necessary. If it is an item that appears frequently on the menu and has good acceptance, repricing may present difficulties. Customers may resent the change and may stop ordering the item to show their displeasure. Sometimes this does not last long, and the item gradually assumes its former importance as a seller.

During periods in which food prices increase rapidly, most customers recognize that the establishment must also increase prices and will accept the price increases. In periods when prices are stable but some other factor makes a change necessary, customers may not be so willing to accept a price change.

Some operators attempt to change a price by removing an item from the menu for a time and then bringing it back at a new price. If it also comes back in a somewhat new form, or in a new manner of serving and with a slightly different menu name, the price change may be less noticeable. Changes also tend to be noticed less if they are made when volume is down.

Prices are frequently changed when a new menu is printed. In fact, the need to change menu prices may sometimes stimulate a menu change more than the need to change items. Some authorities advise against changing format and prices at the same time, saying that customers will notice price changes *because* of the new format.

Using menu clip-ons removes the need to print new menus just to incorporate new prices, and avoids the need to cross out old prices to put new ones in. The latter, especially, can give patrons a negative impression.

In some instances, if a general rise in menu prices must occur, an operation may change several items $0.05 or $0.10 each and then let these remain at this price while changing a few others. In this manner prices are gradually worked up to the desired level. This has the disadvantage of giving customers the impression that the menu and its prices are unstable.

Some announcement, on a clip-on or table tent, can be helpful in indicating to patrons why a change in menu prices in necessary. However, this can also have the undesirable effect of calling attention to changes that some customers might not otherwise notice.

Patrons are likely to be especially aware of price changes for popular items. Also, as noted, they are more aware of a change when the dollar price changes than when the cents price does. Rather than changing a dollar amount, it might be wiser to make a price change in cents, or make it gradually.

Only a few items' prices at a time should be changed if prices are changed frequently. More items can be repriced if the change is less frequent. Some say that only two price changes should be made in a year; others feel that only one is advisable. With some items for which prices may have to change frequently, such as lobster or stone crab, it is advisable to list the price as *seasonable price* or *market price*.

Printed menus that rarely change their items offered or their format are more difficult to change in price than are blackboards, panels, or handwritten menus. Computerized menu programs printed in-house offer versatility in this regard. If a menu has a daily insert in which items change daily or frequently, it is easier to make price changes on this than on the permanent, hardcover menu. It is never a good practice to cross out an old price and write in a new one. Even whiting out a price and writing in a new one is not recommended.

Specials and highly promoted items are difficult to change in price because buyer attention is often centered on them. They should be dropped from promotion for a time and then reinserted with the prices changed. In all cases, price changes on menu items should end up within the range expected by patrons.

## Pricing Pitfalls

All foodservice managers should beware of the following pitfalls committed by inefficient menu planners.

1. Pricing should not be based entirely on just one cost, such as food cost, giving a price that may not reflect actual costs. Other costs vary considerably from a direct relationship with food cost. Thus, pricing only on the basis of food cost can lead to pricing errors.
2. Foodservice pricing should not ignore the economic laws of supply and demand.
3. Value perception on the part of patrons in equating price to value of a menu item should have greater emphasis in pricing.
4. More attention needs to be given to market information in establishing prices.

**SUMMARY**

Pricing a menu is a complex process, and a number of factors need to be considered when doing it. Anyone pricing a menu should know about and experiment with some of the pricing methods, such as marginal analysis pricing and market testing of prices.

Studies have shown that consumers view prices somewhat differently than operators might expect. They want prices grouped within a price range they expect to pay, and resist purchasing items outside this range.

A number of pricing methods are used by the foodservice industry. Probably no one method is used alone, but a combination of them can be used in establishing menu prices.

Patrons will pay only so much for certain kinds of foods in certain kinds of operations, and the menu planner must be sure to meet this restriction. Menu prices are often not easy to change, and planners should be aware of the techniques that make price changes less noticeable by patrons.

**QUESTIONS**

1. What is a *differentiated* product? What is its purpose? How is it used?
2. A foodservice wants a 25 percent food cost. What multiplier or pricing factor should be used?
3. The food cost of a table d'hôte group of foods is $8.20, and a 30 percent food cost is desired. What is the selling price?
4. If raw food cost is $1 and direct labor cost is $0.30, what is the prime cost? If a 40 percent price based on this prime cost is wanted, what is the selling price?
5. Why are guests willing to pay $4 to $6 for a coffee beverage? (Similarly, customers willingly pay over $7.00 for fried onion appetizers, etc.)
6. What pricing principles might be in operation in the case of price value marketing ($1 value menu, 2 for $2, etc.)?

## MINI-CASES

*6.1* Hot Shots is an upscale bar and restaurant targeting a young professional clientele. The bar features premium specialty cocktails that cost a dollar to make and cost the guest $6.00. The bar features premium snack mix with nuts, house-made cheese spread, and whole-grain crackers. The cheese spread is so popular it is available for sale and shipping through the Hot Shots Website. The bar snacks are gratis for bar patrons. Bar business is quite brisk beginning about 4 PM every day.

The restaurant books tables for lunch and dinner and has a private dining room for parties. Restaurant prices range from $8.00 for a grilled vegetable and tofu wrap to $38.00 for antelope tenderloin. The menu is the same for lunch and for dinner. The catering menu has three categories: $15.00 per person, $25.00, or $45.00, with selections available for each category that are judged to be roughly the same food cost. The dining room has struggled, as has the catering business, but both are holding their own for now.

What are some of the underlying principles in operation here?

What strengths need to be preserved?

What changes would you make? Support your answers.

*6.2* Go back to the Children's Menu that you developed in a previous chapter. If you did not do so, create one now. Include the concept of the restaurant for which the menu was designed. If you have not already done so, cost out the recipes and derive portion costs for each menu item. Price out your menu. Evaluate the results. Are the prices in line with what you believe your target market will expect? Are there alterations to the menu or recipes that you might make to bring prices in line with value perceptions?

# 7 MENU MECHANICS

## Outline



## Objectives

After reading this chapter, you should be able to:

1. Identify the basic requirements to make a menu an effective communication and merchandising medium.
2. Describe the services offered by design firms and the considerations associated with determining a good fit.
3. Discuss aspects of using type: typeface, type size, line length, spacing between lines and letters, blank space, weight, and type style.
4. Indicate how to give menu items prominence by using displays in columns, boxes, or clip-ons.
5. Indicate how to best use color in menus.
6. Discuss paper use, construction of covers, and other physical factors.
7. Indicate how menus are commonly printed, how to work with professional menu printers, and the various methods of self-printing.

# Introduction

Certain mechanical factors must be considered in menu planning. No matter how well the menu is planned and priced, it must also be properly presented so that it is understood quickly and leads to satisfactory sales. *Communicating* and *selling* are the main functions of a successful menu. Good use of mechanical factors will enhance a menu's appearance, make a favorable impression on patrons, and advance the overall aims of the operation. Using the services of a professional design firm may be one way to achieve cohesive brand identity

Professional menu printing companies can also be of considerable help in developing a menu that is attractive and achieves its purpose as a merchandising medium. For this reason, the material in this chapter is designed to teach readers how to work with professionals as well as how to do the job without assistance.

# Menu Presentation

How a menu is presented is important to most operations. For commercial establishments the menu does much to convey the type of operation and its food and service. If the menu communicates accurately through design and layout, as well as through the copy, it can sell the items on it.

Most menus are printed on paper and given to patrons to look at, but some are not presented in this way. A cafeteria menu board may show items for sale and list prices. This could be posted so that patrons waiting in line have a preview of their choices. A quick-service operation might have menu boards behind the counter or backlit signage with pictures at the drive-through. Some operations have handwritten menus to give a homey and personal touch. A menu may be made to resemble a small newspaper and list the latest news along with menu items. As long as the primary goal of communicating choices to guests is met, there are a dazzling variety of means to arrive there.

The manner in which menu items are presented should be selected to best meet the needs of the operation. A hospital may have selected menus printed on colored paper, each color indicating a different diet. On some, special instructions concerning selections by patients may be used. Many hospitals have staff visit patients' rooms with handheld computers to take orders and answer questions. The sales department of a hotel or catering department may need a package of menu options to give to people interested in arranging special functions at the hotel.

Some operations need a number of different menus, such as breakfast, brunch, lunch, early bird, dinner, or late night. A country club may need a menu for its bar where steaks, sandwiches, and snack foods are served; another for a coffee shop or game room; a small snack and beverage service near the swimming pool; and another for the main dining room. A hotel or motel might use a special room-service menu. As these menus vary in their purpose and requirements, so must they vary in the manner in which menu items are presented.

The most common menu is the one presented on firm paper, the front being used for some logo, design, or motif. Inside, on the left and right sides of the fold, à la carte offerings (items selected and paid for individually) are listed. The back may also contain à la carte items and alcoholic beverages, or give information about hours of operation and short notes of interest about the operation, locale, or some of the special items served. The items on this heavy paper are permanent.

Often, menu items that change, including table d'hôte meals (foods or meals sold together at one price), are printed on lighter paper and attached to the more rigid menu.

Sometimes menus have items that are à la carte and selections that include a number of side dishes with prices set accordingly. Some or all of the side dishes may be purchased à la carte. (See **Exhibit 7.1**) While this menu shows many à la carte items, the sandwiches are served with fries and slaw and the Specials and Entrée items have choice of potato and (famous) onion bread.

Some menus may offer specials. These can be attached as clip-ons or inserts. If they are used, the basic menu should provide space for them. They should not cover other menu items.

Clip-ons or inserts are used to give greater emphasis to items management wants to push. They should not repeat what is on the menu but offer variety. The use of these clip-ons or inserts makes it possible to change a permanent menu for weekly specials or holidays. Clip-ons and inserts may be of the same color as the menu, but if special effects and heightened patron attention are desired, the use of a different color can help focus attention on them.

Many restaurants list specials on a board as guests enter. Servers then verbally describe them to the guests at the table. When there are more than a few additions, it is helpful to have them available in print form for the customers, including prices.

## Menu Format

Regardless of how a menu is presented to patrons, certain rules in format should be observed. Wording and its arrangement should be such that the reader quickly understands what is offered. If foods are offered in groups, it should be clear what foods are included. Clarity is promoted by making menu items stand out. Simplicity helps avoid clutter. Foods usually should be on the menu in the order in which they are eaten. An exception might be a cafeteria menu board, listing items as they appear in the line. Some offer cold foods first and hot foods last. This avoids the hot foods cooling off while a customer selects cold foods. Some menus also indicate the location of foods, such as in a take-out operation, where different counters offer different foods, or a cafeteria where patrons move from one section to another to get different foods. In this case, the menu board can be helpful by indicating a counter number or using a diagram to show where foods are found. It may prove helpful to have persons representative of the target market provide feedback on how well the menu communicates.

## Production Menus

Some menus may never be seen by patrons. They are written principally for back-of-the-house workers to inform them about what must be produced. This requires a different form and terminology. Selling words and fancy descriptions are not required. The term *carrot pennies,* which sounds good on a menu read by patrons, instead will be *sliced carrots*. Production information is included, such as the amount to prepare, the recipe number to use, preparation time, distribution to service units, designation of the worker to prepare items, and portion sizes. Service instructions, such as the portioning instructions and the dishes and utensils to use, may also be added.

**Exhibit 7.1** Menu with À la Carte and Table d'hôte elements: Twin City Grill

# TWIN CITY GRILL

## STARTERS

**Minnesota Wild Rice Corn Chowder** .......... 3.75 .... 4.75
**Soup of the Day** .......... 3.75 .... 4.75
**Baked French Onion Soup** .......... 5.75
**Beer Battered Button Mushrooms** - Horseradish sauce .......... 6.25
**Spinach and Artichoke Dip** - Garlic toast .......... 7.25
**Coconut Shrimp** - Orange marmalade sauce .......... 7.95
**Beer-Battered Walleye** - Tartar sauce .......... 7.95
**Crispy Calamari** - Lemon mayonnaise .......... 7.95
**Shrimp Cocktail** - Homemade cocktail sauce .......... 8.95

## FLATBREADS

*Our Famous flatbreads are made from fresh unleavened bread that are grilled crisp, brushed with pesto and then topped with mozzarella and your favorite fresh ingredients*

**Oven-dried Tomatoes** - Fresh basil .......... 7.95
**Grilled Chicken** - Oven-dried tomatoes, fresh basil .......... 8.95
**Grilled Vegetables** - Zucchini, yellow squash, red pepper, mushrooms, basil .......... 8.95
**BBQ Chicken** - Oven-dried tomatoes, scallions .......... 8.95
**Grilled Mushrooms** - Shiitake mushrooms, oven-dried tomatoes, scallions .......... 8.95

## SALADS

*All of our salads are freshly tossed and made to share*

**Caesar Salad** - Romaine lettuce, parmesan cheese, homemade croutons .......... 7.95
**With Grilled Chicken or Salmon** - Homemade croutons .......... 8.95
**Mediterranean** - Feta cheese, greek olives, cucumbers, vinaigrette dressing .......... 9.95
**Grill Room** - Grilled chicken, grapes, almonds, blue cheese dressing .......... 9.95
**St. Paul Chopped** - Chicken, pasta, bacon, tomatoes, blue cheese, scallions .......... 9.95
**Asian Salad** - Grilled chicken or salmon, almonds, crispy noodles, oranges, sesame dressing .......... 9.95

## TWIN CITY SIDES

*Plenty to pass*

**Maple Glazed Carrots** .......... 3.95
**Mashed Potatoes** .......... 2.95
**Creamed Spinach** .......... 4.25
**Broccoli** .......... 4.25
**Asparagus** .......... 5.25
**Roasted Mushrooms** .......... 2.95
**Toasted Macaroni & Cheese** .......... 4.50

## CHICKEN & BBQ

*Served with choice of potato and famous onion bread with fresh whipped butter.*

**Garlic Chicken** - Marinated with rosemary & garlic .......... 13.95
**Hickory Smoked BBQ Chicken** - Coleslaw, homemade BBQ sauce .......... 13.95
**Chicken with Mushrooms** - Cabernet-Mushroom demï-glace .......... 14.95
**Hickory Smoked BBQ Chicken & Ribs** - Coleslaw, homemade BBQ sauce .......... 19.95
**Hickory Smoked BBQ Baby Back Ribs** - Coleslaw, homemade BBQ sauce .......... 19.95

10/06

*Courtesy of Lettuce Entertain You Enterprises Inc.*

# Menu Design

The design of a menu contributes greatly to its legibility and patron reaction. Therefore, the design should be well planned. Menus, like individuals, should have personalities. They should reflect the atmosphere and feel of the operation. The eye should be pleased with what it sees on the menu, and patrons should quickly grasp what is being offered at what price. The printing and coloring should blend in with the logo or trademark of the operation, as well as with the type of establishment. The menu should be to a foodservice operation what a program is to a play or opera—an indication of what is to come. It should be an invitation to a pleasing experience and should not promise too much. Patrons should clearly understand what they are to get and the price they are to pay for it.

## Working with Designers

There are firms that specialize in establishing product brand. Services can be as simple as logo design—with the logo carried out on business cards, paper products and menus—or as complex as creating a brand identity and carrying it out in every element of the design of the restaurant. Designers can help create visual solution or total brand identity. Included in services is the specialized area of Web design. A simple yet very professional design might run in the neighborhood of $12,000 to $15,000 from a mid-range professional firm. A Website with *back-end* change capability, whereby the operator can go in and make alterations, may run around $20,000. E-commerce features such as the ability to make reservations online or order (typically for pick-up or delivery) may start at $25,000 and up for a simple system. Complex system capabilities carry a corresponding price tag. Websites are ways for restaurants to market themselves to prospective guests. They generally feature a printable version of the menu. Making sure that the current menu is displayed assists in avoiding disappointment and misrepresentation. Restaurants may wish to link their sites to local convention and visitors bureau sites, healthy choices, ethnic dining, or other pertinent links.

As these sorts of design services are specialized, many operators will not have a local supplier. A poor Website design that originally cost in the tens of thousands may end up costing more than twice that amount if there are serious bugs or the original site has to be scrapped. One reputable firm advised that potential customers do their homework and suggested that buyers look for companies that have in-house designers and programmers. The disadvantage of using companies that use freelancers is that there is less of a team effort, with glitches cropping up due to the work being chopped in pieces. This results in a less-cohesive whole. If bugs appear with an in-house team, it is far more likely that everyone contributing to the project is still present and accountable. Established companies by definition have a track record, resources, and references. Finding a good fit with a professional designer is important. Many design firms may design for a number of types of clients and do not offer specific restaurant expertise in terms of menu content. Others may have access to a consultant in this area or have someone on staff with this expertise. Software can be designed for specific operations that include front- and back-the-house operations, as well as allowing an operator to self-print and self-manipulate professionally designed templates.

## NONTRADITIONAL MENUS

Whatever material is used for the printed menu, it should fit well with the concept of the operation. Modern versions of menu may include Lucite, or a printed menu held onto Plexiglas with colorful rubber bands or creative clipping devices. Natural materials might be utilized to stunning effect if keeping with the theme of the operation.

Computerized screens at the table with descriptions, nutritional analysis, and complementary side and wine-pairing suggestions may be just around the corner so that guests can order from their table. Instantaneous information on item outages could remove the item from view, thus saving the diner disappointment regarding what could have been.

## USING TYPE

### *Typefaces*

One of the most important factors in accomplishing a menu's purpose is the style of type, or *typeface,* used, called *fonts*. There are many different kinds of fonts, and some are more legible and more easily and quickly read than others. The type most often used for menus is a *serif* type, or one in which letters are slightly curved. These are some of the easiest types to read. Letters set in *sans serif* type may look blocky. Today, menu designers can choose from hundreds of readily available fonts. However, any menu should be limited to at most three fonts, or the result will look cluttered. (See **Exhibit 7.2.**)

Studies have been made to ascertain legibility, reading speed, and comprehension using different typefaces. Unfortunately, not all of the fonts that have been studied are used in menus, and some that have considerable popularity in menu use were not included in the studies. Nevertheless, from such studies we can get some idea of the best typefaces to use for menus.

Print comes in plain (regular), bold (heavy print), italics, and script. Any type of italic or script print is more difficult to read than plain type. However, in some cases, these might be preferable to others because of special effects desired. A fine-dining restaurant may want to use these typefaces because they imply elegance in dining. Italic, script, or specialty types can bring on fatigue in reading, but this may not be a factor in short menus. The typeface should reflect the personality of the establishment.

### *Type Size*

The size of type is important to both understanding and speed of reading. Type that is too small makes reading difficult, but type that is too large takes up too much space; it might actually inhibit comprehension because it spreads the words out too much.

Type size is measured in *points.* There are 72 points to an inch. Thus, 18-point type is nearly a fourth of an inch high. Most menu designers use 10- or 12-point type for listing menu items and 18-point type for headings. This can be varied for descriptions.

Readers have ranked their preferences for type size. Three sizes—10-point, 11-point, and

**Exhibit 7.2** Various Font Types

| Faces |
| --- |
| **Serif** |
| Times Roman |
| Braised in butter and then simmered in red wine with shallots and other herbs, this dish has been for centuries one of the most typical of the Bretony area. |
| Palatino |
| Braised in butter and then simmered in red wine with shallots and other herbs, this dish has been for centuries one of the most typical of the Bretony area. |
| Bookman |
| Braised in butter and then simmered in red wine with shallots and other herbs, this dish has been for centuries one of the most typical of the Bretony area. |
| **Sans Serif** |
| Helvetica |
| Braised in butter and then simmered in red wine with shallots and other herbs, this dish has been for centuries one of the most typical of the Bretony area. |
| **Helvetica Black** |
| **Braised in butter and then simmered in red wine with shallots and other herbs, this dish has been for centuries one of the most typical of the Bretony area.** |
| Futura |
| Braised in butter and then simmered in red wine with shallots and other herbs, this dish has |

12-point—ranked together as first choices, followed by 9-point, 8-point, and 6-point. Various type sizes are shown in **Exhibit 7.3.**

Menus printed in a single type size can be monotonous. The sizes are usually varied on a page to give relief. (See **Exhibit 7.4.**) Thus, menu items may be listed in 12-point type, with 9- or 10-point type used for the description just below the item.

Menu headings may be in capital letters in bolder and larger type. Different type from that for the items sometimes may be used, but some mixtures may give an undesirable effect.

At times, to draw attention to an item, larger type is used than that used for regular items. For instance, the menu may use normal-weight 12-point type rather than heavy or light printing for regular items, and then change to a 14-point boldface (heavy) type to give emphasis to a special item. Additional emphasis can be given to an item by placing it in a box and putting an ornamental border around it. (See **Exhibit 7.5.**) If the box is in a different color from the regular menu, the emphasis may be greater.

**Exhibit 7.3** Type Sizes

| Point Size | Name |
|---|---|
| 3½ point | Brilliant |
| 4½ point | Diamond |
| 5 point | Pearl |
| 5½ point | Agate |
| 6 point | Nonpareil |
| 7 point | Minion |
| 8 point | Brevier |
| 9 point | Bourgeois |
| 10 point | Long Primer |
| 11 point | Small Pica |
| 12 point | Pica |
| 14 point | English |
| 16 point | Columbian |
| 18 point | Great Primer |

### *Spacing of Type*

Another factor affecting ease of reading and comprehension of menus is the spacing between letters in a word and between words. If letters and words are set too close together or too far apart, reading is difficult. Associated with this horizontal dimension in typography is the width of the individual characters in a particular font. They may be condensed (narrow), regular, or extended (wide), and this quality has its effect on readability and the scanning rate. Thus, a condensed word would be printed "condensed," a regular would be printed "regular," and an extended would be printed "extended."

Vertical spacing between lines of type, called *leading* (pronounced "ledding") is also important to ease of reading. If no leading is used, the type is said to be set *solid.* The thickness of these leads is also measured in points. **Exhibit 7.6** shows lines set solid and lines with 1-, 2-, 3-, and 5-point leading. A general rule for easy reading is that leading should be two to four points larger than the typeface being used.

### *Weight*

*Weight* is a term used to indicate the heaviness or lightness of print. Light print may appear gray, rather than black, and does not stand out well; normal or *medium* is what is normally used on menus; *bold* or heavy print is quite dense; sometimes even extra bold is used. Bold or extra bold weight helps give emphasis, but too much can lead to a cluttered look.

Emphasis can be given and items made to stand out by the wise use of light and bold type. Very light type will not give good emphasis and is sometimes difficult to read. Extra bold type is extremely dark and black. Bold or heavy may be desirable to draw attention but should be used sparingly. Too bold a typeface is not suitable for the menu of a refined, quiet dining room. (Sometimes a printer may refer to the weight of the type as its color, but this is a term used mostly by professionals.) The light level of an operation must be considered. A restaurant with a low level of light should use a slightly larger, bolder type to keep the menus readable. Light type is best used in an operation with normal or higher than normal levels of light.

### *Use of Uppercase and Lowercase Letters*

Another factor that influences ease of reading and comprehension is effectively combining the use of *uppercase* (capitals) and *lowercase* (small) letters. Uppercase gives emphasis and can set words out more clearly. Lowercase is easier to read. Uppercase is used with lowercase to begin sentences, or for proper nouns. It is usually desirable to capitalize all first letters of proper names and main words in item titles on the menu. For instance, the following would be normal: "Hanger Steak Sandwich." Words such as *or, the, a la, in, and,* or *with* are usually not capitalized. Descriptive

**Exhibit 7.4** Menu Showing Type Size Contrast

**Appetizers**

**Shrimp Cocktail** ______ 9.50
Six ice-cold jumbo shrimp served with tangy cocktail sauce.

**Cold Seafood Platter** ______ 13.50
Oysters, shrimp, lobster, and clams served with
wasabi mustard sauce or drawn butter.

**Calamari** ______ 7.50
Lightly breaded, served with marinara sauce.

**Chicken Sate** ______ 6.95
Marinated grilled chicken tenders with peanut dipping sauce.

**Arugula Salad** ______ 7.95
Crunchy arugula with mushrooms, red onions, hearts of palm,
and cranberry vinaigrette dressing.

**Mozzarella Salad** ______ 8.95
Fresh Buffalo mozzarella, fresh basil, and sliced tomatoes
drizzled with extra virgin olive oil.

information, such as "A combination of shrimp, scallops, halibut, and oysters in Newburg sauce," will not have capitalized letters, except for the name of the sauce, because it is a proper noun.

Often, menu items are put in large, bold caps to stand out, and lowercase type is used for descriptive material below the name of the item, capitalizing only proper nouns and first letters of sentences in the descriptive material. Special effects can be obtained at times by setting all descriptive words in small caps.

**Exhibit 7.5** Menu Item Emphasis

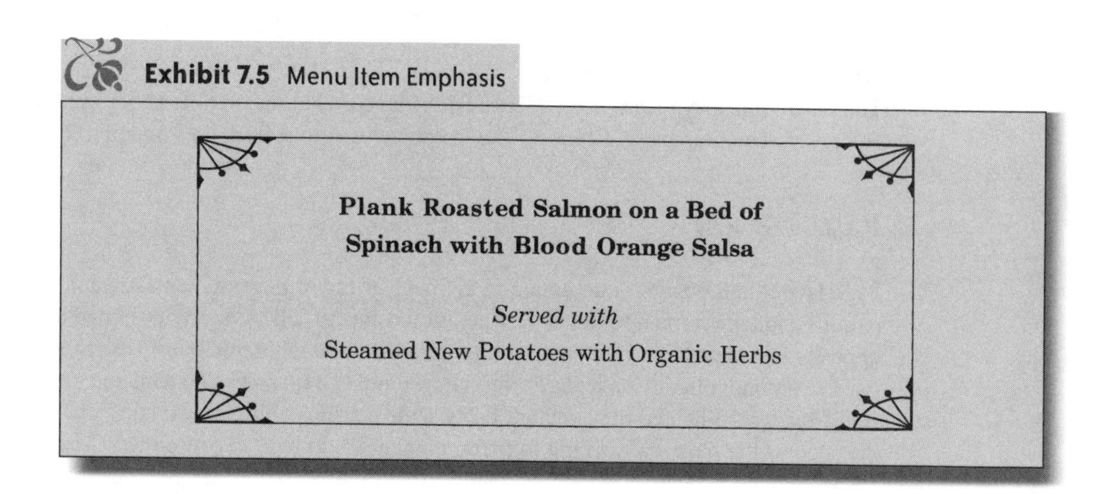

 **Exhibit 7.6** Samples of Leading

**Solid**

A treat to the palate sends taste buds soaring. A treat to the palate sends taste buds soaring. A treat to the palate sends taste buds soaring. A treat to the palate sends taste buds soaring.

**1-Point Leading**

A treat to the palate sends taste buds soaring. A treat to the palate sends taste buds soaring. A treat to the palate sends taste buds soaring. A treat to the palate sends taste buds soaring.

**2-Point Leading**

A treat to the palate sends taste buds soaring. A treat to the palate sends taste buds soaring. A treat to the palate sends taste buds soaring. A treat to the palate sends taste buds soaring.

**3-Point Leading**

A treat to the palate sends taste buds soaring. A treat to the palate sends taste buds soaring. A treat to the palate sends taste buds soaring. A treat to the palate sends taste buds soaring.

**5-Point Leading**

A treat to the palate sends taste buds soaring. A treat to the palate sends taste buds soaring. A treat to the palate sends taste buds soaring. A treat to the palate sends taste buds soaring.

Descriptions should be in keeping with the menu theme and set in a typeface compatible with other type on the page. Descriptions should inform as well as entice. All elements should blend together if a maximum effect is to be achieved.

### *Special Effects Using Type*

The use of type may give special effects. For instance, a nation's script—Javanese, Russian, Greek, Hebrew, Arabic, Japanese, Chinese, Thai—can be used to reflect a restaurant's cuisine.

## Page Design

Page layout and design is an essential element of menu development. A good menu will grab patrons and attract them to items the operation wants to sell. A poorly designed menu will do the opposite—lead patrons into a maze with more than enough items to confuse them.

The amount of copy on a page affects how quickly a menu can be read and understood. In one study, readers indicated they wanted fairly wide margins and disliked copy that ran too close to the edge of the paper, a warning to menu planners who tend to overcrowd areas. Normally, just

Exhibit 7.7 Spacing for Margins

slightly more than 52 percent of a printed page has print on it, and slightly over 47 percent is margins. **Exhibit 7.7** shows a page with a black area in the center indicating print area and margins around it. One would not suspect that the white margin area accounts for nearly half the page. Perhaps, in menu copy, slightly more than this could be covered, but not much more. Providing as much margin space as possible without squeezing menu copy should be the objective.

If space is a problem, extra pages should be added rather than crowding a single menu page. A good margin should remain on the left and right sides and on the top and bottom. **Exhibit 7.8** is an example of good use of space in the menu.

Avoid making menus so large that patrons have difficulty holding them. Some operations make menus very large to give the feeling of luxury, but many guests find them difficult to handle. An operation should give strong consideration to the menu size before deciding on a large one. If a large menu must be used, extra panels that fold open should be considered. Separate menus may also be used. For instance, if desserts are not given with the table d'hôte dinner, or if there is an additional group of à la carte desserts, a special dessert menu may be set up, thus saving space on the main menu. Likewise, alcoholic beverages and wines can be on separate menus. A special fountain menu may be placed on tables and counters where customers can find them, leaving the main menu free to offer a more standard list of items.

### *Line Width*

The width of a line affects reading comprehension. One study found that about two-thirds of menu readers preferred a double-column page to a single-column page. It has also been found that students learn better when reading two-column pages. Most readers like a line length of about 22 *picas* (about 3 ⅔ inches—there are approximately 6 picas to an inch). Long lines may cause readers to lose their place. **Exhibit 7.9** shows a line length of 24 picas. Most books are printed in columns of line length around 25 picas (about 4 3/16 inches), which gives the page only one column.

The eye does not normally flow evenly across a page. It grasps a certain group of words, comprehends these, and then jumps to the next group. If the line is too long, there is a chance of losing the reading place when the eye jumps. Also, too long a line may cause a reader to lose his or her place, since the eyes must refocus as the gaze passes from the end of one long line to the beginning of the next line. One menu authority recommends that, rather than putting prices a distance from menu items in a column to the right, prices should be put right next to or directly under the item.

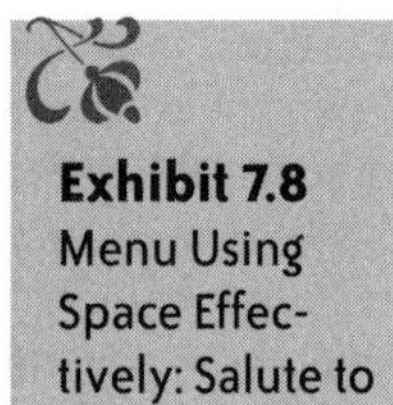

**Exhibit 7.8**
Menu Using Space Effectively: Salute to Excellence, May 2006

*Salute to Excellence*

*First Course*

Mache Lettuce in a Radicchio Cup

*served with*

Grilled Asparagus, Oven Roasted Herb Marinated Roma Tomato, Feta Cheese,
Tabbouleh Salad in an Endive Cup,
Citrus Vinaigrette

---

*Assorted Artisan*
*Dinner Rolls and Flatbreads*
*with Sweet Butter*

---

*Intermezzo*

Mango and Pomegranate Sorbets

*Main Course*

Grilled Petite Filet Mignon topped with Balsamic and Thyme Roasted Red Onion Marmalade,
presented on a bed of Sautéed Spinach,
Red Wine Reduction

*paired with*

Homemade Lump Crab Cake,
Stone Ground Mustard Sauce,
Buckingham's Potatoes, Carrot Mousse,
Bundled Green Beans

*Wine Sponsored by*

E.&J. Gallo Winery

*Grand Dessert Buffet Reception*

featuring
Sweet Street Desserts
and
Rich Products Corporation

*Courtesy of Chicago Hilton Hotel and Towers*

**Exhibit 7.9** Line Length of 24 Picas

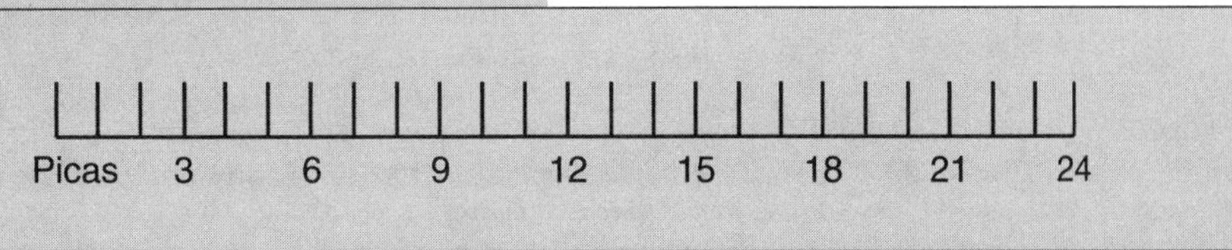


### *Emphasizing Menu Items*

There are a number of ways to emphasize menu items that management wants to sell. An item is set apart by separating it by a bit of space and special type, and then giving some description of the item, as shown in **Exhibit 7.10.** This description can help give emphasis, especially if it is italicized or set in different type. Items can also be shifted slightly for emphasis. Note that when this is done in a column, items will stand out well, as in **Exhibit 7.11.** In both exhibits, items on the top row are capitalized properly, while items in the second row are not, further emphasizing the first line.

In spite of the fact that menu space is valuable, some blank space is needed to set items apart and to avoid having the menu so crowded that the reader is confused. From one-fourth to one-third of the printed area should be blank, in addition to margin space. Headings and lines can be used to indicate separations and help draw attention to items.

The first and last items in a column are seen first and best. This is the place to put menu items one wishes to sell. It is possible to lose items in the middle of a column, so items that management may be less interested in selling, but that have to be on the menu, might be placed here.

Readers tend to skip items. Indenting items presented in a column can help make readers look at all the items in the column. Items too deeply indented, however, are often lost; this is where items are put when one does not wish to give them emphasis. Arranging items in a column from highest in price to lowest, or vice versa, is usually not desirable. People looking for price tend to go to the least expensive one and never look at the others. Mixing up prices makes people look through all the items to see what is there and what the prices are. This is more likely to make even price-conscious buyers see something they find very desirable, and that they will purchase not on the basis of price but on the desirability of the product.

Where the eyes focus is also important in menu design. When patrons open up a two-page menu, their eyes usually go to the right, often to the center of the page or, if not there, to the upper-right hand corner of the page and then counterclockwise to the right bottom. If a menu is a single page, readers will tend to go to the middle of the page and then to the upper right, left and down, then across to the lower right, and then up again. **Exhibit 7.12** indicates this eye movement. It is important that items management most wants to sell be put into those positions where readers look first, known as *emphasis areas.* These items need not necessarily be specials, since many people will come in for specials and will hunt for them on the menu.

**Exhibit 7.10** Item Emphasis

**Coconut Green Curry Mussel Bowl**
*aromatic green curry broth, lemon, fresh mint and sesame crostini 7*

*Courtesy of Deluxe Restaurant; Wilmington, NC; Chef Keith Rhodes*

**Exhibit 7.11** Emphasis on Columns

**Skillet Seared Exotic Mushrooms paired with Six-cheese Tortellini**
in rich herb butter with a drizzle of white truffle and balsamic vinegar reduction 8

**Grilled Beef Tips with Shiraz Bordelaise on Herb Potato Cake**
with shiraz syrup and white truffle oil 8

**Skewered Japanese Tempura Shrimp with Cashew Dipping Sauce**
served with wasabi dressing and ginger shoyu glaze 8.5

*Courtesy of Deluxe Restaurant; Wilmington, NC; Chef Keith Rhodes*

**Exhibit 7.12** Menu View

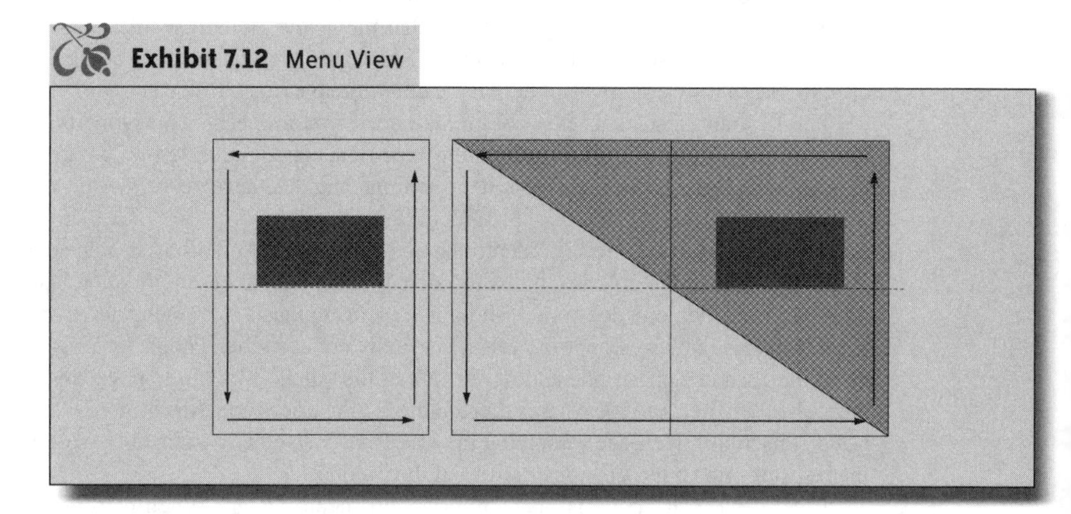

## Color

Besides making an artistic contribution, color can affect legibility and speed of reading. The use of white print on a black background may get more attention than black print on a white background, but it is harder to read. Black on white is read 42 percent more rapidly than white on dark gray, and black type on some colors is read less easily than black on white. However, black on some light colors (mostly creams) is about equal to black on white. The results of a test with var-

**Exhibit 7.13** Color Test Results

| Quite Easy to Read | Fairly Easy to Read | Poorly Read |
|---|---|---|
| Black on light cream | Black on light yellowish green | Black on fairly saturated yellowish red |
| Black on light sepia cream | Black on light blue green | Black on reddish orange |
| Black on deep cream | Black on yellowish red | |
| Black on very light buff | | |
| Black on fairly saturated yellow | | |

ious colors of paper stock are summarized in **Exhibit 7.13.** The order is from easiest to most difficult in each of the three classifications.

In another test, to ascertain how well different colored inks stood out against various paper color tints, the results shown in **Exhibit 7.14** were obtained. Again the order is from easiest to most difficult to read.

Color and design can enhance a menu and make it a better merchandising tool. Color and design are as important to menus as they are to dinnerware. A large amount of color can run into a sizable cost. Yet, as with china, considerable decorative effect can be obtained with only a small amount of color and design. Similarly, plain-colored paper can achieve good effect and is not too expensive. Some paper has color on one side and white or another color on the reverse. As special effects are added, costs increase. Adding silver or gold to a menu can be expensive.

Too much color and design can distract a customer's attention from the menu. Colors that are intense should be avoided unless some special effect is desired. Perhaps a bold color could be used in a club wishing to project a bold atmosphere, but it might not be appropriate in other foodservices.

Clip-ons can be used to give a different design and color. Some menus may be given additional color by using ribbons, silken cords, or tassels, but these are usually restricted to menus for special dining areas.

The variety of colors in paper suitable for menus is wide, including metallic papers. Since

**Exhibit 7.14** Colored Ink Test

| Quite Easy to Read | Fairly Easy to Read | Poorly Read |
|---|---|---|
| Black print on yellow | Blue print on white | White print on black |
| Green print on white | Yellow print on blue | Red print on yellow |
| Red print on white | White print on red | Green print on red |
| Black print on white | White print on green | Red print on green |
| White print on blue | | |

the basic color must serve as a background for the print that describes menu items and lists prices, the print stands out very clearly. Dark browns, blacks, reds, and other deep colors should not be used with dark-colored type, nor should light-colored print be used with light-colored paper.

Color can be used for special events and holidays. Mandarin red with black print is appropriate to commemorate a Chinese New Year; a Halloween menu might use orange and black; black type on emerald green is suitable for St. Patrick's Day.

All colors and designs on the menu should blend with and complement the operation's decor. However, they also can be contrasted with good effect. For example, an outstanding color may be used to give a vivid color splash, just as an interior decorator might use a vividly colored vase to highlight a more subtle color scheme in a room.

Complementary colors are usually those that come from the same primary colors (red, yellow, and blue), but some contrasts between colors coming from different primaries also can be pleasing. For instance, some shades of red and green go well together. While some greens and blues clash, others do not. Knowing how to use colors in design is inherent in some people, but not all. Some menus may seek to achieve a desirable color effect by using colors in stripes, squares, triangles, or circles in some unique pattern. Obtaining a balance takes an ability to blend colors, lightness, and darkness. The services of a menu designer can do much to produce a menu with striking color effects that are in good taste. Menu-publishing software may have built-in complementary color combinations. Some may offer the option of custom color choices.

Color in pictures results from colored inks used in different combinations. The basic four-color-process inks—black, yellow, magenta, and cyan (bright blue)—will give almost any combination of color desired in printing. A fifth color is sometimes added for special effect.

To reproduce a picture, at least four pieces of film, called *separations,* must be made. This is done by taking all yellow tones in one separation. All cyan (blue) tones are taken from the picture in another separation. All shades of magenta and all blacks are taken in other separations. If a special color effect is wanted, other color separations will be made. When these are printed, the human eye will combine and reproduce all colors in the picture. Sophisticated machines now do separations by laser beam, which gives almost perfect placement of one color over another.

A four-color separation can cost more than $500, depending on size, paper, and quantity ordered. However, obtaining the high quality might be worth the expense. Also, buying a large quantity of printed menus can reduce color costs considerably, because the cost is spread, or *prorated,* over time to bring down the unit cost.

There frequently are other costs in using colored pictures. A photograph must be of good quality—a pulled-pork sandwich must look like a *good* pulled-pork sandwich. It should also closely resemble the item actually served in the establishment in order to avoid charges of fraudulent misrepresentation. It is usually worthwhile to pay a professional photographer and perhaps even a food stylist to take photographs.

There are companies that maintain libraries of stock photos. It may be possible to find an appropriate photograph and save the cost of taking a picture. Clip-out sheets can also be used. Be mindful of being able to meet patron expectations.

Color separations can be "doctored" to enhance colors and make them appear as close as possible to the color of the actual food. Color *proofs* must be checked carefully before any printing is done. Poor color reproductions of food items will ruin an otherwise well-developed menu.

Sketches and line drawings, which may or may not have a background, can also be effective. It is possible to add a color within a line picture to help give color contrast, rather than have the line drawing appear filled with the basic color of the paper. Further emphasis can be obtained by putting the picture in a box. Additional color and design can be obtained with decorative borders.

## PAPER

### *The Menu Cover*

The paper used for a menu cover should be chosen carefully. First, consider how often the menu will be used. If it is designed to be disposable, such as a menu on a placemat, then the paper can be a lightweight stock. If it is going to be used regularly, a coated, grease-resistant stock is important. Texture is also a factor to consider, since customers hold the menu in their hands, which can soil menus.

A heavy paper, *cover stock,* is used for most menu covers. This is usually a paper stiff enough to be held in the hands without bending. It may be laminated with a soil-resistant material, such as plastic. The surface is frequently shiny and smooth, but cover materials may have different textures. Laminated covers last for a longer time than untreated ones, and they are easily wiped clean. The weight of cover stock should be in heavy cover, bristol, or tag stock at least .006 inch thick.

Some operations may use a hard-cover folder in which a menu is placed. Often these are highly decorative and represent a considerable investment. Some may be padded, to give a softness in the hand, and covered with strong plastic or other materials that can give the feeling of silk, linen, or leather. These are often laminated onto base materials. Sometimes a foil inlay is stamped on the front. This is done using heat and pressure to set the inlay in the cover material. Gold, silver, and other colors may be stamped onto covers. Using inlays can be expensive.

When these heavy covers are used, it is possible to use a printed menu inside that will not be too expensive. With such a cover, the menu is usually printed on lighter-weight paper. These covers are particularly useful when self-printing.

### *Characteristics of Paper*

The weight of paper for interior pages can be of a lighter weight than that for covers. Usually, strong, heavy book paper is used. Its finish should be such that it resists dirt. Paper can be given different finishes to make it more suitable for menu use. If novelty or striking effects are desired, some specialty papers can be used, as described in **Exhibit 7.15.** The type of ink and printing process used will also help determine the type of paper used. The operator should also investigate the texture and the opacity of the paper.

### *Paper Textures*

Paper textures can range from slight rises, such as is seen in a wood grain, to a rough, coarse surface. It is possible today to give paper almost any texture desired, even that of velvet or suede. *Opacity* of paper (inability to see through it) may depend on the strength of the ink or the use of color. Heavy ink should be used on highly opaque paper. Some transparency may be desirable for some artistic effects. Most often, a maximum opacity is desirable.

**Exhibit 7.15** Basic Paper Terms

| | |
|---|---|
| **Bond paper** | Used for letterheads, forms, and business uses |
| **Book paper** | Has characteristics suitable for books, magazines, brochures, etc. |
| **Bristol** | Cardboard of .006 of an inch or more in thickness (*index, mill,* and *wedding* are types of bristol) |
| **Coated** | Paper and paperboard whose surface has been treated with clay or some other pigment |
| **Cover stock** | Variety of papers used for menu covers, catalogs, booklets, magazines, etc. |
| **Deckle edge** | Paper with a feathered, uneven edge |
| **Dull-coat** | Coated paper with a low-gloss surface |
| **Enamel** | Coated paper with a high-gloss surface |
| **English finish** | Book paper with a machine finish and uniform surface |
| **Grain** | Weakness along one dimension of the paper—paper should be folded with the grain |
| **Machine finish** | Book paper with a medium finish—rougher than English finish but smoother than eggshell |
| **Matte-coat** | Coated paper with a little- or no-gloss surface |
| **Offset paper** | Coated or uncoated paper suitable for offset lithography printing |
| **Vellum finish** | Similar to eggshell but from harder stock, with a finer grained surface |

## Menu Shape and Form

The shape and form of the menu can help create interest and sales appeal. A wine menu may be in the shape of a bottle, while one featuring seafood can be in the shape of a crab, lobster, or fish. When using food shapes, however, the shape chosen must be one that is distinctly recognizable to the average diner. A steakhouse bill of fare may feature a menu in the shape and coloring of a Black Angus steer. A child's menu can be in the shape of a clown whose picture is on the front cover. A pancake house can have its menu shaped like a pancake or waffle.

The fold given the menu may also create an effect. Instead of having a right and left side to a cover, a fold may be used that gives a half page on the right and a half page on the left, so that the menu opens up like a gate. If folds are used, the "foldability" of the paper should be investigated; most coated papers crack easily at a fold.

If a special fold or shape is required, a special *die* (a form used to cut out shapes, such as in jigsaw puzzles) may have to be made. This costs money and should be the property of the one who pays for making the die. Also, the film and/or plates used for printing the menu should become the property of the individual paying for them. This must be clearly understood in making the original agreement. In some instances, printers may not be required to give them up unless such an agreement is made beforehand. Because of the expense and ownership issues, dies are seldom used. Longer or larger paper such as legal or 11- × 17-inch paper can be attractive.

## Printing the Menu

### *Development of Typesetting and Printing*

Around 700 A.D., the Chinese began the process of setting type. They carved each block of type from blocks of wood, rubbing ink over the top surfaces and blotting them on cloth or paper. In Europe, it was not until the early 1400s when Gutenberg invented movable type on the printing press. Letters were cast in metals and put together to make words, then blotted much like their wood counterparts. This method of hand typing was long and laborious.

The *linotype,* a machine that enabled a person to stroke a keyboard that dropped carved letters into place, made the process faster and easier.

### *The Modern Printing Process*

Today, handset and linotype is seldom done except for special purposes. Typesetting has become highly computerized, and in many instances copy is typed into a computer system. The typesetting program reproduces the type either on transparent film or on photographic paper. A copy of the type called a *galley proof* or *reader proof* is made. It is usually in long sheets. This is used to proofread and note corrections. The corrections are made on the computer, and either a new galley proof or a finish proof is made. An artist will either take the finish proof and make up the menu pages or they may be made up in the computer with special programs. *Page proofs* with the illustrations and everything else in place, just as it would be in the final form, must be checked. The page proof is usually a print (similar to a photograph) made from a film. A photographic process is used to make a printing plate from the film once the proof has been okayed. The plate is put onto a press that produces the printed material. Most printing nowadays is produced by the offset method. Ink is applied to the plate, the image is then transferred to another roller, and this image is then *offset* onto a piece of paper. The paper never touches the plate.

*Silk-screening* is another method of printing. A stencil is made, either photographically or by hand, and applied to a silk screen. Ink is forced through the silk onto the paper. This method produces a very intense, dense image and is often used for metallic inks (gold, silver, and copper). It is usually a hand process and is consequently expensive, but the effect may be worth the cost. It is most often used on covers.

Hot foil stamping, *embossing* (creating a raised image with a stamp), and *diecutting* (by which shapes are cut into paper) are other methods of printing and enhancing a menu. As with silk-screening, they are all more expensive to do, and the cost must be weighed against the effect produced.

Permanent menus are usually printed in sizable lots. Some large chains using the same menu can have printings in the hundreds of thousands. A single, small restaurant may print only 500, having only 100 with prices and leaving the remainder with the printer, who can add new prices later when more menus are needed. This takes care of price changes. Most printers want a mini-

**Exhibit 7.16** Self-Printed Menu Using *Soft Café Menu Pro Software*

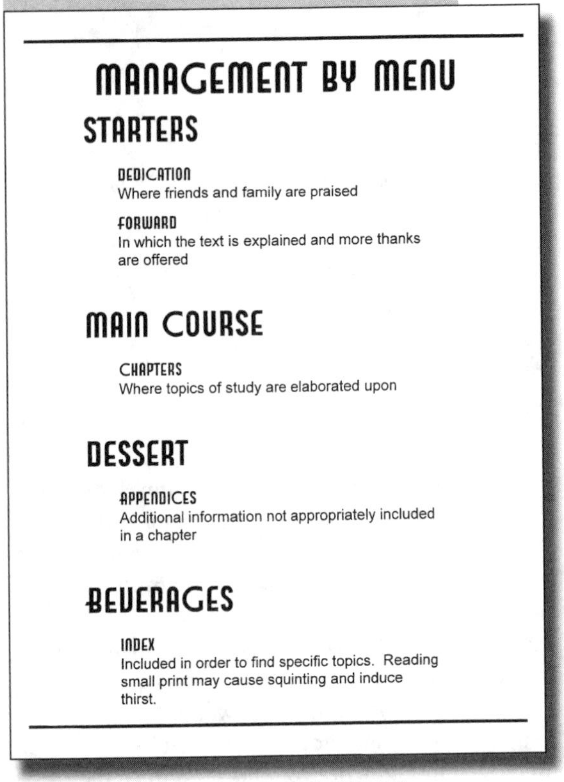

MANAGEMENT BY MENU

STARTERS

DEDICATION
Where friends and family are praised

FORWARD
In which the text is explained and more thanks are offered

MAIN COURSE

CHAPTERS
Where topics of study are elaborated upon

DESSERT

APPENDICES
Additional information not appropriately included in a chapter

BEVERAGES

INDEX
Included in order to find specific topics. Reading small print may cause squinting and induce thirst.

*MenuPro is a registered trademark of SOFTCAFE, LLC.*

**Exhibit 7.17** Self-Printed Menu Using *Soft Café Menu Pro Software*

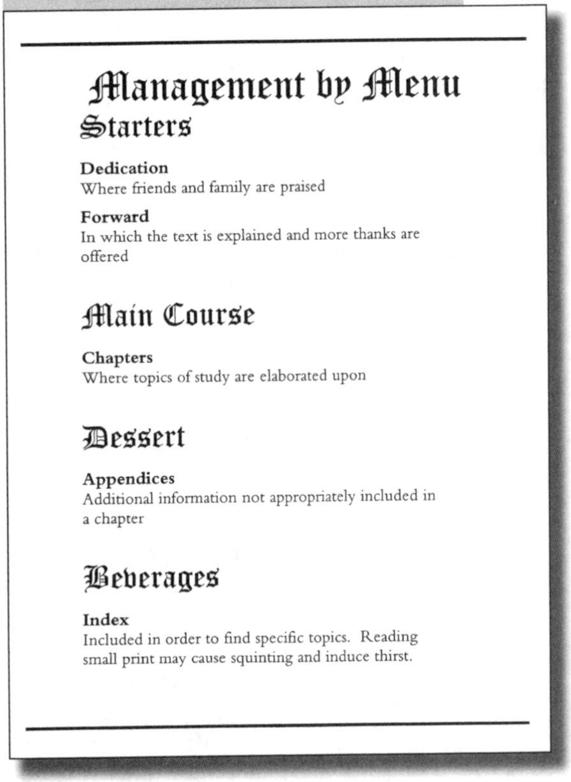

Management by Menu

Starters

**Dedication**
Where friends and family are praised

**Forward**
In which the text is explained and more thanks are offered

Main Course

**Chapters**
Where topics of study are elaborated upon

Dessert

**Appendices**
Additional information not appropriately included in a chapter

Beverages

**Index**
Included in order to find specific topics. Reading small print may cause squinting and induce thirst.

*MenuPro is a registered trademark of SOFTCAFE, LLC.*

mum order of 500. The need to change menu offerings due to seasonal popularity and in response to ingredient price changes should be considered prior to making a decision to stockpile menus.

### *Self-Printing*

Many operations today print their own menus. This can be done almost as well as a professional printing company and often at a huge savings. Having personal control of menu production has some advantages also. Some large operations set up sophisticated systems and produce remarkable results. However, even with much less equipment and facilities, it is possible to print one's own menus, which are adequate for use in the facility. The minimum equipment is usually a computer and a desktop-publishing program with a number of different typefaces and design features. A good laser, color printer is also desirable on which to print camera-ready pages to send to a printer. Software specifically designed to produce menus is available for purchase, as are templates. Care should be taken to avoid a canned appearance that fails to capture the identity of the estab-

**Exhibit 7.18** Management by Menu

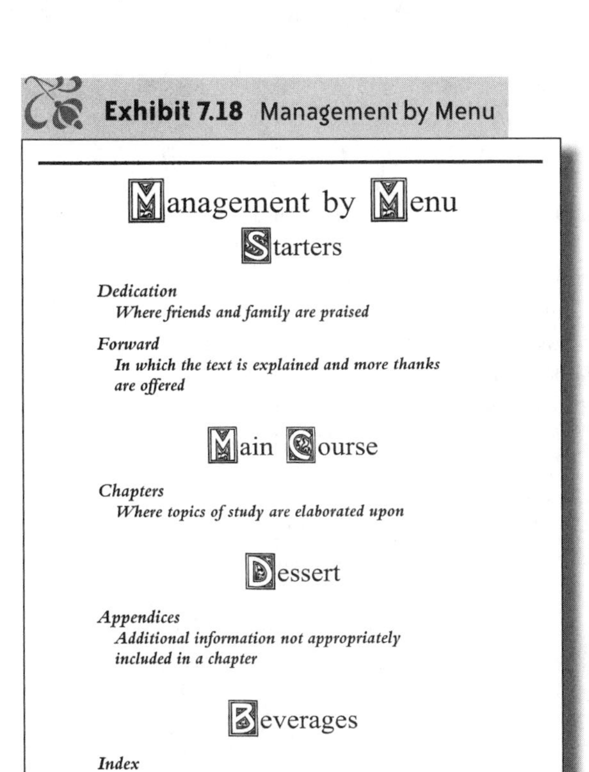


MenuPro is a registered trademark of SOFTCAFE, LLC.

lishment, **Exhibits 7.16, 7.17,** and **7.18** show the same menu self-printed using software in 3 of over 50 styles.

### *Working with Professional Printers*

Certain companies specialize in assisting foodservices in planning, developing ideas for, and printing menus. They usually can be extremely helpful in producing an effective menu. Their experience in setting up menus to do the best merchandising job will be greater than that of the individual menu planner, and this experience can be helpful in avoiding mistakes.

Professional menu printing companies exist to do special art and design work. They may blend printing techniques with special effects to make the best possible menu. If special artwork is needed, either the company will have a staff artist to do it or they will know where to find freelance artists. Often, using a professional menu printer is just one more facet of producing a successful menu. Design firms may work in conjunction with a professional printer. State restaurant associations may be able to provide regional contact information and member references on these firms. An online search will produce numerous results. Caution should be used when contracting with any service provider without an established reputation.

## Summary

Mechanical factors can be extremely important in making a menu an effective communication and merchandising medium. This, of course, can translate into a profit for the operation.

Dark, simple typefaces are easiest to read, but italic and adorned prints can be used for special effects. The best size print for regular menu items is 10 to 12 points, and headings are usually 18 points.

Usually menu items are typeset in 10- to 12-point type with smaller type right below giving the description. The price should be close to the item, either immediately to the right or underneath, so there is no confusion.

Menus should have about two-thirds to three-fourths of their space covered with print. Some blank space is desirable on a menu. The size of a menu should allow it to be handled easily by patrons. If more space is needed, additional pages should be used, or separate menus might be utilized for such items as wines, alcoholic beverages, or desserts.

There are many ways to give emphasis to menu items—a change from one size of type to a larger size, for example. Using bold type can make a menu item stand out. Using all uppercase letters will give emphasis. Setting items in a box with an ornamental border and giving enough space around the box to make it stand out can attract patrons to items, as can clip-ons.

Placement of menu items is also important. The most prominent place on a two-page menu is the middle of the right-hand side. The most prominent place on a single sheet is the middle. The top and bottom are the most prominent places in a column. Items can be lost in the middle of a column. It is not desirable to offer menu items in order of price, because price-conscious patrons will quickly go to the lowest-priced items and not look at other offerings. Mixing prices requires price-conscious patrons to hunt and may lead to their selection of higher-priced items.

The printing of menus has become quite sophisticated. The offset method is quite often used to print menus. The use of full-color pictures requires making four-color separations. This can be expensive, but color is a good selling medium.

The contrast between the color of the type used and the background should be considered carefully. If either the type or background does not give a good contrast, the menu is not read as easily and may be confusing to patrons. Color is also important in giving good decorative effect.

Type, color, and other factors designed to achieve harmony must not be mixed together haphazardly.

The papers used for covers should be either bristol, cover, or tag stock and should be surfaced with some kind of soil-resisting substance. Covers should be made of heavy paper covered with a durable coating. Covers can be given very effective decoration by stamping or high-pressure lamination.

The shape and form of a menu should be carefully considered. Often, a desirable effect can be obtained by working in the logo of the operation, a trademark, or a major food item sold.

Many operations today have computers and laser printers to print menus. This makes the process much faster as well as giving operators more control over design and content.

## Questions

1. Look at any menu and note its typeface, size of printing, legibility, spacing, use of color, and leading. How effectively are these elements used?
2. Look at the same menu. Where is your eye drawn first? Which items does management want to sell?
3. On the same menu, what do its colors tell you? Does its design match the operation's theme and clientele?
4. What is the difference between manual and cast typesetting/printing, offset printing, block printing, and silk-screen printing?
5. What are the advantages of desktop publishing over traditional printing in regard to menus? What are the advantages of traditional printing?
6. What factors should an operator consider when utilizing a professional service provider such as designer or printer?

# MINI-CASES

*7.1* Using a desktop-publishing or document program, create a menu that "breaks the rules" and print it. Alternatively, locate a menu that you believe displays poor menu mechanics. In either case, use the appropriate computer program to "fix" the problems. Enlist the assistance of two persons and have them order off of the "bad" and the "good" menu. What are the results of this experiment? Do they support the principles you have learned to this point? Did you discover anything new?

*7.2* Another approach would be to find two menus, one representing good design and the other poor design. Critique the poorly designed menu, suggesting specific improvements and naming the problems that you expect the changes to address. In the case of the well-designed menu, enumerate specific elements of the menu that compel the diner. Try not to be swayed by your preference for menu offerings.

# 8
# MENU ANALYSIS

## Outline



## Objectives

After reading this chapter, you should be able to:

1. Identify the need for menu analysis and highlight criteria used prior to utilizing the menu and after putting it into effect.
2. Explain the value of subjective evaluation in menu analysis and describe the level of knowledge desired in the performer.
3. Demonstrate how the popularity index or sales ratio is used in menu analysis.
4. Demonstrate how to use menu factor analysis to judge menu item popularity.
5. Identify the Hurst method of menu analysis and describe the benefits.
6. Describe the value of the break-even method as an indicator of what a menu must do to be profitable and demonstrate its use.
7. Identify other methods used by foodservice operations to analyze menus and describe their usage.

# Introduction

After a menu is planned, priced, and set into the form in which it is to be presented, it should be analyzed to see if it: (1) meets the needs and desires of patrons in both the kinds of menu items offered and in price, and (2) meets the needs of the operation from the standpoint of being feasible, profitable, and in line with the goals of the operation. Once the menu items are selected, the menu cannot be considered final until its performance is measured. It is best to do this analysis before printing, since after printing little can be done. Self-printing allows for more frequent adjustments and fine tuning. However, much analysis depends on operating data the menu generates after use. It is up to managers to strike this balance.

It is necessary to check often to see whether menu items are selling, even in institutional operations. Items should be changed if they do not produce sales.

# Common Methods of Menu Analysis

There are seven common ways to analyze menus, each of which will be discussed in detail.

1. A count is kept of items sold per period.
2. A subjective evaluation is conducted, in which the menu is examined by management or by others, such as consultants, skilled in making such judgments to see if it meets certain criteria and standards.
3. A popularity index is developed.
4. A menu factor analysis is created, in which the performance of menu items or groups of items can be judged on the basis of popularity, revenue, food cost, and gross profit margin.
5. A break-even analysis determines at what point a menu will move from a loss to a profit, thus indicating how much must be accomplished in dollar volume and customer count.
6. The Hurst method of menu scoring determines how well a menu scores in sales, food costs, gross profit, and other factors.
7. Goal value analysis evaluates the effectiveness of menu items toward total sales and profits.

Each method can be tailored to meet the needs of individual operations. Any menu can be subjected to all methods, and from each method valuable information may be obtained to make management decisions.

## Menu Counts

One method of keeping track of menu item popularity is simply to make a count of items sold. In some menu analysis methods, a count of the actual number of items sold is made so the sales mix is known.

**Exhibit 8.1**
Monthly Menu Analysis

**MONTHLY MENU ANALYSIS**

Month of ____________________

| Date | | | | | | | | | | | | | | | | | | | | |
|---|---|---|---|---|---|---|---|---|---|---|---|---|---|---|---|---|---|---|---|---|
| 1 | | | | | | | | | | | | | | | | | | | | |
| 2 | | | | | | | | | | | | | | | | | | | | |
| 3 | | | | | | | | | | | | | | | | | | | | |
| 4 | | | | | | | | | | | | | | | | | | | | |
| 5 | | | | | | | | | | | | | | | | | | | | |
| 6 | | | | | | | | | | | | | | | | | | | | |
| 7 | | | | | | | | | | | | | | | | | | | | |
| 8 | | | | | | | | | | | | | | | | | | | | |
| 9 | | | | | | | | | | | | | | | | | | | | |
| 10 | | | | | | | | | | | | | | | | | | | | |
| 11 | | | | | | | | | | | | | | | | | | | | |
| 12 | | | | | | | | | | | | | | | | | | | | |
| 13 | | | | | | | | | | | | | | | | | | | | |
| 14 | | | | | | | | | | | | | | | | | | | | |
| 15 | | | | | | | | | | | | | | | | | | | | |
| 16 | | | | | | | | | | | | | | | | | | | | |
| 17 | | | | | | | | | | | | | | | | | | | | |
| 18 | | | | | | | | | | | | | | | | | | | | |
| 19 | | | | | | | | | | | | | | | | | | | | |
| 20 | | | | | | | | | | | | | | | | | | | | |
| 21 | | | | | | | | | | | | | | | | | | | | |
| 22 | | | | | | | | | | | | | | | | | | | | |
| 23 | | | | | | | | | | | | | | | | | | | | |
| 24 | | | | | | | | | | | | | | | | | | | | |
| 25 | | | | | | | | | | | | | | | | | | | | |
| 26 | | | | | | | | | | | | | | | | | | | | |
| 27 | | | | | | | | | | | | | | | | | | | | |
| 28 | | | | | | | | | | | | | | | | | | | | |
| 29 | | | | | | | | | | | | | | | | | | | | |
| 30 | | | | | | | | | | | | | | | | | | | | |
| 31 | | | | | | | | | | | | | | | | | | | | |
| TOTAL | | | | | | | | | | | | | | | | | | | | |

This count was historically done by having a clerk, cashier, or another individual take the sales checks and make a hand count tabulation. **Exhibit 8.1** is a sheet used for keeping the sales counts of various menu items for one month. This sheet can indicate to management how well certain items are selling. It also summarizes the sales mix, so that it is available for further menu analysis.

Manual counts have become obsolete in operations using electronic cash registers or computers at the point of sale. By punching a preset key, specific menu items can be recorded during the day and then later tabulated for a total count. Thus, in a matter of a few seconds, a complete printout of the number of items sold is possible. If the food cost is also put into these machines, the food cost for each item sold, the total food cost, the gross profit total for each item sold, and the total gross profit are readily available. Thus, much of the basic work required in some of the menu analysis methods is done by the computer or register. The computer can also do a complete menu factor analysis, calculate a break-even point, or analyze a menu according to the Hurst method, methods that will be discussed later.

## Subjective Evaluation

The simplest way to see whether a menu is effective is to evaluate it subjectively by having an independent expert review it to see how well it transmits its message to patrons and predict how well it will perform. If those doing the analysis are expert in menu construction, the opinions will be valuable. If, however, analysis is done by someone who has only a bit of knowledge and who makes judgments purely on the basis of personal views, the evaluation will be of little use.

Menu authorities have published menu evaluation forms that can be used to evaluate various menu factors important to menu success. (A menu evaluation form is presented in Appendix C. It is designed for a hospital or industrial feeding unit, but it could serve as a model for many commercial restaurants.) Managers should set up a menu evaluation form suited to their specific operation.

## The Popularity Index

It may be desirable to know how well menu items or groups of items do in popularity (number of items sold), in generating dollars, in covering food costs, and in the costs of profit.
Menu items within a specific food group can compete against each other for patron selection. One or more items in a group can completely kill the sale of others because of their high popularity. Groups of foods, such as salads and sandwiches, can compete against each other. The relative popularity of separate items and of different food groups should be known so their contribution to the sales mix can be estimated. The potential of the menu as a revenue-producing item and as a means of satisfying customers' desires also can be estimated. A *popularity index* will provide information on this.

Even a noncommercial foodservice needs to study carefully the popularity of its menu items and groups. Both commercial and noncommercial foodservices should also study trends in patron selections. For this a continuing record of selections is needed, developed over a period of time. Not all menu items should be expected to have a high popularity. Some items may be placed on the menu because management feels they must be there even though only a few patrons select them.

Some operations compile a popularity index for menu items as follows: (1) a count of separate items is made within the group to be studied; (2) all selections within the group are totaled; and (3) the percentage for each item of the total sold is calculated. The percentage obtained for each item indicates its popularity when competing with the other items *of its group*. **Exhibit 8.2** indicates the results of a popularity index calculation.

A similar ratio can be calculated for groups of foods, such as sandwiches against salads or frozen desserts against bakery desserts. Such information can be revealing. For instance, a calcula-

**Exhibit 8.2** Popularity Index

| Item | No. Sold | Popularity, % |
|---|---|---|
| Lemon Grilled Free Range Chicken Breast | 44 | 24 |
| Tri Pepper Grilled New York Strip | 38 | 21 |
| Golden Risotto Paella | 14 | 8 |
| Georgia Pecan Dusted Lump Crab Cakes | 54 | 30 |
| Imperial Vegetable Noodle Bowl | 32 | 17 |
| | 182 | 100 |

tion may show that the sandwich group is far more popular than salads or that frozen desserts are destroying the sales of bakery desserts.

A popularity index for a single day is usually not informative. It is more valuable when calculations cover a period of 30 days or longer. Item groups may also vary in popularity over a period of time according to how they are combined with others.

The results of popularity index studies should be analyzed carefully. A continued low ratio for an item should be checked. In some cases, a low popularity can be expected. Items are put on the menu even though they are known to have a low or moderate popularity. Thus, herbed rice might have a relatively low popularity index, but if it is selected enough times, it might be well to leave it on the menu for the few patrons who appreciate it. It is also advisable to check the popularity index with the menu price. If lower-priced items are much higher in popularity than the high-priced ones, patrons may be showing price consciousness. A review of pricing and the type of items offered may be advisable.

Popularity figures can be misleading. Any analysis should look beyond the popularity index to see whether some hidden factors are at work, causing an adverse ratio. Most forms used by foodservices for calculating the sales ratios leave a space for comments so that such factors may be weighed when evaluating the popularity of items. For instance, selections can differ, depending on the day of the week they are offered. Sundays are quite different from other days; Mondays may also be.

Special events in a location near the foodservice operation, perhaps in the same building, can influence selections. A concert or convention will affect the selection of items as well as customer volume.

Selling out of items may force patrons to take different selections than they otherwise would have. Therefore, some record of runouts and times should be maintained.

Weather can affect selections; on a cold day, soup may have a higher selection compared to cold appetizers. Seasons are also influential; fresh strawberry shortcake at a nominal price may have a higher popularity when fresh strawberries are first on the market. A sudden shift in popularity might also mean that a menu item has suffered a variation in quality. Menu placement, presentation on the menu, descriptions, and many other factors must be weighed. **Exhibit 8.3** shows how a record might be kept on one menu item—grilled tuna and couscous, for example.

**Exhibit 8.4** shows a form for recording data to develop a popularity index for five items served both à la carte and table d'hôte. Total contribution of each to sales, food cost, and gross profit are given. The entree salad, pulled pork, and turkey did fairly well, and the chiliburger held its own. Tuna had an unfortunate day. When records such as this are maintained, it is possible to see how well the five items do when competing with each other, and also to check the accumulated popularity index in column 7.

**Exhibit 8.3** Popularity Index Record for Menu Item

**Grilled Tuna and Cous Cous**
(Item)

| Date Served | Forecast | Amount Sold | Sales Ratio | Accumulated Ratio to Date | Contribution to: Sales % | Food Cost % | Gross Profit % |
|---|---|---|---|---|---|---|---|
| 1/25 | 45 | 30 | (5) 7.6% | 7.6% | 7.0 | 7.2 | 7.6 |
| 2/12 | 40 | 38 | (5) 12.4% | 10.0% | 8.1 | 7.8 | 7.9 |
| 4/1 | 40 | 34 | (5) 10.4% | 10.1% | 8.1 | 7.8 | 7.8 |
| 5/20 | 40 | 40 | (5) 14.2% | 11.4% | 8.2 | 7.9 | 8.0 |
| | | | | | | | |
| | | | | | | | |
| | | | | | | | |

Some differentiation in counts may be desirable for à la carte and table d'hôte items. This may be done by indicating the number of à la carte and table d'hôte items sold for a single menu item, such as wild caught salmon sold alone and sold with accompaniments, as shown in **Exhibit 8.5.**

Some operations keep a history of costs and pricing for various items. This record indicates by date the food cost, selling price, and the basis for each, and could be tied in with a popularity index. Changes in any of these factors are recorded. This record allows for an evaluation of costs and selling prices and shows how changes in these factors affected popularity. It also can show if prices are appropriate in various dining areas of a large operation. In some large operations where the same item is sold in different dining areas, prices, and thus popularity, may differ.

The magnitude of a popularity index depends on the number of items against which an item competes. If an item is one of four, and has equal popularity with the others, it would have an expected index of 0.25 (1 ÷ 4). If it is one of five, it would an expected index of 0.20 (1 ÷ 5). Thus, if it is one in four and has a rating of 0.42, 0.17 points above average, it is more than holding its own against the other items. If it has a rating of 0.12, or 0.13 below average, it is not.

### *The Popularity Factor*

It is difficult to compare the popularity indexes of items when they come from groups that do not have the same number of items but are in competition with each other. For instance, if an item that is one of a group of four with a popularity index of 0.25 is compared with an item that is one of a group of five with a popularity index of 0.20, one might conclude that the first is more popular than the second because it has a higher popularity index; this is not true.

To remove this difficulty, the actual popularity index (A) of an item can be divided by its expected popularity index (E) to get what is called a *popularity factor.*

Thus, an item that is one of five, if competing equally well, would have an expected popularity index (E) of 0.20. If its actual popularity index (A) were 0.20, then the popularity factor would be 1.00 (0.20 ÷ 0.20 = 1.00), or what was expected.

## Exhibit 8.4 Form for Recording Data for Popularity Index

Date 9/27 Day Friday Dining Area Cobra Cafe Weather Rain/Sleet Special Events ______
Meal Lunch Total Covers 443 Items Covered Entrees

| (1) ITEM | (2) FORE-CAST | (3) À LA CARTE Price $ | FC* $ | %FC* | (4) TABLE D'HOTE Price $ | FC* $ | %FC* | (5) PORTIONS SERVED | (6) % TO TOTAL FC* | Sales | Gross Profit | (7) POPULARITY INDEX | (8) INDEX TO DATE | (9) QUALITY OF ITEM | (10) COMMENTS |
|---|---|---|---|---|---|---|---|---|---|---|---|---|---|---|---|
| Entree Salad | 60 | 7.30 | 2.40 | 32.0 | 8.50 | 2.54 | 30.0 | 21/48 | 23.3 | 24.7 | 25.4 | 17.7 | 16.0 | Good | — |
| Chili burger | 80 | 4.00 | 1.50 | 37.5 | 5.00 | 1.60 | 32.0 | 38/52 | 21.1 | 20.4 | 20.0 | 25.7 | 22.6 | Fair—little runny | Sold out 12:42 |
| Grilled Tuna | 45 | 6.40 | 1.70 | 26.6 | 7.40 | 1.96 | 26.5 | 9/21 | 7.6 | 9.4 | 10.2 | 7.7 | 12.4 | Poor—too dry | Usually better quality |
| Pulled Pork | 80 | 50 | 2.00 | 30.8 | 6.50 | 3.48 | 38.1 | 22/49 | 22.3 | 20.3 | 19.3 | 18.3 | 16.0 | Good | Sold out 12:50 |
| Smoked Turkey Sandwich | 100 | 4.40 | 1.50 | 35.7 | 5.20 | 1.64 | 31.5 | 48/71 | 25.5 | 25.2 | 25.0 | 30.6 | 27.9 | Good | Price helps sales |
| TOTALS | 365 | — | — | — | — | — | — | 138/251 | 99.8 | 100 | 99.9 | 100.00 | 94.9 | — | — |

*Food Cost

Contributions of items to:

| (11) ITEM | (12) SALES À la Carte | Table d'Hôte | Total $ | (13) FOOD COST À la Carte | Table d'Hôte | Total $ | (14) GROSS PROFIT À la Carte | Table d'Hôte | Total $ |
|---|---|---|---|---|---|---|---|---|---|
| Salad | 153.30 | 408.00 | 561.30 | 50.40 | 121.92 | 172.32 | 102.90 | 286.08 | 368.18 |
| Burger | 152.00 | 310.00 | 462.00 | 57.00 | 48.80 | 156.20 | 95.00 | 210.86 | 305.80 |
| Tuna | 57.60 | 155.40 | 213.00 | 15.30 | 41.16 | 56.46 | 42.30 | 114.24 | 156.54 |
| Pork | 143.00 | 318.50 | 461.50 | 44.00 | 121.52 | 165.52 | 99.00 | 196.98 | 295.98 |
| Turkey | 201.60 | 369.20 | 570.80 | 72.00 | 116.44 | 188.44 | 129.60 | 252.76 | 282.36 |
| TOTALS | 707.50 | 1461.10 | 2,268.60 | 238.70 | 500.24 | 738.94 | 468.80 | 1060.86 | 1529.66 |

**General Comments:**
After four complaints on the tuna, it was taken off the menu.
Compiled by: DSG

However, if A = 0.25, the popularity factor would be 1.25 (0.25 ÷ 0.20 = 1.25), showing it is more popular than expected. If A = 0.15, the factor would be 0.75 (0.15 ÷ 0.20 = 0.75), less popular than expected.

Using the popularity factor makes it possible now to compare the popularity of items from different-size groups. For instance, we may want to compare the popularity of a sandwich with that of a salad. The sandwich is one of eight items and has an expected popularity index of 0.125, with an actual popularity index of 0.15. The salad, one of a group of five items, with an expected popularity index of 0.20, also has an actual popularity index of 0.15. The popularity factor of the sandwich is 1.20 (0.15 ÷ 0.125 = 1.20) and that of the salad is 0.75 (0.15 ÷ 0.20 = 0.75), showing

**Exhibit 8.5** Price Comparison Card

Price Comparison Card

Item Wild Caught Salmon Portion Size 6 oz. À la carte; 4 oz. Table d'hôte

| Where Served | Date | Selling Price À la Carte/Table d'hôte | Food Cost À la Carte/Table d'hôte | Accompaniments |
|---|---|---|---|---|
| Leaf Room | 3/8 | $17.90 $29.90 | $4.88 27% $6.16 21% | À la carte—bread & butter (b&b)<br>Table d'hôte—puree vegetable soup, salad, b&b |
| Café | 3/12 | $11.90 $15.90 | $4.88 41% $6.16 39% | À la carte—b&b<br>Table d'hôte—puree vegetable soup, salad, b&b |
| Café | 3/20 | $13.90 $17.90 | $4.88 35% $6.16 34% | À la carte—b&b<br>Table d'hôte—puree vegetable soup, salad, b&b |
| Gold Room | 3/24 | — $35.90 | — — $6.78 19% | Table d'hôte—puree vegetable soup, salad, wheat berries, coffee or tea |
| Leaf Party Room | 4/20 | $19.90 $27.90 | $5.78 29% $6.48 23% | À la carte—b&b<br>Table d'hôte—puree vegetable soup, salad, roast potatoes, b&b |
| Café | 4/24 | $14.50 $16.50 | $5.74 40% $6.48 39% | À la carte—b&b<br>Table d'hôte—puree vegetable soup, salad, roast potatoes, b&b |

that while both items have the same popularity index within their groups, the popularity factor shows they differ in actual popularity. Calculating the popularity factor puts each item on the same basis for comparison.

Generally, a popularity factor of more than 1.00 shows the item more than holds its own against others; a factor of less than 1.00 indicates it does not. (There is an exception to this rule—food cost—that is explained in the next section on menu factor analysis.)

## Menu Factor Analysis

*Menu factor analysis* is a way of manipulating data to ascertain how well menu items are doing. It develops *factors* to indicate how an item is doing in: (1) popularity; (2) creating revenue or sales; (3) influencing food cost; and (4) how much it contributes to gross profit. Management can use these factors to make decisions on whether to retain the item on the menu, lower its food cost, or increase its price. Menu factor analysis also permits comparison with other factors.

Previously we saw how a popularity index was calculated as a percent of total selections for a group of menu items. The same kind of percentage can be made for indexing dollar sales (revenue), food cost, or an item's gross profit contribution. Thus, if one menu item has 120 orders out of 480 total sold, has $100 sales out of total sales of $500, a food cost of $40 out of $180, and a gross profit of $60 out of $325, the following indexes can be calculated.

| | |
|---|---|
| Popularity | 120 ÷ 480 = 0.25 |
| Sales | $100 ÷ $500 = 0.20 |
| Food cost | $40 ÷ $180 = 0.22 |
| Gross profit | $60 ÷ $325 = 0.18 |

In menu factor analysis, these indexes are called *actual indexes* since they are derived from actual sales and profit figures and not projections.

The *expected* index is a hypothetical index based on what management expects or projects an item to do in terms of popularity, dollar sales, food cost, or gross profit. Thus, if an item is one of a group of four menu items and management expects all to compete equally in popularity, sales, food cost, and gross profit, all the expected indexes would be 0.25. Menu items in a group, however, are usually not equally popular, nor do they bring in the same percent of sales dollars or have the same percent of food cost or gross profit. Some menu items are generally more popular than others. In menu factor analysis, therefore, management usually assigns what it expects an item to do in these categories. Thus, if there is a group of four sandwiches, management might think one sandwich should bring in 34 percent of sales, another 24 percent, a third 22 percent, and a fourth 20 percent, rather than each contributing equally at 25 percent of sales. Thus, it is possible to have an *equal* expected index and a *variable* expected index.

To illustrate how such indexes would be compiled, **Exhibit 8.6** gives the *actual* number of sales with its index, the *equal* expected sold with its index, and the *variable* expected sold with its index. Similar data is then given for dollar sales, food cost, and gross profit.

Although close study of these indexes could reveal much helpful information to management, it is even better to use them to create a factor that will make their relationship stand out more clearly. To calculate this factor, the actual index is divided by the expected index (A ÷ E). If this is done by dividing the actual index by the equal expected index, and the actual index by the variable expected index, we get the figures shown in **Exhibit 8.7.**

In menu factor analysis, a factor of 1.00 indicates an item is doing exactly what is expected; a factor of over 1.00 means it is doing better than expected; and a factor under 1.00 means that it is doing worse than expected. However, with regard to food cost, a factor of 1.00 still means cost is as expected, but a factor under 1.00 is good, and over 1.00 is bad. We can now look at these factors and make a quick summary of what management might see in them.

First, for popularity: If we think that all four menu items should compete equally, the vegetable pita sandwich seems to be doing well (1.17), tuna salad and grilled chicken about as expected (1.00), and smoked turkey not as well as expected (0.83).

However, when we look at the variable factors, or what management expects the items to do, we get a different view. Vegetable pita is not doing quite as well as expected (0.93) but it is not so far off for management to be too concerned. Tuna salad is doing what it should (1.00), and grilled chicken is selling better than management expected (1.09). Smoked turkey is doing just what management thought it should (1.00).

Now, for sales: If we look at the equal dollar sales factors, we see vegetable pita is doing

**Exhibit 8.6** Food Sales Index

| | Actual | | Equal Expected | | Variable Expected | |
|---|---|---|---|---|---|---|
| **Menu Item** | **Sold** | **Index** | **Sold** | **Index** | **Sold** | **Index** |
| | | | **Popularity** | | | |
| Vegetable Pita | 140 | 0.292 | 120 | 0.250 | 150 | 0.313 |
| Tuna Salad | 120 | 0.250 | 120 | 0.250 | 120 | 0.250 |
| Grilled Chicken | 120 | 0.250 | 120 | 0.250 | 110 | 0.229 |
| Smoked Turkey | 100 | 0.208 | 120 | 0.250 | 100 | 0.208 |
| Total | 480 | 1.000 | 480 | 1.000 | 480 | 1.000 |
| | | | **Sales** | | | |
| Vegetable Pita | $154 | 0.312 | $125 | 0.250 | $168 | 0.340 |
| Tuna Salad | 120 | 0.243 | 125 | 0.250 | 119 | 0.241 |
| Grilled Chicken | 120 | 0.243 | 125 | 0.250 | 109 | 0.221 |
| Smoked Turkey | 100 | 0.202 | 125 | 0.250 | 98 | 0.198 |
| Total | $494 | 1.000 | $500 | 1.000 | $494 | 1.000 |
| | | | **Food Cost** | | | |
| Vegetable Pita | $80 | 0.444 | $45 | 0.250 | $90 | 0.500 |
| Tuna Salad | 40 | 0.222 | 45 | 0.250 | 40 | 0.222 |
| Grilled Chicken | 30 | 0.167 | 45 | 0.250 | 25 | 0.138 |
| Smoked Turkey | 30 | 0.167 | 45 | 0.250 | 25 | 0.138 |
| Total | $180 | 1.000 | $180 | 1.000 | $180 | 0.998 |
| | | | **Gross Profit** | | | |
| Vegetable Pita | $74 | 23.6 | $78.50 | 0.250 | $100 | 0.378 |
| Tuna Salad | 80 | 25.5 | 78.50 | 0.250 | 74 | 23.6 |
| Grilled Chicken | 90 | 28.7 | 78.50 | 0.250 | 70 | 0.223 |
| Smoked Turkey | 70 | 0.223 | 78.50 | 0.250 | 70 | 0.223 |
| Total | $314 | 100.1 | $314.00 | 1.000 | $314 | 0.998 |

better than expected (1.25) and tuna salad and grilled chicken (0.97) are not doing too badly. It is the smoked turkey that is in trouble (0.81). However, this all clears up when we get to the variable factors. Vegetable pita now is not doing what management expected (0.92), while the others are doing as well or slightly better (1.00, 1.10, and 1.02). Perhaps the vegetable pita is not bringing in enough dollars because it is not priced high enough, but a look at its variable popularity factor reveals that perhaps not enough sales are occurring. Could this be because the price is too high? Management would have to consider this.

In equal food cost factors, vegetable pita is clearly not doing what it should (1.78) and is driving food costs too high. The others look good, however; in fact, grilled chicken and smoked turkey are excellent (0.67). When we go over to the variable factors, however, the true story comes out. Management has not expected vegetable pita to have the same dollar food cost as the others. Its 0.89 variable factor indicates this. While tuna salad does all right (1.00), grilled chicken and smoked turkey are now down in food cost (1.21). Perhaps management needs to increase their prices to get a better food cost ratio.

**Exhibit 8.7** Menu Factor Analysis

| Menu Item | Equal Factor | Variable Factor |
|---|---|---|
| | **Popularity** | |
| Vegetable Pita | 1.17 | 0.93 |
| Tuna Salad | 1.00 | 1.00 |
| Grilled Chicken | 1.00 | 1.09 |
| Smoked Turkey | 0.83 | 1.00 |
| | **Sales** | |
| Vegetable Pita | 1.25 | 0.92 |
| Tuna Salad | 0.97 | 1.00 |
| Grilled Chicken | 0.97 | 1.10 |
| Smoked Turkey | 0.81 | 1.02 |
| | **Food Cost** | |
| Vegetable Pita | 1.78 | 0.89 |
| Tuna Salad | 0.89 | 1.00 |
| Grilled Chicken | 0.67 | 1.21 |
| Smoked Turkey | 0.67 | 1.21 |
| | **Gross Profit** | |
| Vegetable Pita | 0.94 | 0.62 |
| Tuna Salad | 1.02 | 1.08 |
| Grilled Chicken | 1.15 | 1.29 |
| Smoked Turkey | 0.89 | 1.00 |

Gross profit is the last comparative group. In the equal factor, vegetable pita could do a bit better. Tuna salad and grilled chicken are over 1.00, which means they are doing well. Smoked turkey is not doing as well (0.89), and its *profit margin* or contribution to profit will probably not be as satisfactory. Again, some adjustment takes place in the variable factors. Vegetable pita is now not doing what it should in generating gross profit (0.74). Smoked turkey is doing all right. It is tuna salad and grilled chicken that are carrying the load here, with variable factors of 1.08 and 1.29.

Menu factor analysis can be a revealing and helpful method for analyzing a menu and indicating, perhaps, what should be done in case of problems with menu items. However, it has its drawbacks. The variable factors are based on what management thinks should be true. Management has to be fairly close in its estimates to the actual situation to make the method valuable. Guesses that are too far from the mark can cause problems. One also needs to remember that these are factors based on percentages that do not always reflect all the conditions. Management must look beyond these factors and weigh other things that should be considered. A study problem using menu factor analysis is given in Appendix D.

## THE HURST METHOD

The Hurst method of analyzing a menu considers the effects on sales of pricing, food cost, item popularity, gross profit contribution, and other factors.[1] It is a tool used by management to evaluate menu changes, such as changes in price, items, and food costs. It is a fairly easy method to use and can be quickly performed from a few statistics easily obtained from accounting data.

To use the Hurst method, management should decide on the period for which the scoring will be done. It can be a single meal or a series of meals for which the same menu is used. The period should be a typical one and not one in which unusual events, such as bad weather or a holiday, would keep patrons away.

Eighteen steps are required to make a Hurst menu analysis. **Exhibit 8.8** indicates where data are placed on the form. The numbers (1) to (18) indicate data for each of the 18 steps in the analysis. (Columns and spaces in **Exhibit 8.8** are referred to by "column" or "space" numbers.) Although the number of steps may seem formidable, each step is easily done, and they follow each other in a natural sequence.

The exhibit contains figures of a Hurst study for three menu items. (Normally, there would be more, but to simplify the study, only three were used.) The meal covers lunch for two days. Sales in this example are 100 orders of shrimp at $8.00 with a food cost of $2 per serving, 500 orders

**Exhibit 8.8** Hurst Menu Analysis

**HURST MENU ANALYSIS**

Time Period 2 days (1)

Meal Lunch

Number of Items Evaluated 3

| Columns | 1<br>Menu Item (2) | 2<br>Number Sold (3) | 3<br>Selling Price (6) | 4<br>Total Sales (7) | 5<br>Food Cost (9) | 6<br>Total $ Food Cost (10) |
|---|---|---|---|---|---|---|
| | Shrimp | 100 | $ 8.00 | $ 800 | $2.00 | $ 200 |
| | Beef Ribs | 500 | $13.00 | $6,500 | $3.28 | $1,640 |
| | Turkey | 400 | $ 9.00 | $3,600 | $2.40 | $ 960 |
| **Totals** | | 1,000 (4) | | $10,900 (8) | | $2,800 (11) |

Total Items Sold (Including study items) 1,500 (5)

**Spaces**

| 7<br>Meal Average (12)<br>(total column 4 ÷ total column 2) | 8<br>Gross Profit (13)<br>(total column 4 – total column 6) | 9<br>Gross Profit Percentage (13)<br>(space 8 ÷ total column 4) | 10<br>Gross Profit Average (14)<br>(space 7 × space 9) | 11<br>Total Served (5)<br>(total column 2 + others served) |
|---|---|---|---|---|
| $10,900 ÷ 1,000 = | $10,900 – $2,800 = | $8,100 ÷ 10,900 = | $10.90 × 74% = | |
| $10.90 | $8,100 | 74% | $8.06 | 1,500 |

| 12<br>Percentage of Patrons (15)<br>(total column 2 ÷ total items sold) | 13<br>Menu Score (16)<br>(space 10 × space 12) | 14<br>Comments |
|---|---|---|
| 1,000 ÷ 1,500 = | $8.06 × 66.7% = | Good weather. Many went out of the building to eat elsewhere. (17) |
| 66.7% | $5.37 | Compiler LHK (18) |

of beef ribs priced at $13.00 with a food cost of $3.28 per serving, and 400 orders of roast turkey priced at $9.00 with a food cost of $2.40 per serving. This comes to 1,000 orders for the study group items, but total orders for the period studied were 1,500. By following through on the 18 steps outlined, we can see how a Hurst menu score is compiled.

### *Steps in Performing the Hurst Menu Analysis*

The steps in making a Hurst menu analysis are as follows:

1. Decide on the period and meal to be covered, and the number of items to be evaluated, and fill in the necessary data.
2. Select items that contribute to a major portion of the revenue—all can be included or only the part to be studied. The items selected for special study are called the *study group* in this discussion. Place these items in column 1.

3. Make a count of the number sold for each of the items in the study group. Place the results in column 2.
4. Add entries in column 2 to get the total number of study group items sold.
5. Add to the number of study group items sold (step 4) and all other (nonstudy) menu items sold and place the total on the line for Total Items Sold and in space 11, Total Served.
6. Record the selling price for each item in the study group in column 3.
7. Multiply the number of each study group item sold (from column 2) by its selling price (from column 3) to get the total dollar sales for each item. Record these in column 4.
8. Add column 4 to get the total dollar sales of the study group.
9. Calculate the dollar food cost for each item in the study group. Record this in column 5.
10. Get the total food cost of each sale by multiplying the number of each item sold by its food cost. Record this in column 6.
11. Add column 6 to get the total dollar food cost for all items.
12. Obtain the meal average—or average selling price—of all items in the study group by dividing the total dollars in sales (the sum of column 4) by the total number of study group items sold (the sum of column 2). Place the result in space 7.
13. Calculate the gross profit and gross profit percentage. Subtract the total food costs for all study group items (sum of column 6) from the total dollar sales of study group items (sum of column 4). Record the difference in space 8. Then divide this difference by the total dollar sales (sum of column 4) to get the gross profit percentage. Record this figure in space 9.
14. Calculate the gross profit average by multiplying the check average (average selling price) in space 7 by the gross profit percentage in space 9. Record this in space 10.
15. Calculate the percentage of customers that select the study group items by dividing the sum of the study group items sold (total of column 2) by the total items sold as compiled in step 5 and recorded in space 11. Record the result in space 12.
16. Calculate the menu score by multiplying the gross profit average (space 10) by the percentage of patrons selecting the study group items (space 12). Record the result in space 13.
17. Make any comments in space 14.
18. Sign and date the form.

### *Evaluating the Menu Score*

One menu score by itself means little. It merely indicates how numbers sold interact with prices, sales mix, gross profit, and food costs. A number of menu scores kept over a period of time must be compiled and analyzed to obtain desirable information on how well a menu is doing. (See **Exhibit 8.9.**) Also, comparing scores with those of other operations is not helpful. There are too many varying factors. Only scores generated internally are good for comparison.

Often a low score is the result of too low a selling price resulting in a low check average, or some items not eliciting patron selection. Whatever the cause, management should take steps to correct the situation. The manager may try to reduce food costs and thus raise gross profit, use merchandising to elicit more patron selections, or increase volume. There are many possibilities, and managers should investigate carefully to identify the cause of a low menu score.

**Exhibit 8.9** Effect of Price Changes on Menu Score

| | Item | Selling Price | $ Sales | Total $ Food Cost |
|---|---|---|---|---|
| | 100 Shrimp | $7.50 | $750 | $225 |
| | 500 Beef ribs | 10.00 | 5,000 | 2,000 |
| | 400 Turkey | 6.00 | 2,400 | 600 |
| Totals | 1,000 | | $8,150 | $2,825 |

| | | |
|---|---|---|
| Meal average ($8,150 ÷ 1,000) | = | $8.15 |
| Gross profit % ($8,150 − $2,825 = $5,325 ÷ $8,150) | = | 65% |
| Gross profit average ($8.15 × 65%) | = | $5.30 |
| % Meals of total meals (1,000 ÷ 1,500) | = | 66.7% |
| Menu score (66.7% × $5.30) | = | $3.53 |

A menu planner can use a computer to simulate menu changes before actually putting them into effect. Putting the Hurst method into the computer can make simulation easy, quick, and quite accurate—as long as accurate data are fed into the computer.

## Goal Value Analysis

David Hayes and Lynn Huffman have reviewed menu analysis methods using matrix systems, and have proposed the *goal value method* of menu analysis.[2] It is largely a quantitative method. They use a mathematical method of A × B × (C × D) = Goal value, where A = 1 − Food cost percentage, B = Volume or number sold, C = Selling price, and D = 1 − (Variable food cost percentage + Food cost percentage) to arrive at a goal index they call the *numerical target* or score. They then calculate individual values for each menu item and compare each value with the standard to see if it is above or below standard. If it is above standard, it is doing better than average; if below, it is considered poor as a menu item. **Exhibit 8.10** gives the information needed to set up a goal value system for four menu items—chicken, steak, shrimp, and sole.

**Exhibit 8.10** Data for Goal Value Analysis

| Menu Item | Number Sold | Item Food Cost | Total Food Cost | Selling Price | Total Sales | Food Cost Percentage |
|---|---|---|---|---|---|---|
| Chicken | 24 | $1.75 | $42.00 | $6.95 | $166.80 | 25% |
| Steak | 20 | 4.75 | 95.00 | 11.90 | 238.00 | 40 |
| Shrimp | 16 | 2.60 | 41.60 | 7.50 | 120.00 | 35 |
| Sole | 9 | 3.65 | 32.85 | 8.70 | 78.30 | 40 |
| Total | 69 | 12.75 | $211.45 | | $603.10 | |
| Average | 17.25 | 3.19 | 52.86 | 8.75 | 150.78 | 35% |

**Exhibit 8.11** Achievement Values

| Menu Item | A (1 – food cost percentage) | × | B Number Sold | × | C Selling Price | × | D 1 – (food cost percentage + variable cost percentage) | = | Item Value |
|---|---|---|---|---|---|---|---|---|---|
| Chicken | 0.75 | | 24 | | $6.95 | | 0.43 | = | 53.8 |
| Steak | 0.60 | | 20 | | 11.90 | | 0.28 | = | 40.0 |
| Shrimp | 0.65 | | 16 | | 7.50 | | 0.33 | = | 25.7 |
| Sole | 0.58 | | 9 | | 8.70 | | 0.26 | = | 11.8 |

The following results are obtained on chicken, steak, shrimp, and sole: The average food cost is 35 percent, the average number sold is 17.25, and the average selling price is $8.75. Setting the variable cost at 32 percent, the combined food and variable cost is 67 percent (35% + 32%). Thus, the *goal value standard* for the four items is (1 – 0.35) × 17.25 × ($8.74 × [1 – 67%]), or 32.3. The individual achievement values are shown in **Exhibit 8.11.** Thus, steak and chicken do well, being above the goal value standard of 32.3, while shrimp is below standard and sole does poorly. Chicken is the best performer.

Let us say management wants to remove the sole and replace it with a pasta dish with a selling price of $6.50, a food cost of $1.43 (22 percent), and a variable cost of 32 percent. It expects 20 sales. The total number of covers does not change, so the pasta steals 11 selections from the others. **Exhibit 8.12** shows what might be found. The average food cost is 31.4 percent ($177.70 ÷ $565.70), the average number sold is 17.3 (69 ÷ 4), and the average selling price is $8.20 ($565.70 ÷ 69). The combined variable and food cost is 63.4 percent (31.4% + 32%). The goal value standard is (1 – 31.4) × 17.3 × ($8.20 × [1 – (31.4% + 32%)]) = 35.62.

The individual menu item evaluation is shown in **Exhibit 8.13.** Now pasta becomes the best-selling menu item, and steak drops to a level slightly above the standard. Shrimp does poorly.

The goal value method of analysis can be very useful to foodservice operators. However, in

**Exhibit 8.12** Goal Analysis

| Menu Item | Number Sold | Food Cost Percentage | Total Food Food Cost (in dollars) | Cost Percentage | Selling Price | Total Sales |
|---|---|---|---|---|---|---|
| Pasta | 20 | 22 | $1.43 | $28.60 | $6.50 | $130.00 |
| Chicken | 20 | 25 | 1.75 | 35.00 | 6.95 | 139.00 |
| Steak | 18 | 40 | 4.75 | 85.50 | 11.90 | 214.20 |
| Shrimp | 11 | 35 | 2.60 | 28.60 | 7.50 | 82.50 |
| Total | 69 | | | $177.70 | | $565.70 |

**Exhibit 8.13** Menu Item Evaluation

| Menu Item | A (1 – fixed cost percentage) | × | B Number Sold | × | C Selling Price | × | D 1 – (fixed cost percentage + 32%) | = | Item Value |
|---|---|---|---|---|---|---|---|---|---|
| Pasta | 0.78 | | 20 | | $6.50 | | 1 – (0.22 + 0.32) | = | 46.6 |
| Chicken | 0.75 | | 20 | | 6.95 | | 1 – (0.25 + 0.32) | = | 44.8 |
| Steak | 0.60 | | 18 | | 11.90 | | 1 – (0.40 + 0.32) | = | 36.0 |
| Shrimp | 0.65 | | 11 | | 7.50 | | 1 – (0.35 + 0.32) | = | 17.7 |

these calculations, variable cost remains fixed. In actual practice, it does not. The method can be made more effective if a high, medium, and low variable cost is assigned.

## The Break-Even Method

Another way to analyze a menu is to see whether it *breaks even,* or covers all the costs of doing business. This method can be used by both for-profit and not-for-profit operations. If the success of an operation is generally measured in terms of cost control and budget, the break-even method is a good approach. Commercial operations just getting started might use the break-even point. Later, when costs are being covered and profit is steady, the operator may switch to another type of menu analysis.

If the costs of doing business are not covered, there is a loss; if they are more than covered, there is a profit. To know when a menu will break even, one can total all the costs of producing and serving the expected volume of meals and deduct this from the expected income. Thus, if a menu in one day is expected to bring in $6,000 and costs are $5,500, the profit is $500.

At times it is desirable to estimate a break-even point prior to putting a menu into effect. The menu planner may want to know how many dollars in sales a menu must generate before it breaks even and how many guests it must attract to do so; or the planner may want to know how much money the operation must take in to make a desired profit, or how many guests will be required to make this profit.

To calculate a break-even point, three things must be known: (1) the average check; (2) fixed costs in dollars; and (3) the fixed cost percent.

The *average check* is the total dollar sales divided by the number of guests served in a particular period. For example, if sales for a certain period were $100,000 and 10,000 guests were served, the average check was $10 ($100,000 ÷ 10,000 = $10).

*Fixed costs* are costs incurred regardless of whether a single sale is made. These include rent, administrative expenses, license fees, depreciation, insurance on equipment and other capital values, some labor costs and employee benefits, heat, electrical power, and advertising. They are sometimes called *turn-key expenses* because as soon as the key is turned in the door in the morning, and before any business is done, they are incurred. The *fixed cost percentage* is the dollar value of fixed costs divided by sales.

Several formulas are used to ascertain at what point break-even occurs. The following symbols are often used.

| | | |
|---|---|---|
| BE | = | Break-even point |
| FC% | = | Fixed cost percentage |
| FC | = | Fixed cost in dollars |
| AC | = | Average check in dollars |

A simple formula to determine the break-even point in dollars is BE = FC ÷ FC%. If fixed costs are $232,367 and the fixed cost percentage is 46 percent, the formula is BE = $232,367 ÷ 0.46 = $505,146.65.

The break-even method can also be used to calculate the number of customers needed to cover costs. The formula used to obtain this is BE = FC ÷ (AC × FC%). If the average check is $6.54, BE = $232,367 ÷ ($6.54 × 0.46) = $232,367 ÷ $.015 = 77,198 customers.

To find the dollar sales needed to cover fixed costs and give a desired profit, add into the dollar fixed costs the amount of profit desired, using the following formula: BE = FC + P (profit) ÷ FC%. If the desired profit is $15,000, the calculation is $232,367 + $15,000 ÷ 0.46 = $537,754.34.

To find out how many customers would give this profit, the formula would be BE = (FC + P) ÷ (AC × FC%). This would translate into ($232,367 + $15,000) ÷ ($6.54 × 0.46) = 82,182 customers.

## Other Methods of Menu Analysis

Operations often use data and analytical methods of their own to calculate how well a menu might do or is doing. Some like to obtain a computer printout immediately after a meal to see how well certain menu items did. A printout may also be obtained of how many of certain items each server sold—often called a *productivity report.* Some operations collect the gross dollar sales per seat. Another method might be to check operating figures against a profit, or to look at the profit and loss statement.

A daily food cost report, as shown in Chapter 5, acts as a check on how well a menu is doing. It indicates total sales and food cost, and often makes comparisons with other periods so management can see how well the current menu is generating business.

## Summary

The final step in menu development is analysis of the menu's results. Menu development is actually a continuous process, and most menu items and prices can never be considered final. Menu analysis provides the operator with an approach for improving a menu's popularity and, in commercial operations, making a profit.

It is important that menus meet cost restraints and make a profit for commercial operations. Menu analysis methods available include making menu counts, using subjective evaluation, the popularity index, menu factor analysis, the Hurst method, and break-even analysis. These methods can be helpful in indicating menu item profitability and item popularity. The use of an outside consultant may also provide this information, as well as indicate how well menu items are presented. Besides the methods detailed here for menu analysis, different operations may use their own individual analytical methods, including analysis of daily computer printouts of sales.

## QUESTIONS

1. A vendor offering three main items—hot dog, hamburger, and fried fish sandwich—wishes to make a menu factor analysis to see how well these items are doing. Management expects the following indexes.

| Item | Popularity Index | $ Sales Index | $ Food Cost Index | $ Gross Profit Index |
|---|---|---|---|---|
| Hot dog | 0.35 | 0.38 | 0.37 | 0.40 |
| Hamburger | 0.40 | 0.38 | 0.38 | 0.35 |
| Fish sandwich | 0.25 | 0.24 | 0.25 | 0.26 |

The following information is available.

| Item | Number Sold | Selling Price | $ Food Cost |
|---|---|---|---|
| Hot dog | 162 | $1.59 | $0.21 |
| Hamburger | 286 | 1.89 | 0.31 |
| Fish sandwich | 97 | 1.79 | 0.17 |
| Total | 545 | $5.27 | $0.69 |

Complete the menu factor analysis, and then analyze the findings. (Remember that dollars of food cost are compounded with number of sales of an item, and if the number of sales of an item is high this tends to inflate food cost.)

2. If 123 salmon steaks, 171 beef steaks, 86 broiled chickens, 36 lobster tails, 192 ribs of beef, and 204 seafood casseroles are sold, what is the popularity index for each?

3. Fixed costs are 45 percent of $3,904. Average check is $9.95. What must the dollar volume be to break even? How many customers must be served?

4. Using the Hurst method of analysis, complete the following.

| 1<br>Item | 2<br>No. Sold | 3<br>Selling Price | 4<br>Item Food Cost | 5<br>Total Food Cost | 6<br>Total Sales |
|---|---|---|---|---|---|
| Chicken | 15 | $ 9.50 | $2.86 | | |
| Tuna | 16 | 12.00 | 3.20 | | |
| Fettuccine | 14 | 8.75 | 2.10 | | |
| Scallops | 15 | 11.00 | 3.07 | | |
| Pot pie | 13 | 10.75 | 2.03 | | |
| Tilefish | 11 | 12.50 | 2.40 | | |
| Pork chop | 12 | 11.25 | 2.94 | | |
| Calf's liver | 14 | 10.50 | 2.88 | | |
| Lamb chops | 12 | 14.00 | 4.53 | ________ | ________ |

The number of items sold is the total count of entree items for this lunch sold to club members, that is, 122.

**7**
**Meal Average**
**$__________**

**8**
**Gross Profit**
**$__________**

**9**
**Gross profit %**
**__________%**

**10**
**Gross Profit Average**
**$__________**

**11**
**Total Meals Served**
**__________**

**12**
**% of Customers**
**__________%**

**13**
**Menu Score**
**$__________**

**14**
**Comments**

## MINI-CASE

**8.1 Regarding the Management by Menu Analysis Café:**

1. In each category, calculate expected and actual popularity factors.
2. What conclusions do you draw from these numbers? Explain your reasoning.
3. Calculate contribution margins of each item.
4. What conclusions do you draw from these figures? Explain your reasoning.
5. Does the new information, coupled with data obtained from performing popularity factor analysis, change any of your earlier conclusions? Explain.
6. Numbers aside, give your subjective evaluation of the menu. What market might this menu appeal to? What changes might you make to improve it? Are ingredients cross-utilized efficiently? What other factors might have gone into the development of this menu?

### MANAGEMENT BY MENU ANALYSIS CAFE

#### SOUP

Tofu Chili .......... $4.95
- 20 orders sold-20% food cost

Cream of Roasted Garlic .......... $3.95
- 15 orders sold-30% food cost

Gumbo with Seafood and Sausage .......... $4.95
- 25 orders sold-35% food cost

#### SALADS

Chilled Thai Noodles with julienne vegetables .......... $5.95
- 10 orders sold-25% food cost

Fried Chicken Salad .......... $6.95
- 15 orders sold-35% food cost

Anti pasta with sausage, marinated vegetables and cheese .......... $7.95
- 25 orders sold-40% food cost

Squid Salad- Marinated with Onions and Peppers .......... $6.95
- 5 orders sold-20% food cost

Nicoise Salad- Seared Tuna, Green Beans, Hard Boiled Eggs, Black Olives, Red Onions and Steamed Potato Cubes .......... $8.95
- 20 orders sold-25% food cost

#### PASTA

Seared Tuna with tiny new peas in 3-cheese sauce over fettuccine .......... $8.95
- 30 orders sold-25% food cost

Calamari Diablo over Spaghetti .......... $7.95
- 25 orders sold-30% food cost

Eggplant and vegetables Parmesan over Fettuccine .......... $7.95
- 35 orders sold-30% food cost

Spaghetti Marinara with Tofu "Meat"balls .......... $8.95
- 30 orders sold- 25% food cost

# 9
# THE BEVERAGE MENU

## Outline



## Objectives

After reading this chapter, you should be able to:

1. Identify the basic requirements for planning an alcoholic beverage menu.
2. Indicate how the beverage menu can be properly implemented through skilled merchandising and service.
3. Describe how to institute controls to ensure that the beverage menu satisfies guests and meets cost and/or profit goals.

# Introduction

No other phase of foodservice has changed as much in recent years than that of the service of alcoholic beverages. An organization called MADD (Mothers Against Drunk Driving) has been a big factor in bringing about this change. Irresponsible consumption of alcoholic beverages resulting in social disturbances, deaths, and damage to property, plus the danger heavy drinking can impose on individuals and families, were factors cited by this organization in asking that heavier penalties and more control be put on the sale of alcoholic beverages. As a result, all 50 states passed laws increasing responsibility for damages and injuries related to alcohol sales. Lawsuits that resulted in extremely heavy financial and other penalties caused insurance companies to raise their rates to a point that some operators closed. Third-party responsibility (*dramshop laws*) were established so that someone serving alcohol could be held responsible for the results. Many foodservices today have training programs to teach employees how to meet the restrictions of the laws and follow strict operating procedures. The NRAEF's Serv Safe Alcohol, TIPS, and the AHLAEIS CARE programs are all recognized training programs.

Simultaneously, with this trend (and perhaps aided by it) came a great change in people's drinking patterns. A much larger number of people stopped drinking alcoholic beverages entirely, while others changed to lighter drinks such as beer and wine instead of spirits. It is estimated that 40 percent of adult Americans do not drink at all, although that percentage is thought to be shrinking. Between 1980 and 1998, there was a steep decline in wine, beer, and spirit consumption in the United States. Wine is predicted to remain the largest growth market, assisted in part by demographics. Half a million potential new consumers come of age each year. Vinexpo predicts that by 2008, the United States will be spending 20 percent of the world's total expenditures on wine. The emphasis seems to be on quality, as opposed to quantity. There is a decline in beer consumption, but there is an increase in niche market beers. Spirits are expected to see steady, not remarkable, growth. Remarkable is the 19 percent increase in vodka sales. This may be explained by the introduction of numerous flavored vodkas on the market, as well as the popularity of specialty martinis.

Medical experts have presented much evidence on the dangers of alcoholic abuse, but studies on light to moderate alcohol consumption, particularly regarding red wine, have cited both harmful and helpful effects.

The requirements for an adequate beverage menu, therefore, have changed. Menus now often feature nonalcoholic drinks and offer a wider selection of beers and wines. A reception may offer no spirits or mixed drinks. Instead, beer, wine, and nonalcoholic drinks will be offered. Even at banquets the elaborate use of wines has decreased.

A good beverage menu can help an operation's profits considerably if: (1) merchandising and service personnel promote the sales of drinks, responsibly; (2) procedures are established to control costs and proper inventory levels are maintained; and (3) the menu includes choices that complement the food menu and meet patron expectations.

Much of what has been discussed regarding the planning, pricing, and analysis of food menus can also be applied to beverage menus. However, there are some differences. Usually, alcoholic beverage prices are based on a higher markup than food prices. Also, alcoholic beverages are often really an accompaniment or an accessory to food, and so more emphasis must be given to merchandising and selling. Finally, control procedures must be tighter in the handling of alcoholic beverage and income obtained from it.

The factors that affect food menu development (personnel, physical facilities, patron expectations, and costs) also apply to beverage menus, but there is an additional controlling factor. Alcoholic beverages are regulated by law, and the operation *must observe state regulations for alcohol service.* Certain kinds of beverages can be sold only in certain kinds of places, and a liquor license must be obtained by operators wishing to sell. Purchasing may also be controlled to a point. Taxes may be higher, and certain selling hours must be observed. The law on serving minors is strict and penalties can be severe.

# Presenting the Beverage Menu

Beverage menus may be presented in several ways. It is not uncommon to see alcohol listed on the regular (food) menu, when only a few beverage choices are offered. Some institutional menus may offer cocktails and wines this way. Hospitals and nursing homes, for example, may provide wine to patients whose physicians allow it. These may be offered on separate lists or on the regular menu.

Some establishments keep separate beverage menus. A number of these may be used. One may cover drinks such as mixed drinks and cocktails; another will present the wine list; another will offer after-dinner drinks. Yet another will list afternoon beverages, including cocktails; mixed drinks; tall, cool drinks; and dessert-type wines, such as port, Madeira, and muscat. The sales department may have still another—a banquet wine and drink list for use in planning special events with patrons.

No matter what the presentation, it is desirable that the regular menu *announce* the availability of alcoholic beverages. A considerable amount of revenue and profit can be lost if guests don't know about the availability of alcohol as a choice.

Some menus can be quite elaborate and contain a large number of items. If the establishment has a substantial wine cellar, the wine list will naturally be extensive—it may run several pages. If it does, the pages should be marked so that patrons can quickly find the type of beverage desired. Tabulations on the margins can indicate what each page offers.

It is not usually necessary to offer an exhaustive assortment of drinks or wines to meet patrons' needs. Simple, well-thought-out choices can do the job. Just as food menus should be analyzed, liquor menus should be reviewed regularly to determine which items are selling. Those items that do not move should be eliminated. A sizable amount of money can be tied up in a beverage inventory, and unless these inventories are turned over fairly often, they can be a hidden expense.

Generally, simplicity is a goal when planning beverage menus. However, this is usually easier said than done. There are a certain number of cocktails, highballs, and other mixed drinks that patrons will want *and* expect. If a wine list is to cover dry white wines and red wines (both domestics and imports), sparkling wines, apéritif wines, and dessert wines, the list will be extensive. However, it is best not to offer too much—only those items that meet the expectations of guests should be offered.

Since menu planning for alcoholic beverages is a highly specialized area, it is often advisable to have an expert's advice. Most wine purveyors have individuals on their staffs who are knowledgeable in planning wine lists. Some of the big brewers, distillers, and liquor distributors can also give assistance. Each will have a considerable amount of written material for use in designing an attractive menu. Films and other educational materials may be available from these companies. Some states restrict the use of purveyor-supplied promotional material that is related to the merchandising of alcoholic beverage.

The presentation of specialty drinks may take skill to set up. Color may be necessary to enhance the menu, and this may require the advice of an expert in menu planning.

If only one beverage menu is used, careful consideration must be given to presentation of the various kinds of spirits. If only a single hardback card is used, the operation's logo can appear on the front with perhaps a presentation of drinks that the establishment especially wishes to sell. **Exhibit 9.1** shows the alcoholic beverages should be classified on a menu.

Because requirements vary greatly, beverage lists should be specifically designed for the individual enterprise. Operations with an ethnic theme or clientele should feature a beverage list consistent with that theme. A German restaurant needs a different list of wines and beers from those offered by a French or Italian restaurant. A menu featuring Mexican foods may present margaritas prominently and perhaps make a specialty of them. Some rich Spanish red wines, such as the full-bodied Marques del Lagar from the Rioja hills or Sangre de Toro (blood of the bull), can be offered, along with one or more of the soft white wines of northern Spain. *Cerveza* (beer) will almost certainly be served. Sangria in glasses or small pitchers is often available. French restaurants will need a good list of quality French wines. Italian restaurants may feature Italian wines, and may also include a few French and California wines. Follow-through is important in a beverage menu, and some research is necessary to make sure that an item is authentic when it is put on a menu. This also includes offering drinks in traditional glasses or steins.

Prices should be presented clearly for each item or group of drinks. The descriptions, type,

**Exhibit 9.1** Liquor Menu Classifications

| Main Classifications | Order within the Classification |
|---|---|
| Wines | Usually, the order of consumption is the order of the listing: aperitifs; dry dinner wines, red or whit; rosés; sparkling wines; dessert wines. Descriptive words such as dry, crisp, fruity, cherry, melon, and oak help guests make a pleasing choice. |
| Beers and Ales | List those available in draft or in bottles. Imports are usually listed separately from domestic items. Including descriptions is helpful, particularly when describing niche beers |
| Cocktails | Martinis, specialty and blended drinks, etc. (The order here is not especially important; merchandising may dictate which order is best.) |
| Mixed Drinks | Vodka, gin, rum, and other spirits in mixed drinks, such as with water, soda, tonic water, or juices. (Again, the order is not too important, but it should be dictated by preferences of guests.) |
| After-Dinner Drinks | Frequently, brandies and some other liqueurs are listed, followed by coffee drinks. |

and use of color should promote clarity and legibility. (See Chapter 7 for information on menu mechanics.) Effective merchandising is vital.

## The Wine List

Eight to ten wines on the menu may be sufficient to satisfy patron desires and meet operational goals at a very casual dining establishment. The connoisseur will wish to have a longer list, but some operations find that around four wines are enough. One expert has said that if a wine list is well compiled, a list of 50 to 75 wines should satisfy even the most discriminating individuals in their selection of wines for a meal. More wines than ever before are available. Guests anticipate a selection that is as exciting as the food choices.

Most operations find that premium wines by the glass are very popular with patrons. In other operations a list of bottled wines may be offered, in addition to wine by the glass. (See **Exhibits 9.2 and 9.3.**) Due to the size of red wine glasses, some upscale restaurants offer the portion in a small carafe so that guests are secure in their perception of value. *Wine flights* are very small "tastes" of wine amounting to about a glass.

Wine lists are difficult to design. Wine menus should be set up by someone who knows a lot about wines. Certain wines *must* be on the menu, such as white or red dinner wines. Dry wines may have to be from both foreign and domestic sources, depending on the patronage. There is a large variety of wines, such as those from the Rheingau and Rheinhesse in Germany; from the Alsace and Moselle areas nearby; from the Burgundy, Bordeaux, and other wine districts of France; and from many other countries. Domestic wines are dominated by California vineyards but include an ever-growing list of regional wineries that are popular with patrons. A complete list can be very extensive, and the individual who is making up the list should know how to reduce it without failing to meet guest expectations.

A distinction should be made between dinner wines, apéritif wines, sweet wines, and effervescent wines on the menu.

Dinner wines should be divided between dry reds, whites, and less dry rosés. Most patrons prefer dry wines, but some may like the small degree of sweetness in a rosé. (It is important that the word *dry* be used to describe a *lack of sweetness,* rather than a quality of sourness, in a wine.)

Whether the wines are domestic (produced in the United States) or imported will depend on patron preferences and availability of items. Preferably, the list should include a selection of both.

Some domestic wines are named after imports they are meant to resemble. One may find on a list of domestic wines such names as Burgundy, Beaujolais, and Chablis. These may not be good representations of the European wines they are supposed to imitate. This will mainly depend on the type of grape and processing used. A domestic wine bearing a generic name—meaning a specific *kind* of wine from a district or area, such as Burgundy—may or may not come from the pinot noir grape grown in Burgundy, France. If the buyer wishes a wine that more nearly resembles the foreign wine, he or she should order a domestic wine bearing the name of the *varietal* (variety of grape) used for that wine. Thus, if one orders a domestic wine named "Pinot Noir," a varietal name rather than a generic name, such as Burgundy, the wine will more nearly resemble a French Burgundy because U.S. regulations require that, when the varietal name is used, 75 percent of the wine must come from that specific type of grape. Similarly, if one wishes a domestic wine resembling a true Chablis, the order should be for a wine labeled "Pinot Chardonnay," the grape used to produce this wine in France. The price of a domestic varietal wine should be higher than a domestic generic one, other factors being equal. Master sommelier Ronn

**Exhibit 9.2**
Wine Menu: Tucci Benucch

# VINI TUCCI

**EVERYDAY WINES AVAILABLE BY THE GLASS**

| | *GLASS* |
|---|---|
| **CIELO PINOT GRIGIO**<br>It's light, refreshing, and easy to drink - what's not to love! | **$5.00** |
| **ARANCIO GRILLO**<br>If you love the texture of Chardonnay, but prefer the freshness of Sauvignon Blanc, try this Grillo. | **$5.00** |
| **RUFFINO LIBAIO CHARDONNAY**<br>An Italian interpretation of an American classic! | **$5.00** |
| **STELLA MONTEPULCIANO D' ABRUZZO**<br>If there ever was a wine tailor made for Marinara dishes - this is it. | **$6.00** |
| **AVALON CABERNET SAUVIGNON**<br>This California stunner proves you don't have to pay a king's fortune to enjoy a great cabernet. | **$6.00** |
| **DA VINCI CHIANTI**<br>Did Leonardo make wine too? No, but these Sangiovese grapes are from his home town. | **$6.00** |

**SPUMANTI E VINI PER DOLCE**

| | *GLASS* | *BOTTLE* |
|---|---|---|
| **PISANI PROSECCO**<br>Do as the Italians would do - begin your meal with a delicious glass of Italian bubbly! | **$7.00** | **$26.00** |
| **MOSCATO D' ASTI**<br>A gently sparkling dessert wine | **$6.00** | **$22.00** |

**BIANCHI**

| | *GLASS* | *BOTTLE* |
|---|---|---|
| **SANTA RITA SAUVIGNON BLANC**<br>Sure, Chile produces great red wines, but they are equally famous for their citrusy Sauvignon Blancs. | **$6.00** | **$22.00** |
| **COOPERIDGE WHITE ZINFANDEL**<br>Crisp and refreshing with strawberry and raspberry fruit flavors. | **$5.00** | **$20.00** |
| **HIRSCHBACH RIESLING**<br>Well balanced German style with just the right amount of sweetness. Great with spicier dishes. | **$6.00** | **$22.00** |
| **ESTANCIA CHARDONNAY**<br>Fuller bodied Chardonnay with ripe pear and melted butter flavors. | **$8.00** | **$30.00** |
| **TAMELLINI SOAVE SUPERIORE**<br>This venetian white will remind you of a Pinot Grigio, crisp and refreshing! | **$8.00** | **$30.00** |
| **FRANCO FIORINA GAVI**<br>This complex white features pear and citrus fruits. | **$8.00** | **$30.00** |

**ROSSI**

| | *GLASS* | *BOTTLE* |
|---|---|---|
| **STEFANO FARINA DOLCETTO D' ALBA**<br>A light, easy to drink dry red wine that is great with pizza. | **$7.00** | **$26.00** |
| **AGOSTINO PAVIA BARBERA D' ASTI**<br>Can't decide between a fruity Pinot Noir or an earthy Chianti? We suggest you try this Barbera instead. | **$7.00** | **$26.00** |
| **CUSAMO NERO D'AVOLA**<br>The ripe blackberry and spice notes will remind you of a Zinfandel but in a more food friendly way. | **$7.00** | **$26.00** |
| **GABBIANO CHIANTI CLASSICO**<br>Lush texture with flavors of dried black cherries. A can't miss Chianti Classico! | **$8.00** | **$30.00** |
| **MC WILLIAMS ESTATE SHIRAZ**<br>Shiraz fans will not be disappointed with this Aussie classic. | **$7.00** | **$26.00** |
| **RANCHO ZABACO ZINFANDEL**<br>This Zinfandel is robust, intense, spicy, and just plain fun to drink. | **$8.00** | **$30.00** |
| **RED ROCK MERLOT**<br>Perfect for those in search of a smooth red. | **$6.00** | **$22.00** |
| **IRONY VINEYARDS PINOT NOIR**<br>The lighter texture and soft bing cherry flavors will pair with anything on the menu. | **$8.00** | **$30.00** |
| **HAYMAN & HILL NAPA CANERNET SAUVIGNON**<br>Everything you would expect from a great Napa Cab. For a classic pairing - try it with our Filet Trio. | **$10.00** | **$38.00** |
| **BUGLIONI VALPOLICELLA CLASSICO**<br>A zesty wine with intense flavors. Think baby Amarone! | | **$40.00** |
| **VITANZA ROSSO DI MONTALCINO**<br>This Rosso is the fresher and easier to drink version of Brunello di Montalcino. | | **$50.00** |

*Courtesy of Lettuce Entertain You Enterprises Inc.*

**Exhibit 9.3**
Drink Menu featuring Wine Listing: Mon Ami Gabi

# Mon Ami GABI

## COCKTAILS

GABITINI - Grey Goose Vodka straight up with blue cheese olives, Gabi's favorite! 9.95

C'est Si Bon! - It's so good! - Grey Goose L'orange, vermouth, watermelon syrup and lime 9.95

THE FRENCH 75. . . . . . . . 9.95
Bombay Sapphire and orange juice, topped with champagne

CALVADOS SIDE CAR . . . . . . . . 9.95
Daron Calvados, Cointreau in a sugar rimmed glass

LEMON DROP MARTINI . . . . . . . . 9.95
Grey Goose Le Citron, squeeze of lemon, and a sugar rim

THE PARIS ROMANCE . . . . . . . . 9.95
Amaretto Di Saronno, Malibu Rum and fruit juices

GABI'S COCKTAIL . . . . . . . . 9.95
Red Lillet, filled to the top with champagne

MANHATTAN POMEGRANATE . . . . . . . . 9.95
Jim Beam, sweet vermouth and pomegranate juice with a splash of orange

MON AMI GABI COSMO . . . . . . . . 9.95
Grey Goose L'Orange, Cointreau, cranberry juice and a squeeze of fresh lime

MOJITO GABI . . . . . . . . 9.95
Bacardi, pomegranate juice, mint leaves, lime and soda garnished with mint

CLASSIC MARTINI . . . . . . . . 9.95
Bombay Sapphire and dry vermouth with a splash of orange bitters

ESPRESSO MARTINI . . . . . . . . 9.95
Stolichnaya Vanil Vodka, with a splash of Baileys, Kahlua and a shot of espresso

## BIÈRE

GOOSE ISLAND (SEASONAL) . . . . 5.95
CHIMAY TRIPLE ALE . . . . . . . 8.95
TROIS PISTOLES . . . . . . . . . 8.95
HEINEKEN . . . . . . . . . . . . 4.95
AMSTEL LIGHT . . . . . . . . . . 4.95
STELLA ARTOIS . . . . . . . . . 6.95
BUCKLER N/A . . . . . . . . . . 4.95
PILSNER URQUELL . . . . . . . . 4.95
FISCHER AMBER . . . . . . . . . 5.95
HOEGAARDEN . . . . . . . . . . 6.95
FAT TIRE 22 OZ. . . . . . . . . . 10.95

## LES VINS BLANCS BOTTLED WHITE WINES

### CHAMPAGNE/SPARKLING WINES

01 Langlois Chateau, Cremant de Loire, Brut, N. V. . . . . . . . 40
02 Joseph Perrier, Cuvée Royal, N.V. . . . . . . . . . 65
03 Tattinger, Brut, N.V. . . . . . . . . . . 85
04 Nicolas Feuillatte, Brut Rose, N.V. . . . . . . . . . 95
05 Perrier Jouet, Grand Brut, N.V. . . . . . . . . . 115
06 Bollinger, Special Cuvée, N.V. . . . . . . . . . 125
07 Veuve Clicquot Ponsardin, Vintage Reserve, 98 . . . . . . . . 165

### BOURGOGNE *CHARDONNAY*

23 Bourgogne, Chardonnay, Henri Darnat, La Jumalie, 04 . . . 47
24 Saint Veran, Verget, 04 . . . . . . . . . . 48
25 Montagny, 1er Cru, Alain Roy, 04 . . . . . . . . . . 49
26 Pouilly-Fuissé, Domaine Les Vieux Murs, 04 . . . . . . . . . . 50
27 Chablis 1er Cru, Vaillons, Jean Paul & Benoit Droin, 03 . . 74
28 Meursault, Domaine Latour-Giraud, Les Narvaux, 01 . . . . 90

### ROSÉ

30 Domaine Massamier, Cuvee Oliver, 04 . . . . . . . . . . 28

### ALSACE

50 Pinot Blanc, Gustave Lorentz 05 . . . . . . . . . . 36
51 Pinot Gris, Gustave Lorentz, 05 . . . . . . . . . . 38
52 Gewurztraminer, Gustave Lorentz, 04 . . . . . . . . . . 40

### LOIRE *SAUVIGNON BLANC / CHENIN BLANC*

56 Muscadet, Serve & Maine, Domaine de la Quilla, 04 . . . . 26
57 Vouvray, Château Moncontour, 04 . . . . . . . . . . 29
58 Quincy, Jacques Rouze, 04 . . . . . . . . . . 39
59 Sancerre, Henri Bourgeois, La Vigne Blanche, 04 . . . . . . . 48
60 Pouilly Fumé, Henri Bourgeois, 04 . . . . . . . . . . 58

## LES VINS ROUGES BOTTLED RED WINES

### BORDEAUX *MERLOT / CABERNET SAUVIGNON*

80 Bordeaux Superior, Moulin D' Issan, 04 . . . . . . . . . . 36
81 Graves, Chateau de Maine, 03 . . . . . . . . . . 43
83 Lalande-de-Pomerol, Château Haut-Chatain, 01 . . . . . . . 49
84 St Emilion, Château Guibot, La Fourvieille, 01 . . . . . . . . 54
85 Pessac Leognan, Château Gazin Rocquencourt, 99 . . . . . . 57
87 St Emilion, Château St. George, 00 . . . . . . . . . . 67
89 Saint Julien, Château Moulin Riche, 99 . . . . . . . . . . 72
90 Margaux, Blasson D' Issan, 03. . . . . . . . . . 75
91 Lanlande-de-Pomerol, Pascal Chatonnet, La Sergue, 00 . 76
93 Margaux, Château Tayac, Cru Bourgeois, 00 . . . . . . . . . . 83

### BEAUJOLAIS *GAMAY*

101 Beaujolais, Château de Lacarelle, 03 . . . . . . . . . . 32
102 Cotes de Brouilly, Château Thivin, 04 . . . . . . . . . . 43
103 Morgon, Marcel Lapierre, 05 . . . . . . . . . . 45

### LOIRE *CABERNET FRANC*

105 Saumur-Champigny, Dom. Filliatreau, La Grande Vignolle, 05 . 35

### BOURGOGNE *PINOT NOIR*

110 Bourgogne, Domaine Parent, Pinot Noir, 03 . . . . . . . . . . 36
111 Mercurey, Domaine de la Croix Jacquelet, Faively, 02 . . . . 45
112 Monthelie, Jacques Parent, 03 . . . . . . . . . . 53
113 Ladoix, Domaine de Merode, Les Chaillots, 04 . . . . . . . . 55
114 Nuits-St-George 1er Cru, Laboure Roi, Les Damodes, 00 . 58
115 Marsannay, Domaine St. Martin, Les Eschezeaux, 03 . . . . . . . . 60
116 Santenay, 1er Cru, A. Marie et J. Marc Vincent, Les Gravieres, 03 . 71
117 Gevrey Chambertin, Joseph Drouhin, 04 . . . . . . . . . . 81
118 Gevrey Chambertin, Domaine Rene LeClerc, 04 . . . . . . 92
120 Clos de la Roche, Domaine Pierre Amiot et Fils, 01 . . . 170

### RHÔNE/Southern France *GRENACHE / SYRAH*

130 Côtes du Rhône, Perrin Reserve, 04 . . . . . . . . . . 36
131 Costieres de Nimes, Chateau de Campuget, '1753,' 04 . . . 38
132 Côtes du Rhône, Domaine de la Janasse, 05 . . . . . . . . . . 40
133 Gigondas, E. Guigal, 03 . . . . . . . . . . 64
134 Châteauneuf-du-Pâpe, Perrin, Les Sinards, 03 . . . . . . . . 73
135 Gigondas, Tardieu-Laurent, 01 . . . . . . . . . . 79
136 Cornas, Domaine Courbis, Les Eygats, 01 . . . . . . . . . . 115

Vintages subject to availability

*Courtesy of Lettuce Entertain You Enterprises Inc.*

Wiegand estimates that 65 percent of all wines in the United States come from California, 10 percent come from other states, and 25 percent are imports.

For specialty restaurants, either the menu or the service personnel will indicate the wines offered. A seafood restaurant, for example, might feature white wines, while listing a few reds and rosés. A steakhouse might reverse this, offering predominately red wines but including whites with depth such as chardonnay.

Normally, red dinner wines are served at room temperature (about 60°F to 65°F, or 15.6°C to 18.3°C) with red meats; white dinner wines are served chilled (about 45°F to 50°F, or 7.2°C) with seafood or poultry. The custom of serving red wines with red meats and white wines with white meats has all but faded away. Many authorities today say that patrons should select the wines they like. Still, guidelines exist for restaurants that want to stay with the traditional etiquette, as shown in **Exhibit 9.4**.

If there is a doubt about which wine should go with which food, a rosé can be offered, since it is suitable for use in place of either a red or white wine. Many times, patrons who know very little about wines prefer a rosé with just a touch of sweetness. White wines generally have less body and less full flavor than reds, although chardonnay has been referred to as a red wine masquerading as white. Reds are also slightly more astringent. Some whites have a tart quality, as evidenced in a Graves from Bordeaux or in some German dry white wines. This in no way detracts from their quality, but some patrons, not appreciating these qualities, may prefer a softer, fruitier wine.

Sweet wines, such as muscat, Barsac, sparkling wines, or sweet Rieslings, complement desserts. Port and some other quite sweet, fortified wines are proper after dinner, and many Americans still drink brandy or sweet liqueur in place of dessert.

**Exhibit 9.4** Matching Wine with Food

| Menu Item | Wine Suggestion |
|---|---|
| Appetizer | Champagne, dry white wine, lighter reds |
| Salad | Generally no wine (wine does not typically complement acidic dressings) however cheese dressings, such as roquefort, may be well suited to wine) |
| Fish or seafood | Dry or medium-dry white |
| Beef | Hearty red or full whites |
| Lamb | Hearty red |
| Veal | Light red or full-bodied white |
| Ham or pork | Dry or medium-dry white or rosé |
| Turkey, duck, chicken | Full-bodied white or light red |
| Game (venison, pheasant, wild duck) | Hearty red |
| Lasagna, spaghetti, pizza | Hearty red |
| Cheeses, full-flavored | Full dry white (with roquefort) |
| Cheeses, mild | Mild table wines of any type |
| Desserts, pastries, fruits mousses | Semisweet sparkling wine, sweet white table wine |

If the management knows its wines, it can feature a wine that is not well known but is of fine quality. Spanish Garnacha is a contemporary, little-known wine. Aszu Tokay from Hungary is a fine offering as a sweet dessert wine. A Barbera from the Piedmont district in Italy is an excellent alternative to some of the higher-priced French red dinner varieties.

### *Merchandising Distinction and Variety*

Wines are distinctive in numerous ways: they are produced from different varieties of grapes, from different vineyards, and in particular vintage years; they come from various districts, parishes, chateaux, or schlosses (castles); and they are shipped by many different vintners. These factors offer many merchandising opportunities to foodservice operators.

Most wines have a story behind them, and good merchandising should make use of interesting wine lore. A good example of one of the romantic stories of wine that can be used is the famous remark of Dom Pérignon when he first tasted champagne, which he had invented accidentally. "I am tasting stars!" he exclaimed. The fact that the Aszu Tokay wine from Hungary has been called the "king of wines and the wine of kings" also makes good copy. Est! Est! Est!, a famous wine from Italy, derives its name from the exclamation made by the servant of a bishop who was sent ahead to sample the wine before his master came. The servant, who had heard of the wine's fame, said when he tasted it: "It is! It is! It is!" Another story that makes for interesting menu reading is that one of Napoleon's colonels thought so highly of the wines of the famous Clos de Vougeot vineyards that he ordered his soldiers to salute when they marched past the vineyard. A mention that Zinfandel (a red dinner wine) comes only from grapes grown in California can be used to increase this wine's prestige.

Use of proper terminology is crucial. Knowledgeable patrons will quickly catch the error if the server calls a Moselle wine, which historically comes in tall green bottles, a Rheingau or Rheinhesse wine, which come in darker-colored bottles. Similarly, labeling a German Sekt or an Italian Asti Spumante as "champagne" is deceptive. Wine descriptions can help sell, but they must truly represent the qualities and other factors found in the wine.

### *Merchandising the Formal or Catering Wine List*

Operations such as clubs, hotels, restaurants, and others serving fine foods may wish to present the particular name of the wine served with a printed menu. This can be done in various ways. The U.S. State Department sets up the White House menus; often, when one wine is served, it will be listed at the bottom of the menu after the foods, as shown in **Exhibit 9.5.** At other times, the particular wine served with several courses will be listed to the left of the courses with which it is served.

**Exhibit 9.5** Formal Wine Listing

Coquille of Seafood Neptune
Cheese Straws
Piccata of Veal Luganese
Saffron Rice
Asparagus Tips
Champagne
Mousse

*Bernkasteler Doktor 1969*

The menu shown in **Exhibit 9.6** is typical of those used for formal meals today. An appetizer course or a light soup might be used to start the meal and a dry white complementing it is listed underneath. Perhaps the most elaborate kind of a menu would be the seven- or eight-course meal served at very formal banquets. **Exhibit 9.7** shows a way of listing the wines for a formal meal. (If an eight-course meal were served, there would be a poultry course following the fish course.

**Exhibit 9.6** Wine Dinner: Wildfire Wattle Creek Wine Dinner

# WILDFIRE®

**STEAKS, CHOPS & SEAFOOD**

## WATTLE CREEK WINE DINNER

*Reception*

**Lobster Wrapped Shrimp**
pineapple, star fruit chutney

***Wattle Creek Viognier '04***

*First Course*

**Wildfire Wood Roasted Fish & Chips**
halibut, truffle steak fries

***Wattle Creek Chardonnay '03***

*Second Course*

**Roasted & Sliced Beef Tenderloin**
morel mushroom hash, cipolline onion rings, bordelaise sauce

***Wattle Creek Cabernet Sauvignon '02***

*Third Course*

**Selection of Artisan Cheeses**

***Wattle Creek Shiraz '01***

*Dessert*

**Classic Chocolate Tart**
dark chocolate sauce, blackberry ice cream

***Cockburn 1998 LBV Port***

Chef, Steven Lukis
Winemaker, Michael Scholz
Wine Director, Brad
Wermager

Wildfire Chicago
Tuesday, April 18th, 2006

*Courtesy of Lettuce Entertain You Enterprises Inc.*

**Exhibit 9.7** Formal Wine Menu from Everest

## Table of Contents



## Table of Contents Red Wines



## Table of Contents Dessert Wines



## Table of Contents Les Digestifs du Maison



*Courtesy of Lettuce Entertain You Enterprises Inc.*

Exhibit 9.7 *(Continued)*

## Cabernet Sauvignon

### California

| | | | |
|---|---|---|---|
| 2342 | Andrew Geoffrey. Diamond Mountain. Napa | 2002 | 180.00 |
| 2466 | Andrew Geoffrey. Diamond Mountain. Napa | 2001 | 190.00 |
| 2486 | Andrew Will. Champoux Vineyard. WA | 2002 | 140.00 |
| 2416 | Andrew Will. Kilpsun Vineyard. WA | 2002 | 140.00 |
| 2419 | Araujo. Eisele Vineyard. Napa | 2001 | 390.00 |
| 2305 | Araujo. Eisele Vineyard. Napa | 1999 | 440.00 |
| 2300 | Araujo. Eisele Vineyard. Napa | 1996 | 480.00 |
| 2398 | Archipel. Napa | 2002 | 89.00 |
| 2446 | Arrowood. Reserve Speciale | 2000 | 160.00 |
| 2453 | Atlas Peak. Howell Mountain. Napa | 2003 | 220.00 |
| 2487 | Barnett Vineyards. Rattlesnake Hill. Napa | 2000 | 290.00 |
| 2414 | Barnett Vineyards. Rattlesnake Hill. Napa | 1996 | 290.00 |
| 2488 | Behrens & Hitchcock. Ink. Napa | 2000 | 290.00 |
| 2475 | Behrens & Hitchcock. Kenefick Ranch Cuvée. Napa | 2000 | 290.00 |
| 2417 | Behrens & Hitchcock. Ode to Picasso | 2000 | 390.00 |
| 2439 | Benziger. Tribute. Sonoma Mountain Estate. Sonoma | 2003 | 190.00 |
| 2347 | Bond. St. Eden. Napa | 2002 | 690.00 |
| 2444 | Bond. St. Eden. Napa | 2001 | 590.00 |
| 2493 | Bond. Melbury. Napa | 2002 | 690.00 |
| 2484 | Bond. Melbury. Napa | 2001 | 590.00 |
| 2471 | Bond. Vecina. Napa | 2001 | 490.00 |
| 2472 | Bond. Vecina. Napa | 2000 | 440.00 |
| 2325 | Brochelle. Napa | 2002 | 120.00 |
| 2458 | Bryant Family Vineyard. Napa | 2003 | 1400.00 |
| 2435 | Bryant Family Vineyard. Napa | 2001 | 1400.00 |
| 2423 | Bryant Family Vineyard. Napa | 2000 | 1100.00 |
| 2302 | Bryant Family Vineyard. Napa | 1999 | 960.00 |
| 2301 | Bryant Family Vineyard. Napa | 1997 | 1100.00 |
| 2358 | Beaulieu Vineyard. G. Latour. Private Reserve. Napa | 2001 | 190.00 |
| 2304 | Beaulieu Vineyard. G. Latour. Private Reserve. Napa | 1974 | 190.00 |
| 2306 | Beaulieu Vineyard. G. Latour. Private Reserve. Napa | 1966 | 240.00 |
| 2307 | Beaulieu Vineyard. G. Latour. Private Reserve. Napa | 1964 | 320.00 |

*Courtesy of Lettuce Entertain You Enterprises Inc.*

Another method of presenting the wines served at a formal meal is to list the wines separately from the food. This would be done with a folded menu, which, when opened, has the wines listed on the left half and the food courses on the right half.

## The Spirit List

In simplistic terms, *spirits* are alcoholic beverages that are not wines, liqueurs, or beers. Spirits include distilled beverages, such as bourbon, rum, scotch whiskey, tequila, gin, and vodka. Spirits generally have a higher alcoholic content than wines, beers, and liqueurs.

The list of menu spirits is rather traditional, and less variation is found on it than on a wine list. Only a few drinks may be on the menu, it being understood that the bar can furnish others if desired. Thus, a menu might list only those drinks that management wishes to push, as shown in **Exhibit 9.2**.

### *Stocking Spirits*

The items on a spirit list should be balanced to provide what management feels will satisfy guests' needs. Normally, individuals ordering highballs, cocktails, or similar drinks know what they want, and a list is not required. In this case, the purpose of a menu is more to inform the patron about the prices rather than to spell out what is available. Most operations will find that they need a good list of *call liquors,* or brand names. Wild Turkey Bourbon, Grey Goose Vodka, Chivas Regal Scotch, or Tanqueray Gin are examples. The number of these should be limited as much as possible. Checks should be kept on how frequently requests are made for call items. Those that move slowly should be eliminated.

If a couple of good call scotches, bourbons, gins, and, rum are stocked, most requests can be filled. Often, if a patron who orders one high-quality spirit that is not in stock is offered another of equal popularity, the patron will accept the substitution. If bartenders and service personnel keep track of how many call items are *not* requested and what they are, a better list can be built. Local tastes and value perceptions must be considered.

### *Proof*

Alcoholic content of spirits is measured in *proof,* expressed as a percentage of volume of water to alcohol.

The term was originated by the British, who found that if gunpowder was mixed with alcohol and water, it would burn, but only if a specific amount of alcohol was mixed with the water *and no less.* If even the slightest amount of water above the limit was added, the gunpowder would not burn. The British used this test as a means of checking the alcoholic content of spirits. If the spirit burned, they said it was "proof" that the spirit contained an adequate quantity of alcohol. **Exhibit 9.8** indicates the equivalent proofs of British, European, and American standards.

In the United States, proof of a spirit is two times the percent of alcohol by volume or weight. Thus, if a spirit contains 50 percent alcohol, the proof is 100 ($2 \times 50$). It is important to know whether alcohol content is based on alcohol by volume or alcohol by weight. A spirit that is 50 percent alcohol by weight contains more alcohol than a spirit that is 50 percent alcohol by volume. This is because alcohol is lighter than water, so it takes more alcohol to equal water weight.

**Exhibit 9.8** Comparison of Proofs

| | British | American | European (Gay-Lussac) |
|---|---|---|---|
| *100% alcohol* | 175 proof | 190 proof | 100 proof |
| | 100 | 114 | 57 |
| | 88 | 100 | 50 |
| | 85 | 98 | 49 |
| | 80 | 90 | 45 |
| | 75 | 86* | 43 |
| | 70† | 80† | 40 |
| | 65 | 74 | 37 |
| *Water* | 0 | 0 | 0 |

*Many blends and lower-priced whiskies are 80 or 86 proof.
†The British and American governments do not allow spirits such as gin, rum, brandy, or whiskey to be less than this in proof.

Wines have less alcohol (7 to 14 percent) than spirits, and beers generally have the lowest alcoholic content (from 2.5 to 8 percent). Spirits range widely, from 50 to 190 proof, depending on the type of spirit and the brand.

## Beers and Ales

On the beverage menu the list of beers, ales, and similar drinks should be limited. A modest list would feature three or four domestic beers and ales and three or four imports. Ethnic restaurants often include beers of that country on the menu. Of course, if the operation features a beer list, more beers should be offered. If several good draft beers are served as house beverages, the need for a bigger list, even in one of these restaurants, may not be necessary. On some beverage menus, one or two draft beers *must* be offered. Light beers are also popular, and a small selection of these should be available.

Local favorites may need to be included on the menu. Chicago beer drinkers, for example, have made Goose Island a local favorite. Coors is practically a mandatory menu item in some western states. A German restaurant will be expected to offer several dark ales from that country. It's a simple matter of knowing the clientele.

## After-dinner Drinks

The list of after-dinner drinks on a beverage menu should also be limited to what the trade requires and no more. Port, brandy, and coffee drinks may be featured. Often, the menu can offer some mixed after-dinner drinks that the establishment feels can be sold at a good profit, and that will please patrons. These can be listed as clip-ons or boxed entries on the beverage menu to call special attention to them, in a way similar to that described for food menu items the management wants to feature. Table tents also work well.

### NONALCOHOLIC DRINKS

Many restaurants offer a wide selection of nonalcoholic beverages for young patrons and those who choose to not drink. Tougher drinking and driving laws also encourage operations to develop interesting drinks that do not contain alcohol. Offering nonalcoholic beverages makes good economic sense because these items can help increase check average.

Restaurant patrons often like to order a pre-dinner drink with appetizers and non-drinkers appreciate choices of attractive, good-tasting beverages that compliment these foods. Popular cocktails such as Margaritas, Daiquiris and many others can be ordered without alcohol. Creative combinations of fruit or vegetable juices with flavorings, cream, or soda can be made in a wide variety of ways. Freshly juiced fruit and produce beverages are also popular and capitalize on the increased desire to ingest healthy food and beverages. Offering soft drinks with a garnish and/or flavoring served in a cocktail glass makes these common drinks more festive. Specialty coffee, tea, and hot chocolate beverages are also popular menu items. After dinner beverages can include these specialty items along with nonalcoholic versions of ice cream drinks. Bottled or flavored waters can be promoted using name brands that have appeal such as Fuji or Perrier.

## Wine Service

Servers must know all of the establishment's wines and which wines might be suitable with certain foods in order to make suggestions. With training, servers can hone in on patron needs or tastes and follow through with proper service.

In large dining rooms, a wine steward or *sommelier* will be on staff. It is this individual's job to discuss wines with patrons, make suggestions, and take the order. Both sommeliers and servers should be able to present a good reason for a wine suggestion, as wine is more important than ever to restaurant patrons. For instance, if lamb and beef make up the entree orders, a red Bordeaux or Burgundy will have flavor qualities that go well with them. If the selection is seafood, a Chablis, Chenin Blanc, or German Reisling would bring out the best flavor qualities of the seafood. Lobster is a slightly sweet meat, and a wine that is crisp and delicate with just a touch of sweetness will be highly desirable. However individual tastes vary, and servers should be prepared to make alternate suggestions.

It is also necessary to enhance the proper wine with the proper service. Some wines should be served at specific temperatures. Some guests may desire wine at a different temperature, and service personnel may ask, when the wine is ordered, if a specific temperature is desired. If the guest prefers that a red dinner wine be chilled, the server should show no evidence that this is not the usual service but go ahead and serve it that way. (Chilling a red wine is thought to harm the flavor and bouquet.) It is proper for the individual ordering a red wine to ask the waiter to pull the cork ahead of service so that the wine can breathe. This allows the wine to oxidize slightly, giving it a fuller flavor and a better bouquet. Older, high-priced red wines should not be allowed to oxidize as much as newer reds.

Proper presentation and pouring are key elements of good service. It is customary for patrons ordering wine to be shown the label on a bottle before the cork is pulled to allow them to see that it is the right wine. The foil around the cork should be removed and the cork pulled following procedures. After the cork is pulled, white wine may be put back into the wine holder to continue chilling: Holders are often chilled marble sleeves. Red wine may be placed directly on the table, the label toward the host, or it may be put into a basket. The server may set the cork down at the host's right so that he or she can examine it. Due to a shortage of cork, plastic "corks" or screw-top bottles are often used instead of the traditional cork stopper.

If there was any doubt about the quality, the server or *sommelier* historically tasted the wine in a silver *tastevin.* (This should be only a tiny sip.) The wine was then poured in a small quantity into the host's glass to judge suitability. These days, as the wine belongs to the guest, it is the host who commonly tastes for acceptability. All of the other guests are served before the host.

The proper glass should be used. (See **Exhibit 9.9.**) Using small glasses that are completely or nearly full is a mistake. Only a part of the wine glass should be filled, "to leave room for the nose." A Burgundy wine requires a large, tulip-shaped glass, a claret needs a narrower one, a Rhine or Moselle should have a glass with a tall stem. The all-purpose wine glass is also proper. Champagne is best served in a tall, narrow tulip glass rather than a flat, open champagne glass. (Brandy should be served in a proper snifter and *not* in a small liqueur glass.)

The reason a wine glass is turned in slightly at the top is to concentrate the wine odor so that it is fuller as the wine is consumed. Before tasting, wine may be swirled in the glass, especially a red wine, to give it some oxidation and assist in building flavor and bouquet. It is proper to hold the wine up to the light to check its clarity and color, or to look at it through the glass against the white tablecloth to note these visual qualities. About two ounces of wine should be poured into the taster's glass so that plenty of space is left for sensing the bouquet. Servers should leave the bottle on the table for the host to pour more wine for the guests if the server is not there to do it. In fine dining establishments, the wine is placed away from the guests due to the belief that the server should do the serving.

**Exhibit 9.9** Wine Glasses

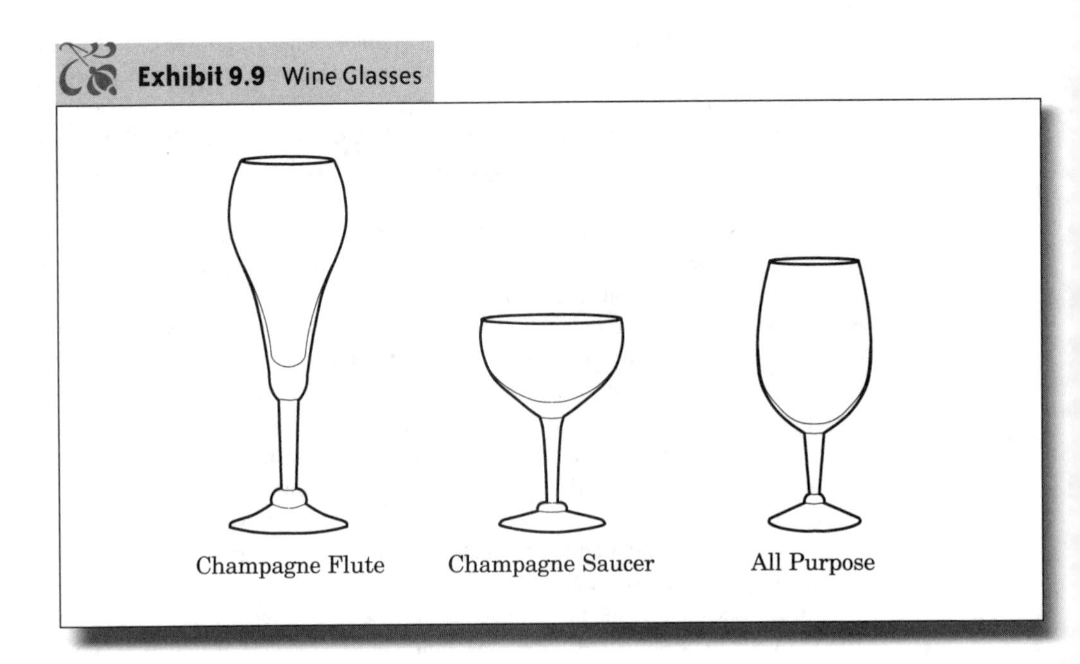

## Beverage Service

Even if a menu has been masterfully planned to please patrons and meet profit needs, it can be completely negated by poor selling and service. It is essential that servers provide the follow-through that will obtain the proper sales mix and check average, and satisfy customers.

Beverage service should be built around the emphasis of relaxation and pleasure. Many people drink at social events, with friends, or to relax and spend a few idle, peaceful moments. Others want entertainment as well as drinks. Servers should be friendly, helpful, courteous, and know how to handle people and be able to sense their moods. Some will want drink recommendations and some conversation. Others will want only the service required and then to be left alone. Servers should know the liquor menu very well, how drinks are made, in which type of glass they are served, how they are garnished, and beverage prices.

The service of beverages in a bar or lounge will be slightly different than that in a dining area. In the bar or lounge, people may be primarily there to drink and not especially to eat, although most operations will serve food in the bar or lounge. In the dining area, the patrons' main purpose is to have food and, possibly, to complement it with beverages. Servers should be trained to quickly determine what guests want and then obtain it for them. Service should be sure, deft, and courteous. It is very important that servers watch their assigned tables for signals for repeat orders, and to take caution not to overserve a guest.

Drinks should be served in attractive, clean, sparkling glassware, be made from standardized formulas, and be attractively garnished. It is important in beverage merchandising to have drinks prepared to attract attention. A specialty martini can be a conversation piece at the table if served with a beautiful garnish of fresh fruit. A draft beer served in a glass so cold it "smokes" will draw notice. Sometimes unusual glassware can be a differentiating factor. Serving drinks in a special glass that can be kept by the patron or in a hollowed-out pineapple or coconut is enjoyable. Beer steins may be used for German beers. Cocktail napkins should be presented before the drinks are served. If snacks or hors-d'oeuvres are served, the proper serving utensils and dishes should be on hand.

When guests enter a dining area, a server should come to the table as soon as they are seated and ask, "What kind of beverage may I bring you?" This question is more direct than, "Can I bring you anything?" Any special promotions should be described. If guests seem as though they would like a drink but do not know what to order, the server should promptly make a suggestion.

## Liquor Pricing

Liquor is priced differently from food. Markups are usually higher, and there is a tendency to *average out* prices more by groups of similar drinks than by individual costs. Thus, a large group of mixed cocktails may be priced at $4.00 each, even though the cost of producing the different drinks may differ somewhat. With food items, individual pricing would likely be used. Many operations attempt to have an average ingredient cost for liquor items of around 20 to 25 percent, while at the same time they may want a food cost of 30 to 40 percent.

As in all menu pricing, a knowledge of cost is still desirable before establishing prices.

## Pricing Wines

Wine prices should reflect the menu price range for food. Patrons like to see a bottle of wine priced somewhere around the price of entrees.

Some operations price wines at 200 percent of their bottle cost—the equivalent of marking them up 100 percent. However, this may increase the price of some rather expensive wines above what the market will bear, or it may bring in less total profit if only one bottle is sold when two could have been. Patrons often compare wine prices seen in stores and resent the higher markup in foodservice operations. Because of this, beverage pricing is more often done on the basis of competition or what the market will bear than food is. Servers working on a wine-selling commission also are happier if more wine is sold because it not only increases their commission, it also increases the amount of tips from the customer because of the larger bill.[1] Furthermore, it is desirable to have wines appear on tables because there is a greater tendency for patrons at adjacent tables to purchase wine if they see others having it. Therefore, some operations may price higher-costing wines at a lower markup to move them. Usually, the lowest limit used for the markup for wines is 50 percent over bottle cost. The reason for this is that costs on wine are usually calculated on the following cost percentages.

| | |
|---|---|
| Interest on inventory[2] | 11% |
| Storage and other space costs | 5 |
| Cooling and handling | 3 |
| Labor | 15 |
| Glassware | 5 |
| Breakage | 2 |
| Total | 41% |

The margin of 9 percent on a 50 percent markup (50% – 41% = 9%) must go for profit and variable and fixed costs. Because 9 percent may be insufficient to cover these, the markup may have to be higher.

## Pricing Spirits

The costs of drinks made from spirits, such as highballs and cocktails, depend on the quantity and cost of the spirits used and the cost of other ingredients in the drink, such as a carbonated beverage, ice, fruit, or vermouth. To control costs, an operation should establish *exact* quantities to be used for all drinks and see that these limits are followed.

Usually, the main ingredient cost of a drink is calculated, and then other costs are added to this. For instance, if two ounces of bourbon costs 1.24 and vermouth $0.16, the total cost to make a Manhattan is 1.40. If the beverage markup is to be based on a 25 percent product cost, the selling price will be about $5.60.

Some operations estimate the added ingredient cost over the main ingredient cost. Often this is considered to be around 20 percent of the main beverage ingredient cost.

In pouring beverages, there can be some miscalculation in determining how much is available for sale. There can be losses from evaporation, overpouring, or spillage. Often an allowance of about 5 percent is allowed for loss in spirit dispensing, and 7 to 10 percent in dispensing tap beer or other tap products.

When the loss factor is figured in beverage yields, the calculation, based on a liter (33.8 oz), used for pouring 1½-ounce drinks would be $33.8 \times 0.05$ = a 1.7-ounce loss $(33.8–1.7) = 32.1 \div 1.5$ oz $= 21.4$. So about 21 drinks could be poured from a liter after the pouring loss is subtracted. Similarly, calculations should be used to find approximate portions obtained when pouring from irregular-sized bottles. In many operations this calculation will be unnecessary, thanks to computerized or electronic dispensing of drinks that count drinks poured from controlled stocks.

As stated earlier, there is a tendency to group prices in beverage menus around certain kinds of drinks. To get a group price, an operation may keep a record of the number of different kinds of drinks sold during a period and then make a study to see what the cost are. This will help determine an average selling price for each item in the group, along with the resulting profit percentage. For instance, **Exhibit 9.10** shows that the total cost of pouring a certain group of drinks is \$308.79. If the cost markup is 25 percent, then \$1,235.16 should be taken in (\$308.79 ÷ 0.25 = \$1,235.16).

**Exhibit 9.10** Record of Drinks Sold

| | Ounce Per Drink | Cost Per Ounce | Cost Per Drink | Total Cost for Sale |
|---|---|---|---|---|
| **DRINKS, PLAIN** | | | | |
| 102 Scotch | 1.5 | \$0.85 | \$1.25 | \$130.05 |
| 84 Bourbon | 1.5 | 0.30 | 0.45 | 37.80 |
| 23 Gin | 1.5 | 0.30 | 0.45 | 10.35 |
| 28 Vodka | 1.5 | 0.30 | 0.45 | 12.60 |
| 6 Tequila | 1.5 | 0.65 | 0.99 | 5.85 |
| 4 Rum | 1.5 | 0.30 | 0.45 | 1.80 |
| Total plain drinks cost | | | | \$198.45 |
| **COCKTAILS, MIXED** | | | | |
| 88 Gin or Vodka | 2.0 | \$0.30 | \$0.60 | \$57.00 |
| 25 Manhattans | 1.5 | 0.30 | 0.45 | 11.25 |
| 66 Other Bourbon | 1.5 | 0.30 | 0.45 | 29.70 |
| Total mixed cocktails cost | | | | 91.95 |
| 424 Total | | | | 290.40 |
| Add 20% Other Material Costs | | | | 58.08 |
| Total Cost of Pouring Drinks | | | | \$308.79 |

If management feels scotch at \$5.00 per 1.5-ounce drink is too high (the markup at a 25 percent food cost) and wants to see what happens if it is sold for \$3.00, and all other drinks and cocktails are sold for \$2.00, the calculation would be: 102 × \$3.00 = \$306.00 and for the others 322 × \$2.00 = \$644.00, for a total of \$950.00. At a cost of \$308.79, this is a percentage cost of 32.5 percent. If management feels that it wants a percentage markup based on a 25 percent food cost, this is not satisfactory. A tight bar control must be practiced so basic ingredients are used in the amounts estimated.

Some operations have a high markup and charge one price, regardless of what the drink is. The price set is sufficient to cover all costs of drinks it produces. If any adjustment must be made, the amount of the main ingredient poured can be varied. This can be done by pouring two ounces of gin or vodka instead of 1½ ounces for a martini, or pouring 1¼ ounces of a call brand of scotch instead of 1½ ounces, thus covering the costs of an expensive item.

### Pricing Tap Products

Draft products are usually priced on the basis of the unit cost. For instance, a 12-ounce pilsner beer glass holds about 10.4 ounces of beer. The amount held in a glass varies with the size of the head of foam—from ½ to 1 inch. Operations should find out their own average serving amount. There is usually a 7 percent pouring loss when beer is poured from the tap. If a half-barrel of beer (1,984 ounces) costing 100.00 is used, the unit cost is calculated as follows:

1,984 oz × (100% − 7%) = 1,845 ounces
1,845 × 10.4 oz = 177 glasses
100.00 ÷ 177 glasses = \$0.565 cost per glass

If the operator wants a 25 percent product cost, the selling price is (\$0.565 ÷ 0.25 = \$2.26). If the operator wants a 20 percent product cost, the price is \$2.83 (\$0.1568 ÷ 0.20 = \$2.825).

Bottled beer is usually sold on a straight markup basis. For instance, if a 12-ounce bottle costs \$0.75 and the desired cost percentage is 25 percent, the price is \$3.00; if it is 20 percent, the price is \$3.75.

Other pricing methods as mentioned in Chapter 6 on pricing could also be used for tap products. Charging only what the market will bear might be a good method for pricing tap products.

## Beverage Control

Control of alcoholic beverages is essential to profitability of the liquor operation. Liquor that is "lost" cannot be sold. If controls are inadequate, potential profit can be lost. Through overpouring, spillage, improper mixing, improper ringing up of sales, customers who refuse drinks or avoid paying, and employee pilferage, liquor profits can be lost. Seeing that alcoholic beverages items are served as planned and that costs are controlled comes with a program that includes good purchasing methods, standardized recipes, proper production, adequate record keeping, and financial controls. It is the purpose here to briefly summarize some of the factors that can lead to serving the right product at the right cost. There are a variety of computer software programs for managing beverages from inventory to profit analysis. Taking advantage of these products can greatly increase efficiency.

## PURCHASING CONTROLS

The menu largely dictates what is to be held in stock for beverage service, and the stock requirements will, in turn, dictate what is to be purchased. The purchasing of alcoholic beverages is somewhat simplified in that items are usually purchased by brand name. Customers will often ask for a particular brand of vodka, a *call brand.* (Such an item is usually sold at a higher price than regular stock.) Thus, a certain number of standard brand items will be required to maintain a stock for patron calls. Wines, beers, and ales are generally requested by brand, also.

Every establishment generally has in stock several *house* or *well* items. These are good-quality brands of spirits that are used for mixed drinks and for other drinks when no call brand is specified. Often these are selected on the basis of local tastes and price. Requests for wines, beers, and ales can also be filled from house stocks. For instance, if a customer orders a draft beer, the house stock is served. Similarly, a customer ordering a glass of dry, white wine will likely get a house brand. Increasingly, guests are brand conscious, which often results in brand loyalty.

Selection of alcoholic beverages must be based on several factors, including price, patron preferences, quality, and proof. Even though a particular item may be of high quality, it should not be stocked if only a few customers order it.

Value in spirits is often based on the quantity one receives plus the proof of the item for the price. A buyer should check the proof of items when purchasing, although if purchasing by brand, this may not be important since patrons who want a brand either know the proof or are not concerned. The buyer needs to be aware of proof, however, in order to price products. A spirit of 100 proof should sell for more than one of 80 proof, all other factors being equal. Buyers should also check the quantity of liquor in the bottle. A 750-ml bottle holds close to the same quantity as a fifth-sized bottle—the fifth holds 756.5 ml. However, in a case (12 bottles) of fifths one gets nearly three ounces more liquor than one does in a case of 750-ml bottles. A quart holds 945.6 ml, while a liter holds 1,000 ml, so when one purchases a case of bottles in liters, one is getting nearly 653 ml, or about 22 ounces, more. Many operations still use ounce measures in pouring. In pricing, it is necessary to convert bottled drink cost into item cost, so whichever form of measure is quickest for the operation should be used. (See **Exhibit 9.11.**)

**Exhibit 9.11** Volume Measures for Standard Bottles*

| Bottle | Ounces | Milliliters |
|---|---|---|
| 750 ml | 25.4 | 750.00 |
| Fifth (⅕ gl) | 25.6 | 756.50 |
| Quart (U.S.) | 32.0 | 945.60 |
| Imperial Quart | 38.4 | 1,133.54 |
| Liter | 33.9 | 1,000.00 |
| 1.75 | 59.3 | 1,750.00 |
| 3.00 | 101.6 | 3,000.00 |
| Gallon | 128.1 | 3,780.00 |
| Imperial Gallon | 153.7 | 4,534.15 |

*In these calculations the metric equivalent of 1 oz was taken to be 29.5 ml. If this had been weight and not volume, an ounce would be equal to 28.4 gm, approximately.

The quantity purchased should be held to the lowest possible figure, with safeguards against running out. Usually an operation calculates its average use of items and establishes both a minimum and maximum level of stock for maintaining stocks between deliveries. The minimum level is set so that the stock will be available during normal usage, plus an additional safety factor. The maximum level is set so too much money is not tied up in inventory. Occasionally, stocks may be raised above a maximum level if use is expected to be high. Seldom should it be allowed to go below the minimum level. Very slow-moving items are usually reordered when the last bottle is issued or the bar runs out.

Buyers should not be misled by discounts into purchasing more than the maximum level, unless the discount is substantial. If the discount is less than the interest of money invested in a large stock, the deal will not be a favorable one for the buyer. Storage, risk of loss, and other factors also must be considered as costs.

Beverage purchase requisitions are usually the basis for purchase orders. Only those authorized to buy beverages should have access to purchase requisitions or purchase orders. They should be signed by management or the highest purchasing authority. A record should be maintained of all purchase orders for liquor. From the purchase record a *daily receiving report* should be set up, indicating what beverage orders are expected for delivery on what days, how much was ordered, and the price. The specification for each item can be consulted to obtain additional information for the receiving report. In some states, liquor may only be purchased at state-run stores by the manager, using a specific payment method.

### *Receiving*

When goods arrive, they should be checked to see whether they are the right item and size ordered and the proper quantity. If the delivery matches both the purchase order and the invoice, the latter can be signed and a copy retained and sent to accounting with the receiving report. If the delivery is not correct, the discrepancy should be noted on the receiving report and on the retained invoice. If the discrepancy is major, delivery should be refused. After receiving, alcoholic beverages should be moved quickly to a locked storage, since security is a factor during transport and until properly stored.

Goods should be stored on shelving in the same order as they appear on the inventory sheet. In this way, the items are easily found and checking is easier during inventory.

### *Issuing*

The issue of bar goods requires special treatment. *Nothing* should be issued to the bar without a properly signed requisition, as shown in **Exhibit 9.12.** The requisition with a duplicate is brought to the issue room and presented to the storeroom clerk if one exists who fills the order. Accompanying the requisition should be one empty bottle for every bottle to be issued. The storeroom clerk should check each empty bottle to note the code before disposing of it. Some states require that empty bottles be retained for inventory by officials. Sometimes bars and lounges might lose track of a bottle. This can happen with full-bottle spirit sales or wine sales, when a guest takes the bottle. In this case a sales slip for the item, signed by the bar manager, is presented to the clerk instead of the empty bottle. Issues for room service and catering operations may have to be handled a bit differently, but good control and security must be part of the procedure.

When the person to whom the items are to be issued receives them, the storeroom clerk and the receiver should each sign both copies of the requisition. The storeroom clerk keeps one copy and the receiver gets the other. Increasingly, paper is being done away with as operations are using handheld computerized devices for record keeping. Signatures can be stored along with the documentation and uploaded into a central computer file. Bars and lounges and other liquor-dispensing units should order on the basis of *par stock,* or an allowable inventory based on what management thinks is needed for the bar to operate over a specific period, usually a single shift

**Exhibit 9.12** Beverage Requisition Form

BEVERAGE REQUISITION

Dept: ______________________ Date: ______________

| Quantity | Item | Quantity | Item |
|---|---|---|---|
| | Liquor | | Beer |
| | | | |
| | | | |
| | | | |
| | | | |
| | | | |
| | | | |
| | | | |
| | | | |
| | | | |
| | | | |
| | | | |
| | | | |
| | | | Soft Drinks |
| | | | |
| | | | |
| | | | |
| | | | |
| | | | |
| | | | |
| | | | |
| | Wine | | Fruit, etc. |
| | | | |
| | | | |
| | | | |
| | | | |
| | | | |
| | | | |
| | | | |
| | | | |
| | | | |

SIGNATURE: ______________________

or a full day. In that case, issues from the storeroom to the various units should take place at the beginning or end of a shift.

A bar stock inventory is done in tenths of a bottle or fourths of a bottle. Each time an inventory is taken, a check is made to see that all bottles bear the coding stamped on them by the storeroom clerk when the cases are opened. After cases are emptied, they should be broken down and folded. Unbroken cases can easily hide pilferage.

*Bin cards* for checking off items removed from storage can act as a perpetual inventory, but it is preferable to keep the inventory on a computer. A regularly scheduled physical inventory should also take place. Historically, two people have worked together to take the inventory. One identifies items and counts while the other checks the bin card and writes in the correct count on the inventory sheet. It is usually better for verification if one of the inventory takers is not a member of the beverage department. If two people are not available, management may elect to perform this function using a handheld device that reads bar codes or labels. Some nightclubs with multiple outlets have found success in employing a single individual whose sole task is to perform alcohol inventory, reconciling actual product on hand with recorded sales. Discrepancies must be explained by unit managers.

## Production Control

Bartenders may like to have the freedom to mix drinks according to their own formulas, but many operations find that consistent drink production ensures proper quantities and quality, keeps costs down, and contributes to responsible service. Standard recipes for drinks should be as well tested and costed out as food recipes. The cost of the garnish is added to the cost of the main ingredients. The standardized recipe should give the drink name, list ingredients in amount in order used, preparation method, portion size, type of glassware, garnish, and any other information to produce the drink. Code numbers should be included on the recipe if the restaurant uses a coding system. (See **Exhibit 9.13.**)

Many bars that want to have control over the production of drinks publish manuals and brochures of recipes that standardize pouring. It is important for managers to see that these recipes are followed. When the standardized recipe is used, the cost and the resulting drink can help ensure patron acceptance and management control.

Some operations allow bartenders to *free pour* without measuring ingredients. With experienced personnel, an almost exact measure may be poured each time. Studies have shown, however, that better consistency in drink quality and better cost control occurs when measures are used.

*Jiggers* and *shot glasses* are often used as measures. A jigger is often a measure with two different size cups, one on either end. The most commonly used has a ¾-ounce and a 1½-ounce cup. Some jiggers are scaled in ¼-inch markings and hold from 1 to 1½ ounces. The standard jigger size is considered 1½ ounces. A shot glass measures from ⅞ ounce to several ounces. The standard is often set at 1 ounce. The jigger is usually used for mixed drinks, while shot glasses are used for making nonmixed or straight-shot drinks.

**Exhibit 9.13** Standardized Recipe for a Manhattan

| Ingredient | Ingredient Unit | Drink Measure | Purchase Unit Cost | Measure Cost |
|---|---|---|---|---|
| Bourbon, whiskey | 32 oz | 2 oz | $23.50 | $1.46 |
| Vermouth, sweet | 32 oz | 1 oz | 6.00 | 0.18 |
| Bitters | 16 oz | dash (½ oz) | 6.50 | 0.20 |
| Cherry (100 count) | 1 gal | 1 cherry | 18.00 | 0.18 |
| | | | | $2.02 |

Method:

On the rocks: Fill an old-fashioned glass (7 oz) with small ice cubes; measure with jigger; add bitters. Stir. Garnish with cherry on bar pick. Add swizzle stick.

Straight or up: Measure with jigger into shaker; add ice and stir. Drain into cocktail glass (4½ oz). Garnish with cherry on bar pick.

The lines on glass measures can be deceiving. Glasses are slightly concave or curved, so the patron seeing the glass sees the spirit above the line when it is actually just on the line. Some glasses are purposely lined just under the proper amount so that when a drink is overpoured, the exact amount is still given.

Caps that measure the exact amount poured can be put on bottles. When the established amount is poured through the cap, pouring automatically stops. Some caps also count the number of pours. Once used, a cap will not release liquid until the bottle is first set upright, then tilted for another drink.

Push-button pouring systems can be used to control drink amounts, and some attached to electronic cash registers or computers are quite sophisticated. Not only is the right drink poured in the right amount, but information on the drink, its price, and place of service is also recorded in the machine. Thus, at the end of the day a complete sales readout can be obtained.

Some patrons do not like to see drinks poured from caps or from push-button units, preferring the free-pour system. For this reason, when patrons can see service, such units might not be used, but they can be used as controls at service bars.

## Service Control

Controlling the service of liquor is the final check in good beverage management. Point-of-sale systems have greatly simplified service control. Orders are entered into the system by server name or code and table number. Checks can easily be separated or split with a record of server and item sales. Those using historical "paper" systems should be aware of the following principles. Bar checks for servers and bartenders should be numbered, and when issued, the check numbers should be recorded and the receiver should sign for the checks. When checks are returned unused, they are checked in. If a check turns up missing, the server who was issued the check is responsible.

The transfer of checks from the bar to the dining service should be worked out well. Usually the check and a transfer log are sent to the dining area by someone at the bar. The host, hostess, or maitre d' receives the check and signs a transfer log, which frees the bar of responsibility to obtain payment. Service personnel should use one check for each table served. (See **Exhibit 9.14.**) The bartender at the service bar should do the same. Sometimes bartenders do not use checks but cash registers instead. When the customer is ready to pay, the total bar bill is rung up and the paid check is then put into a locked box so it cannot be used again. This is done because if the server retains the check and gets a similar order, the check could be reused. The dishonest server could then pocket the extra cash since the recorded check has already been paid and rung up. Such checks held out this way by servers are called *floaters*. Using a locked check box helps prevent this from happening, but only if the check is placed there immediately after payment.

Some systems call for servers to obtain payment upon service for each check. If there is a reorder, the check is recalled, the additional items ordered sent out, and the additional amount totaled. Other systems call for cash to be presented by the server at the end of a shift. Then the receipts are totaled and the server presents the amount of cash matching the total, keeping the extra as tips. It seems that most establishments prefer the former method, where payment is received immediately. The system should serve the guest needs for the type of establishment offering beverage service.

Many guests dislike having to pay each time they order. Guests are less likely than ever to

## Exhibit 9.14 Bar Check

| TABLE | GUESTS | SERVER | DATE | CHECK NO. 981201 |
|---|---|---|---|---|
| 1 | | | | |
| 2 | | | | |
| 3 | | | | |
| 4 | | | | |
| 5 | | | | |
| 6 | | | | |
| 7 | | | | |
| 8 | | | | |
| 9 | | | | |
| 10 | | | | |
| 11 | | | | |
| 12 | | | | |
| 13 | | | | |
| 14 | | | | |
| 15 | | | | |
| 16 | | | | |
| 17 | | | | |
| 18 | | | | |
| 19 | | | | |
| 20 | | | | |
| BAR TOTAL → | | | | |

SIGNATURE ROOM NO.

| CHECK NO. | DATE | GUESTS | CASHIER'S INITIALS | BAR TOTAL |
|---|---|---|---|---|
| 981201 | | | | |

carry cash. Moreover, having patrons settle up at the bar while waiting for a table is generally a convenience for the establishment, rather than a consideration for guest needs. Just as a grocery shopper does not wish to pay separately at produce, dairy, and seafood departments, many diners prefer to be charged only once under one roof.

## Reconciling Sales

Beverage managers often try to establish methods that will allow a check on income received to determine whether income is what it should be. This is called *reconciling sales.* There are a number of methods for doing this. Reconciling sales can start with a daily cost sheet that includes the daily receipts, material use, food cost percentage, accumulated figures for these, and, perhaps, some historical data for comparison purposes. However, this sheet presents what *is* happening but not what *should* happen. Similarly, a profit-and-loss statement tells what *has* happened—not what should happen.

Some operations record inventory issues at retail prices, that is, what the item is to make in income. Theoretically, the cash or income received should then equal the issue price. There are problems in this, however. If only one type of drink is given at one price from the amount in each bottle, the system can work, but this is usually not the case in most beverage operations. Most often, different amounts are poured from one bottle for different-priced drinks. For instance, what should the retail value of a bottle of gin be if a gin fizz of 1½ ounces sells for $4.50, a specialty martini with 2 ounces of gin sells for $6.75, and a Tom Collins using 1½ ounces of gin sells for $4.25? The answer will depend on how many of each are sold from the bottle, and the one drink, one-price method will not work. Computerized point-of-sale systems come to managements aid here, too.

### *Sales Tally*

A more certain way to reconcile sales is to make a *sales tally* from guest checks and check the income against the amount of liquor used per drink for that day or that period. First, the exact amount of liquor used in each kind of drink must be known. Recipes will provide that information. Then multiply the total for the drink sold by the amount of liquor used per drink to get the total amount of liquor used for the day. This total is then multiplied by the selling price to get the actual income. This can be done for each drink made from each kind of spirit to see that sales come fairly close to matching the amount of liquor used. Thus, income can be determined for each type of drink sold, as shown in **Exhibit 9.15.**

The tally method can be cumbersome and time-consuming, but with electronic cash registers or point-of-sale systems, it is considerably simplified.

**Exhibit 9.15** Income from Four One-liter Gin Bottles

| Drink | Amount of Gin Dispensed per Drink, in ounces | Number Sold | Selling Price | Total Amount Used, in ounces | Total Income |
|---|---|---|---|---|---|
| Martini | 2.00 | 34 | $6.00 | 68.0 | $204.00 |
| Gin and Tonic | 1.50 | 12 | 5.00 | 18.0 | 60.00 |
| Tom Collins | 1.75 | 8 | 5.50 | 14.0 | 44.00 |
| Gimlet | 1.50 | 6 | 5.75 | 9.0 | 34.50 |
| Straight Gin | 1.50 | 13 | 5.00 | 19.5 | 65.00 |
| Totals | | 73 | | 128.5 oz | $407.50 |

### *Averaging Drinks*

Some operations make a tally, such as that shown in **Exhibit 9.16** of every kind of drink sold over a period. Then, an average number of drinks sold from a bottle is obtained, and this, instead of a tally, is used to estimate how much income should be obtained for the amount of liquor used.

### *Averaging Percentage*

Other operations use an *average percentage* of total sales instead of averaging the number of drinks. From the breakdown in **Exhibit 9.17** of drinks sold from a liter, the following percentages are obtained: martini, 62.9 percent; gin and tonic, 11.4 percent; Tom Collins, 11.4 percent; gimlet, 8.6 percent; and straight gin, 5.7 percent. The next calculation required is to find out the number of drinks obtained from the bottle for each kind of drink served, allowing for about a 5 percent loss estimate. The results are shown in **Exhibit 9.17.**

Note that the income per liter of gin is about the same when using either the average percentage method or the drink-averaging method.

The advantage in using percentages is that they can be applied to any size bottle of beverage. Thus, it would be possible to calculate the potential value of a fifth of gin in the same manner as

**Exhibit 9.16** Drink Averaging

| Drink | Amount Number of Drinks Sold | Selling Price | Total Oz Dispensed | Income $ |
|---|---|---|---|---|
| Martini | 11.0 | $6.00 | 22.0 | $66.00 |
| Gin and Tonic | 2.0 | 5.00 | 3.00 | 10.00 |
| Tom Collins | 2.0 | 5.50 | 3.50 | 11.00 |
| Gimlet | 1.5 | 5.75 | 2.25 | 8.62 |
| Straight Gin | 1.0 | 5.00 | 1.50 | 5.00 |
| Totals | 17.5 | | 32.25 | $100.62 |

**Exhibit 9.17** Average Percentage Method

| | |
|---|---|
| Martini | 1,000 ml ÷ 60 ml per drink = 16 |
| Gin and Tonic | 1,000 ml ÷ 45 ml per drink = 21 |
| Tom Collins | 1,000 ml ÷ 52 ml per drink = 18 |
| Gimlet | 1,000 ml ÷ 45 ml per drink = 21 |
| Straight Gin | 1,000 ml ÷ 45 ml per drink = 21 |

Next, the percentage of drinks served is multiplied by the number of drinks given per container times the selling price, which equals the contribution of each drinker to the total income per bottle.

| | | |
|---|---|---|
| Martini | 62.9% × 16 × $6.00 = | $60.384 |
| Gin and Tonic | 11.4% × 21 × 5.00 = | 11.97 |
| Tom Collins | 11.4% × 18 × 5.50 = | 11.286 |
| Gimlet | 8.6% × 21 × 5.75 = | 10.3845 |
| Straight Gin | 5.7% × 21 × 5.00 = | 5.985 |
| Income per Bottle | | $100.0095 |

The percentage method can also be used to estimate the total income from different-sized bottles if the total amount dispensed is obtained. This figure is divided by the amount in one container for which the expected income is known.

Suppose for a given period 12 1.75-liter bottles, 4 1-liter bottles, 6 quarts, and 8 750-ml bottles of gin were sold. The total sold would be as follows.

| | |
|---|---|
| 12 × 1,750 ml | = 21,000 ml |
| 4 × 1,000 ml | = 4,000 ml |
| 6 × (32 oz × 29.5 ml) | = 5,664 ml |
| 8 × 750 ml | = 6,000 ml |
| | 36,664 ml |
| | ÷ 1,000 ml |
| | = 36.66 ml |

If each liter is expected to bring in approximately $100.00, then $100.00 × 36.66 = $3,666.00, the expected total income from the sale of this gin.

that of a liter of gin. First, estimate the percentage of drinks that will be obtained per bottle and then use the following formula.

Income per drink = Percentage of drinks sold × Number of drinks per bottle × Selling price of each drink

Then, total the income from each drink to get the expected income.

### *Standard Deviation Method*

Another method used to calculate income expected from a total amount of sales per bottle is called the *standard deviation* method. A calculation is made to ascertain what income would be obtained

**Exhibit 9.18** Standard Deviation Method

| Item | Number Sold | Drinks Sold | Price | Income |
|---|---|---|---|---|
| Scotch | 150 qt (32 oz) | 3,000 | $6.00 | $18,000 |
| Gin | 100 qt (32 oz) | 1,600 | 5.00 | 8,000 |
| Rum | 50 qt (32 oz) | 1,500 | 5.00 | 7,500 |
| Bourbon | 180 qt (32 oz) | 5,400 | 5.50 | 29,700 |
| Total | | | | $63,200 |

if one kind of standard size drink were sold at one price from a bottle. The amount of scotch, gin, rum, and bourbon sold in a period is then obtained, and the expected income ($63,200) is calculated, as shown in **Exhibit 9.18.** However, if sales from this amount of liquor during one period are $57,512, or only 91 percent of expected, a check of the results from other periods may be needed. A study of several other periods may indicate that the difference between the expected income and actual gain may average out at 91 percent. Managers can typically expect a normal variance, and actual sales will usually be a percentage of potential or expected sales. A very low or high variation should be examined.

Potential or expected sales will only be realized if the planned sales mix holds. If the sales from a liter of gin, as shown in **Exhibit 9.15**, are not what is expected, there will be a definite variance in income from that expected. A sudden drop in sales of normally popular drinks should be investigated. There could be any number of reasons for a loss of sales. How large a variation should be permitted before an investigation is made? This depends upon the particular operation. A 1 to 3 percent variation probably does not warrant investigation, but beyond that point it is wise to find out whether something is wrong. Liquor is particularly susceptible to pilferage, theft, and a variety of fraudulent practices.

### *Drink Differential Procedure*

The *drink differential* procedure is sometimes used to allow for the use of one type of liquor in a number of different drinks sold at different prices. Suppose management decides that each ounce of scotch should be sold at $3.00, each ounce of bourbon at $2.00, and each ounce of gin at $1.80. Next, an expected sales value is established for each bottle. If a liter is used, allowing for 5 percent spillage, each liter would give about 37.9 ounces, and the expected sales value per bottle would be as follows.

| | |
|---|---|
| Scotch | $115.80 |
| Bourbon | 75.80 |
| Gin | 72.00 |

However, not all these items are sold straight. For example, 2 ounces of bourbon can be used in Manhattans selling at $2.50, 1.75 ounces in Old-Fashioneds priced at $3.00 each, and 1.5 ounces in Whiskey Sours priced at $5.00 each. The selling price should be Manhattan, $4.00 (2 oz × $2.00); Old-Fashioned, $3.50 (1.75 oz × $2.00); and Whiskey Sour, $3.00 (1.5 oz × $2.00). Thus, there is a $0.50 average under the selling price of Manhattans and Old-Fashioneds $3.50 – $4.00) and ($3.00 – $3.50), respectively. Whiskey sours reap $2.00 over ($5.00 – $3.00). This illustrates some of the challenges of using this method. Adjusting the value per bottle based on a sales tally performed by the point-of-sale system can assist in making an adjusted value per bottle.

There are other methods of trying to check on the ratio of income to inventory used, but these are the basis of ones most commonly practiced historically. As the industry rapidly converts to electronic cash registers and point-of-sale systems, the prevailing method is one in which a readout is obtained that will give the different kinds of drinks sold, the total income derived from each, the amount used, and a comparison with the expected sales income. The totals of actual sales and expected sales should be very close. Any system used should not cost more than the information it supplies is worth.

## Summary

Beverage menus differ in merchandising, pricing, and in other ways from menus offering only food, and they require special planning. Liquor menus can be separate from food menus or combined with them. Sometimes a few alcoholic beverages are offered with the food menu, and a much more extensive alcohol selection is available as well. Sometimes a separate menu is provided for spirits and another for wines. Bear and ale menus may also be needed. It is sometimes desirable in fine-dining operations to have a separate port and dessert wine menu to offer guests at the conclusion of a meal. Beverage menus should be simple and direct. Most patrons know what they want and simply wish to check prices.

It is not enough just to write a good beverage menu. Follow-through is important if the liquor menu is to be profitable. This involves seeing that the right service is given, the quantity and quality of each drink is controlled, and costs are controlled so the menu benefits the operation.

Alcohol service should be prompt and proper for the beverage. Some drinks require special service. Servers should be knowledgeable salespeople, especially when it comes to wine and other alcohol. The procedures used in serving should be those that will help develop sales. Responsible service cannot be overstressed.

Management should see that wine is served at the proper temperature in the right glass and in the proper manner. All of these require skill on the part of servers.

The pricing of liquors is often on the basis of material costs, being about 20 to 25 percent of the price. Spirit pricing is based more on competition and what the market will bear than food pricing is. There is also a tendency to group liquor prices within categories with a single price for all drinks in the same category.

Wine pricing is different from spirit pricing. Often the cost of the wine is doubled to arrive at a selling price. Many patrons like to see a bottle of wine priced at around the price of entrees, though afficiandos will often pay much more.

Control of liquor costs and the control of liquor itself starts with purchasing and follows through receiving, storing, issue, preparation, and service. Purchasing is simplified because buys can be made on the basis of brands both for *call* liquors and for *house* or *well* liquors. The volume of alcohol obtained and the *proof* are other factors to weigh in purchasing. The quantity purchased should be based on usage during a certain time period between orders. The amount on hand should not only provide for the usage during the period but also allow for unexpected use.

Receiving is a critical step in liquor control, because it is here that a check is made to see that the right item is received in good condition. Immediately after receiving, deliveries should be sent to storerooms. As goods come into the storeroom they should be unpacked, marked with an operation code, and stacked properly on shelves or bins. Good security is needed in the storeroom, and only authorized employees should be allowed on the premises. Issuing only on the basis of properly signed requisitions, exchanging new bottles for empty ones, disposing of or promptly removing the empty ones after verification of codes, and having all requisitions signed by the receiver and storeroom clerk are all procedures that help ensure beverage control.

All drinks should be made according to established, standardized recipes. Recipes and bar manuals may be used to standardize quantities and glassware and assure that proper garnishes are served. Recipes should be costed out, and pricing should be based on these recipes. Whether preparation personnel are allowed to free-pour or required to use jiggers, shot glasses, caps, or automatic pushbutton devices will depend on the operation.

Paper bar checks should be carefully controlled, and servers should sign for the ones they are given and account for each check received. Service personnel are usually held responsible for walkouts and lost checks. Point-of-sale systems have simplified controls.

*Sales reconciliation* methods are used to check expected income against actual income to see whether the bar is being operated and controlled correctly. These methods can include counting the drinks that can be sold per bottle, pouring drinks in different amounts at different selling prices, taking a sales tally of the drinks sold, averaging out expected bottle income by the number of drinks sold or percentages of drinks in a bottle, and the standard deviation method. The best record-keeping method is to use an electronic cash register or point-of-sale system that can quickly give a full printout of sales, costs, and number of drinks sold.

## Questions

1. The cost of gin is $0.60 an ounce, the cost of dry vermouth is $0.24 an ounce, and the cost of olives and ice is calculated as being 20 percent of the gin and vermouth costs. Calculate the cost of a martini made by the following recipe.

| | |
|---|---|
| Gin | 2 oz |
| Dry vermouth | ½ oz |

2. What are the selling prices of the martini in question 1 when the price is set at a 20 percent cost ratio? A 25 percent cost ratio? A 30 percent cost ratio?

3. A bottle of wine costs the operation $11.00 and yields four glasses. What is the cost per glass?

4. A 16-gallon beer keg yields about 200 12-oz servings. What would the beverage cost be on a keg costing $85.00? How about premium beer at $150.00? What would the selling prices be with 20 percent and 30 percent cost ratios?

5. The following information is obtained on bourbon sales during one shift. What is the potential or expected income per liter?

| Drink | Number Sold | Ounces Served | Selling Price |
|---|---|---|---|
| Straight | 27 | 1½ | $4.00 |
| Manhattan | 32 | 2 | 5.00 |
| Old-fashioned | 40 | 1¾ | 4.00 |
| Whiskey sour | 12 | 1⅕ | 4.00 |

## Mini-Cases

9.1 Develop a five-course wine dinner for the next appropriate seasonal holiday. Describe the market you will target, price per person, and merchandising plan to obtain reservations.

That was fun, wasn't it? Now develop a three-course beer menu for an upcoming seasonal event. Describe your market, price per person, and merchandising plan.

9.2 In what foodservice establishments would it be inappropriate or unsuitable to serve alcohol? Why?

Consider the advantages and disadvantages of selling a product for which the law holds the operator responsible for overserving an adult or serving an underage minor who may have false identification.

What are your thoughts on why our industry is perceived as having a higher than average problem regarding employees who abuse or are addicted to alcohol?

# 10
# PRODUCING THE MENU

## Outline

Introduction
Securing the Product: Purchasing
- The Role of the Purchaser
- Determining Need
- Searching the Market
- Purchasing Controls
- Value Analysis

Preparing the Food: Production
- Kitchen Personnel
- Control Procedures

Summary
Questions
Mini-Cases

## Objectives

After reading this chapter, you should be able to:

1. Describe the relationship between planning the menu and the role of the purchaser.
2. Explain how purchasing needs are determined and how methods of meeting those needs are found.
3. Compare and contrast the various methods of bidding.
4. Explain the relationship between specifications and quality standards.
5. Identify the purpose and benefits of purchasing controls and be able to give examples of procedures used.
6. Explain the relationship between planning the menu and the role of production.
7. Describe the purpose and benefits of production controls and procedures used to facilitate them.

# Introduction

The menu cannot stand alone. By itself, it is simply a document that has potential for management of a foodservice. To be effective, the menu must be integrated with thorough purchasing and production procedures. No matter how well a menu is planned, if these two major steps are not done properly, the menu will fail.

While the menu is the central document that dictates what is to be done in almost every operating department of a foodservice, it is completely dependent on these departments for its fulfillment. It controls and directs, but it cannot act. In this respect the menu is somewhat like the staff member in an organization who originates directives but has no authority to take action. Instead, the person must make recommendations to a line member, who then translates these into action. Similarly, the menu depends on the purchasing and production departments to implement its recommendations.

Purchasing must correctly interpret what the menu calls for and procure the necessary materials *in time* for the items to be produced. Production must prepare the food in a timely manner with due regard to food cost, quality, safety, and quantity. Both functions are a vital part of a complete foodservice system that revolves around the menu.

# Securing the Product: Purchasing

Purchasing is a complex function. It comprises the following steps:

1. Determine the need for an item, along with the quality, quantity, and other factors required to satisfy that need.
2. Search for the item on the market.
3. Negotiate between the buyer and the seller, ending in a transfer of ownership.
4. Receive and inspect, ending with acceptance or rejection of the item.
5. Storage and issuing items.
6. Evaluate the purchasing tasks, as judged by the performance of the product, and the economy and efficiency with which results were achieved.

In almost every purchasing task, these six steps must occur; only in rare instances can any be circumvented or omitted. The sixth step is extremely important, yet is often overlooked. Unless it takes place, the purchasing task may never be improved.

## The Role of the Purchaser

The job of the purchaser is to interpret properly what is needed to produce the items listed on the menu, and to procure them at prices within constraints established by the operation. This is not easy, and some buyers fail because they do not understand the menu requirements or do not know how to satisfy them. Some buyers may lack sufficient background to search out the required products on the market. Many buyers act merely as order givers and not buyers. Some purchasers may know the market, food production, and service very well, but fail to do a good job because they are market-oriented; that is, they fail to follow through and see that their operation correctly receives and uses products. Some buyers may also be misled by bargain prices, and obtain materials that do not fulfill the menu's promise. Two factors essential to value in any menu are *quality* and *merchandising appeal.* If these are lacking, the price paid by the operation in lost profits may be high. By contrast, a knowledgeable buyer who gets the right item at the right time and at the right price helps boost menu potential.

Expert purchasing requires knowledge of the production, processing, and marketing of products, their use in the foodservice establishment, and menu pricing. The buyer needs to know a lot about how the market operates, about markups, seasonal factors, and where to find specific items of the best quality for the best price. **Exhibit 10.1** shows a produce report obtained from U.S. Foodservice Market Report. A good buyer is constantly in search of products that will simplify preparation and handling, improve quality, and facilitate service. A purchaser should also be active in analyzing cost and performance factors and in sorting out procedures that will improve the purchasing function.

Markets can change quickly, and a buyer must be prepared to move with the changes. A rich background of information and knowledge about the market is needed if a buyer is to function adequately in it. Sharing information with kitchen personnel charged with menu development assists the operation to respond proactively. A buyer must safeguard the interests of the enterprise. The buyer must not waste resources nor involve the enterprise in legal or ethical problems. Buying requires a high level of integrity. The policies of the enterprise must be followed. Purchasing is a management function and, if delegated, must be given only to those who will protect the interests of the operation.

Orders serve the execution of the menu; items must be purchased to meet menu needs. Production notes the menu item and orders either from the storerooms or writes up a purchase/requisition requesting the buyer to obtain it. The buyer receives the purchase requisition and places and receives the order, either sending it directly to the kitchen or to stores. Requisition from production moves items from stores, then into use and service.

## Determining Need

A production department lets the buyer know what products are needed by listing them on a requisition, as shown in **Exhibit 10.2**. Sometimes the amount of an item to order is based on a minimum amount that must be kept in stock (*par stock*), a safety factor, and an average usage amount from the time of ordering to delivery. For instance, if the par stock for tomatoes is four cases, the safety factor is two cases, and two more are needed for usage between order and delivery, then the *reorder point* (ROP) for the item is eight ($4 + 2 + 2 = 8$). **Exhibit 10.3** shows an order list that holds orders to established par stocks.

Sometimes orders are automatic; an established amount is kept in stock, and when replenishment occurs, the maximum level is reached. These are called *standing orders.* This can occur with many dairy, bread, coffee, or tea products where the supplier sees that a par stock level is maintained.

## Exhibit 10.1 Produce Report

**Produce.** The chief California lettuce harvest region is transitioning to Salinas but volumes from the area are limited due to the earlier cool wet weather. Favorable weather should help lettuce shipments improve in the coming weeks but overall volumes may be inconsistent through May. In turn, volatile lettuce prices and suspect product quality are expected to endure possibly into next month. Idaho potato planting is progressing under mostly good conditions. Inflated potato prices are likely through the spring.

| | Market Trend | Supplies | Price vs. Last Year |
|---|---|---|---|
| Limes (150 count) | Decreasing | Good | Lower |
| Lemons (95 count) | Steady | Good | Lower |
| Lemons (200 count) | Increasing | Good | Higher |
| Honeydew (6 count) | Decreasing | Good | Higher |
| Cantaloupe (15 count) | Decreasing | Good | Lower |
| Blueberries (12 count) | Decreasing | Good | Higher |
| Strawberries (12 pints) | Steady | Good | Higher |
| Avocados, Hass (48 count) | Increasing | Good | Lower |
| Bananas (40 lb), term | Increasing | Good | Higher |
| Pineapple (7 count), term | Increasing | Good | Lower |
| Idaho potato (60 count) | Increasing | Good | Lower |
| Idaho potato (70 count) | Increasing | Good | Lower |
| Idaho potato (70 count), term | Increasing | Good | Lower |
| Idaho potato (90 count) | Increasing | Good | Lower |
| Idaho potato #2 (6 oz. min.) | Increasing | Good | Higher |
| Processing potato (100 lb) | Steady | Good | Lower |
| Yellow onions (50 lb) | Increasing | Good | Higher |
| Yellow onions (50 lb), term | Increasing | Good | Lower |
| Red onions (25 lb), term | Decreasing | Good | Higher |
| White onions (50 lb), term | Decreasing | Good | Higher |
| Tomatoes, (large case) | Steady | Good | Higher |
| Tomatoes, (5×6, 25 lb.), term | Decreasing | Good | Higher |
| Tomatoes, vine ripe (4×5) | Increasing | Short | Higher |
| Roma tomatoes (large case) | Decreasing | Good | Higher |
| Green peppers (large case) | Decreasing | Good | Lower |
| Red peppers (large, 15 lb. cs.) | Increasing | Good | Higher |
| Iceberg lettuce (24 count) | Decreasing | Good | Lower |
| Iceberg lettuce (24 count), term | Decreasing | Good | Higher |
| Leaf lettuce (24 count) | Decreasing | Good | Higher |
| Romaine lettuce (24 count) | Increasing | Good | Higher |
| Mesculin mix (3 lb), term | Decreasing | Good | Higher |
| Broccoli (14 cnt.) | Decreasing | Good | Lower |
| Squash (case) | Increasing | Good | Higher |
| Zucchini (case) | Decreasing | Good | Higher |
| Green beans (bushel) | Decreasing | Good | Higher |
| Spinach, Flat 24's | Decreasing | Good | Higher |
| Mushrooms (large), term | Steady | Good | Same |
| Cucumbers, bushel | Increasing | Good | Higher |
| Pickles (200–300 ct.), term | Increasing | Good | Higher |
| Asparagus (small) | Decreasing | Good | Lower |
| Freight (CA to city avg.) | Increasing | Good | Higher |

**Exhibit 10.2** Requisition

## Pantry Requisition

DATE: ______________________ ORDERED BY: ______________________

| Quantity | Size | |
|---|---|---|
| | | MILK |
| | ½ pt | White |
| | ½ pt | Chocolate |
| | ½ pt | Skim |
| | ½ gal. | Soy |
| | gal | 1% milk |
| | each | Yogurt |
| | | CREAM |
| | qt | Whipping cream |
| | gal | Sour cream |
| | | |
| | case | EGGS |
| | | |
| | | CHEESE |
| | case | Bleu |
| | case | Cream |
| | box | Cheddar |
| | | Cottage |
| | box | Feta |
| | box | Pepper Jack |
| | | |
| | | VEGETABLES |
| | case | Cabbage |
| | case | Cherry tomatoes |
| | case | Tomatoes |
| | each | Cucumbers |
| | each | Green peppers |
| | case | Lettuce—Boston |
| | bskt | Radishes |
| | each | Red cabbage |
| | bag | Chives |
| | bunch | Basil |
| | bunch | Cilantro |
| | bunch | Green onions |
| | case | Garlic |
| | | |
| | | FRUITS |
| | pts | Blueberries |
| | pts | Strawberries |
| | each | Peaches |
| | lb | Grapes |
| | each | Apples |
| | each | Oranges |
| | each | Grapefruit |
| | each | Watermelon |
| | doz | Lemons |
| | doz | Limes |
| | each | Pomegranate |
| | each | Avocado |
| | qt | Cherries w/stems |
| | ea | Pineapple |
| | each | Melons |

| Quantity | Size | |
|---|---|---|
| | | PICKLES-OLIVES-SEAFOOD |
| | gal | Dill pickles |
| | gal | Sliced pickles |
| | gal | Olives |
| | qt | Olives; ripe |
| | gal | Sweet relish |
| | case | Anchovies |
| | | Caviar |
| | can | Tunafish |
| | lb | Crabmeat |
| | | |
| | | CONDIMENTS |
| | #10 | Mustard |
| | #10 | Chili Sauce |
| | 14 oz | Catsup-bottles |
| | gal | Horseradish |
| | pt | Lemon juice—bottle |
| | gal | Vinegar, Balsamic |
| | gal | Vinegar, Rice |
| | gal | Canola Oil |
| | gal | Olive oil |
| | drum | Mayonnaise |
| | Qt. | Soy sauce |
| | Jar | Blueberry jam |
| | Jar | Pumpkin butter |
| | qt | Tahini paste |
| | pt | Peanut butter |
| | | |
| | | SEASONINGS |
| | box | Paprika |
| | Jar | Curry paste |
| | box | Peppercorns |
| | box | Salt |
| | box | Season salt |
| | box | Dry mustard |
| | box | Celery seed |
| | | |
| | | JUICES |
| | case | Tomato |
| | Gal | Passionfruit |
| | Gal | Pineapple |
| | | |
| | | MISCELLANEOUS |
| | lb | Sugar |
| | box | Profitorohas |
| | case | White chocolate |
| | #10 | Chocolate sauce |
| | #10 | Caramel sauce |
| | box | Corn chips |
| | | |
| | | PAPER SUPPLIES |
| | box | Disposable gloves—M |
| | each | To go containers |
| | box | Frill toothpicks |
| | box | Doilies |
| | box | Pan liners |
| | roll | Plastic wrap |
| | roll | Foil |

**Exhibit 10.3** Order Sheet

PAGE 1

**Ordering Sheet**

DAY:______________ DATE:______________________ DELIVERY:______________________________

| ON HAND | ORDER | PAR STOCK | SPECIFICATION AND DESCRIPTION | PURVEYOR | UNIT PRICE | EXTENSION TOTAL |
|---|---|---|---|---|---|---|
| | | | **—Meat Order—** | | | |
| | | 12 pcs | Canadian bacon 12-8 lb stick | | | |
| | | 100 lb | Corned beef, Ch. deckle-off 7–8 lb | | | |
| | | 25 pcs | Choice top butts 12–14 lb | | | |
| | | 8 pcs | Choice strips 17–18 lb | | | |
| | | 300 lb | Commercial chuck | | | |
| | | 20 lb | Beef liver | | | |
| | | 15 pcs | Knuckles, choice 8–9 lb | | | |
| | | 8 pcs | Choice insides 16–20 lb | | | |
| | | 4 pcs | Hams, round VC 10–12 lb | | | |
| | | 3 pcs | Hams, pullman canned 6–7 | | | |
| | | 40 lb | Lamb breast | | | |
| | | 12 pcs | Pork loins 10–12 lb | | | |
| | | 12 pcs | Choice beef ribs 28–32 lb | | | |
| | | 30 lb | Spareribs | | | |
| | | 70 lb | Pork sausage 5 lb | | | |
| | | 10 lb | Salt pork | | | |
| | | 24 lb | Fresh brisket, Ch. deckle-off | | | |
| | | 10 lb | Pork tenderloins | | | |
| | | 30 lb | Veal stew | | | |
| | | 100 lb | Beef bones | | | |
| | | 12 lb | Wieners 8/lb | | | |
| | | | **Total Meat** | | | |
| | | | **—Poultry Order—** | | | |
| | | 8 cs | Turkeys 24–26 lb toms | | | |
| | | 4 cs | Chicken 2¼ lb p.c. 65 lb/cs | | | |
| | | 40 lb | Chicken livers, fresh | | | |
| | | | Fowl livers, frozen | | | |
| | | 12 cs | Eggs, Grade A extra large 30 doz/cs | | | |
| | | | **Total Poultry** | | | |
| | | | **—Fish Order—** | | | |
| | | 100 lb | Halibut 40 lb | | | |
| | | 75 lb | Shrimp, green 15–20 ct 10-5 lb | | | |
| | | 140 lb | Grouper | | | |
| | | 4 pcs | Smoked salmon | | | |
| | | 40 lb | Lobster tails, African 10 lb 31–35 ct | | | |
| | | 150 lb | Filet of sole 50 lb | | | |
| | | 60 lb | Crab meat 6-5 lb Backfin | | | |
| | | 5 lb | Scallops 5 lb | | | |
| | | | **Total Fish** | | | |
| | | | **—Produce—** | | | |
| | | 20 cs | Head lettuce 24's | | | |
| | | 12 cs | Tomatoes 20 lb flat 5×6 | | | |
| | | 8 | Potatoes, 100 ct 50 lb bag Yukon gold | | | |
| | | 6 | Onions, 50 lb Vidalia | | | |
| | | 1 | Onions, red 25 lb | | | |
| | | 2 | Cabbage, new green 1 cs | | | |
| | | 1 | Cabbage, red | | | |
| | | 2 | Celery cabbage | | | |
| | | 2 | Celery pascal, Cal. 30/cs | | | |
| | | 2 | Cucumbers 1 lb med. | | | |
| | | 2 | Carrots 50 loose | | | |
| | | 2 | Romaine 24's | | | |
| | | | **Total Produce** | | | |

## Searching the Market

A number of different methods for searching the market and ordering may be used. The most simple is to use a *call sheet* or *quotation sheet*. The list of needs is established and then copied on the sheet. Often the grade, quality, size, and other requirements for each product are noted, and then a number of purveyors are called and asked what their price is for each item on the sheet. This process can be quite complex when new menu items have unusual ingredients for which no specifications have been established. After this process is completed, the quoted prices are compared and the purveyor is chosen. The buyer then places the order. Orders can be placed via telephone, fax, e-mail, online, or personal contact.

### *Bidding*

Bidding is a *formal* method of buying. An operation indicates its needs to a number of purveyors, who quote prices for each item. The buyer then selects the most appropriate purveyor. The most formal bidding procedure is when specifications for items are written, indicating the quantity, quality, and packaging desired, and noting other factors such as billing, delivery periods, and general conditions of performance and responsibilities of the seller and buyer. Then purveyors offer items and prices in written bids based on these requirements. Sometimes samples are submitted by the supplier, along with the offer to sell, and these samples are examined before a decision is made. Some samples may be retained to compare with the goods delivered.

In formal bidding, purveyors may have deadlines for submitting bids. Bids may be publicly opened and awarded. Bidders may be notified in writing of bid acceptance, and they will then have a specified time in which to perform. In many cases, a *bid bond* is required, which guarantees that if an award is made to a purveyor, performance is expected as agreed on the basis of the bid. If the purveyor does not perform as stated, the bonding company must either see that the purveyor does perform or pay the bond forfeit. Usually bidding of this form involves a rather large volume of business and is done by state, federal, or other governmental agencies, and by large enterprises, such as hospitals. Unless volume is large, the cost of administering a bid bond program is not worthwhile.

***Informal (Negotiated) Buying***—The most informal bidding method is called *negotiated* buying, which occurs when perhaps only one or several items are needed—and at least three purveyors are contacted over the phone—much as in call sheet buying. Then a decision is made as to who among the three will get the sale. Sales reps, unless new, generally work on commission and are authorized to determine which price schedule an establishment qualifies for. Volume is the primary determining factor in this decision.

***Cost Plus Buying***—In *cost plus* buying, a purveyor buys items at cost with a specified markup added, part of which is the cost to the operator. For example, an operator agrees with a fresh produce purveyor to send him or her all produce orders in return for the purveyor's seeking out the best quality and price on the market. The buyer pays an agreed-on markup for this service. Such an arrangement may be limited to the purchase of specified products, generally commodity items.

***Blank Check Buying***—Another method called *blank check buying* allows a purveyor to go out on the market and obtain necessary items no matter what the cost; the buyer agrees to pay whatever the market price is plus the purveyor's standard markup. In such cases, a purveyor may be instructed to purchase at the best possible price and bill the enterprise for it. Such an arrangement should be made only with reliable purveyors with whom the establishment has dealt for a long time.

### *Writing Specifications*

The heart and soul of purchasing is the *specification,* which is the delineation of what the buyer wants in an item. It should cover all item characteristics and all other factors needed to get the right product at the right price. Some characteristics can be extremely simple and brief. They may only indicate what is wanted, how much or how many, the brand, and the packaging size. Involved specifications are needed when the item required is not common or well known and there are few quality or other purchase factors established for it. For instance, the purchase of vacuum-dried apricot nuggets may require a buyer to write up a detailed list of quality factors needed in the item. If an enterprise wishes something special that differs from the commonly marketed item, it may be necessary to prepare a detailed specification.

Buying should not take place until management determines exactly what is needed. Even a brief specification must include important item characteristics so that the right item is obtained.

Usually, a specification should include the following:

1. Name of the item
2. Quantity needed
3. Grade of the item, brand, or other quality information
4. Packaging method, size package, and special requirements
5. Basis for price—by the pound, case, piece, dozen, etc.
6. Miscellaneous factors required to get the right item, such as the number of days beef should be aged, the region in which the item is produced, or the requirement that all items be inspected for wholesomeness

When writing specifications, much detail can be eliminated if factors that apply to all items are detailed together in a "general specification" section. This section can carry instructions for delivery, methods of billing, and bid acceptance requirements. Buyers should learn the common procedures in marketing so that they use market terms correctly. Buyers and sellers use a language unique to purchasing. Terms such as *California lug, No. 10, 18, Brix,* and *5 × 5 tomato* have precise meanings that both understand. This code shortens and simplifies buying and communication between buyers and sellers. Some common terms are definitive enough to eliminate the need for further detail. All meat purveyors understand the meaning of the federal government's *Institutional Meat Purchase Specifications* (IMPS) commonly used on the market. The mention of any IMPS number for a meat item in specifications eliminates a great deal of detail that otherwise would have to be written out. If a specification requests a No. 109 rib roast, this number indicates the product's preparation, trim, and distance from the plate and from the ribeye of the meat. The North American Meat Processors Association (NAMP) has published *The Meat Buyers Guide* to assist the operator with information on meat cuts and related food safety and nutrition information. More information is available at *www.namp.com*. **Exhibit 10.4** indicates how a buyer might set up specifications for some market items.

When a considerable amount of market information exists on food products, the writing of a specification is not difficult. When items are not well known, writing specifications becomes

## Exhibit 10.4 Purchase Specifications

| | Hamburger* | Turkeys |
|---|---|---|
| *Name:* | Hamburger, IMPS No. 136 | Turkeys, Beltsville, fresh-killed |
| *Amt Needed:* | 150 lb (1,200 patties) | 80 lb |
| *Grade:* | From U.S. Good (top) | U.S. Grade A, ready-to-cook, young toms |
| *Packaging:* | 2 oz patties, frozen; packed in 25 lb lots, layer packed with wax paper separators | Wrapped in polyethylene and air exhausted, two to a carton, delivered at 40°F (4.4°C) or less but not frozen |
| *Price:* | Price per lb net | Price per lb net |
| *Misc:* | Conform to all IMPS requirements; only from chucks, rounds, flanks, or shanks; deliver at 0° F or lower internal temperature | Minnesota grain-fed birds each between 24 and 26 lb, no tolerance permitted over or under |

| | Fish | Tomatoes |
|---|---|---|
| *Name:* | Cod fillets, boneless, no skin | Tomatoes, canned |
| *Amt Needed:* | 40 lb (about 60 to 70 fillets) | 10 cases |
| *Grade:* | Strictly fresh caught cod processed in plants meeting federal sanitary standards | U.S. Grade B or Choice |
| *Packaging:* | Dry layer packed in 20 lb lots | 6/10's |
| *Price:* | Per lb | Per dozen 6/10's |
| *Misc:* | Shall be treated with no preservatives or seasonings; from Boston docked cod, hake, pollock, cusk or haddock 1½ to 3 lb size (scrod) | Shall be tomatoes with no added juice or other liquid, California pack |

| | Pears | Shrimp |
|---|---|---|
| *Name:* | Pears, halves, Bartlett | Shrimp, headless, frozen, in shell (raw green) |
| *Amt Needed:* | 2 cases | 120 lb (24 5-lb packages) |
| *Grade:* | U.S. Grade A or Fancy | Highest quality, pink shrimp |
| *Packaging:* | 24 2½'s | In 5-lb blocks with no added moisture in wax paper wraps and in cardboard boxes, 8 boxes to the carton |
| *Misc:* | 7 to 9 count per can; minimum drained weight 17 oz; heavy syrup (18 to 22° Brix). | Shall be large size, 21 to 25 per lb |

| | Eggs | Butter |
|---|---|---|
| *Name:* | Eggs, fresh, in shell | Butter, sweet cream |
| *Amt Needed:* | Two cases (1 lot) | 400 lb |
| *Grade:* | U.S. AA | U.S. Grade A (93 score) |
| *Packaging:* | 30 doz paper cartons | 5-lb packs, 72 pats per lb |
| *Price:* | Per dozen | Per pound |
| *Misc:* | Size large, min. wt. net per case 45 lb; no dozen shall weigh more than 25 oz nor less than 23 oz | Pats shall be individually separated by waxed paper and layer packed; deliver over two-month period in lots not under 40 lb each |

| | Apples | Milk |
|---|---|---|
| *Name:* | Apples, fresh, Rome Beauty | Milk, fresh, homogenized, 1% |
| *Amt Needed:* | 20 Washington cases | 60,300 half pints |
| *Grade:* | U.S. Extra Fancy or Washington Extra Fancy | U.S. Grade A, pasteurized |
| *Packaging:* | 100 size, minimum net weight per box 45 lb | In sealed ½ pt paper containers; 64 containers per carton |
| *Misc:* | Paper cartons; apples shall have been stored in federally supervised environment controlled warehouses | Flash pasteurized; milk shall not be over two days old from milking time; shall conform in all respects to local and state ordinances |

*Institutional Meat Purchase Specifications of the USDA*

more difficult. Buyers have been especially hampered in writing specifications for many convenience foods because there are few factors in the market that differentiate item characteristics, the quality, and other elements needed for their purchase. However, many federal and state agencies have begun to develop specifications for these items, which can be of considerable assistance to enterprises writing similar specifications. Others may take a recipe for a product, and from this prepare a specification indicating the ingredients that must be in it. Government agencies or other organizations may have good information on the quality factors for some items that are very much like the one to be purchased. For instance, there are some good standards and purchasing criteria for canned meats. If these are studied, perhaps the necessary standards for the frozen equivalent can be developed.

Two code numbers are often used in purchasing, and foodservices may use these as their code numbers for items. The two are the *Institutional Meat Purchases Specifications* (IMPS) and the Universal Product Code (UPC).

The IMPS are code numbers used to indicate specific kinds and cuts of meat. For example, numbers in the 100s indicate beef, the 200s indicate lamb and mutton, 300s veal and calf, and 400s pork. Portion cuts are identified by placing a 1 in front of the cut number. Thus, a 109 rib is a beef rib ready for the oven, and 1109 is a rib steak ready to cook. Using the number makes it simple and easy for buyers to indicate exactly what they want without a lot of writing.

The *Universal Product Code* (UPC) is another method used to standardize purchasing. It consists of a series of numbers with lines of different width and boldness, which indicate specific food items. We see these used at supermarket checkout counters, where the bar code is drawn across a light, which reads the code and records the sale, indicating the product and price. UPCs also have specific numbers related to the lines and, if the light will not read the lines, the operator can punch in the number and the machine will record the information in the same way it would if it read the lines. UPCs are increasingly being used to perform inventory using handheld scanners. Palm-sized PDAs can be programmed with bar-code sending capabilities using software produced by companies such as Chef Tec. Purveyors are generally happy to consult regarding the development of specifications as part of their customer service.

### Quality Standards

Buying quality can be defined in different ways. Many buyers use a brand name to assure consistent quality. Thus, a buyer may order Heinz catsup because it is a recognized item of a certain quality. Brands, however, are only as good as the manufacturer that makes them. Their quality is generally not based on any standard other than that observed or established by the manufacturer.

Quality definition can also be established by *grade.* Federal standards are usually the basis for these grades. Many food grades exist. Grading is the separation of a product into different quality levels. Thus, for canned fruits there may be three quality levels: Grade A (highest), Grade B, and Grade C. Buyers and sellers need to know what quality levels these grades represent. Grades are usually established on known quality factors and do not change. **Exhibit 10.5** shows quality grade symbols for several items.

*Trade grades* may be used that have recognition only in specific markets. Thus, trade grades have been established for fresh fruits and vegetables, dried fruits, eggs, and poultry, but use of federal grades has all but replaced these.

For instance, very few markets now use the trade grades for eggs because federal grades are more universally known. At times the federal government may adopt a trade grade and copy it.

**Exhibit 10.5**
Quality Grade Symbols

U.S. Department of Agriculture

This has happened with meat and canned items. A federal grade is established after consulting with industry and is issued as a *tentative grade.* It is tested for a time to see how it works, and buyers and sellers can suggest changes. After a period of testing, the grade may be revised and then be made official. Federal grades are usually established for different levels of the market—the *consumer,* the *wholesale,* and the *manufacturing* levels. A food buyer may use all three. However, consumer or retail grades dominate the foodservice market because these standards are how most foods are graded, and processors and producers tend to follow these standards more than any others. Thus, eggs today move almost totally on the basis of consumer grades and not wholesale ones.

The federal government's grading system for processed foods is based on the development of scores for certain items. Then, based on the total score of an item, a grade is assigned to it. Thus, an item scoring anywhere from 90 to 100 may be Grade A; another, scoring from 80 to 89, Grade B; and another, from 70 to 79, Grade C. Anything below a score of 70 is said to be below standard. This does not mean that the food is inedible or not suitable for some uses, but that it is below the lowest standard.

Various methods of scoring have been developed for different groups of foods. The scoring for meat varies considerably from the scoring of fruits and vegetables. Buyers must learn the basis for scoring different food products, the value given different quality factors, and what the scores mean in item quality.

A very important standard for food buyers is called *standard of identity.* This is a statement by the federal government that defines exactly what an item is by its characteristics. Thus, no manufacturer of egg noodles can use the term *egg* to describe noodles unless the noodles contain 5½ percent dry egg solids. Unless a product comes from a specific species and variety recognized in the standard of identity, it may not be called by the name of that particular species and variety. For example, juice from anything other than *citris paradisi* cannot be called grapefruit juice. Standards of identity also indicate what is meant by terms such as *diced, salt-free, shoestring,* and *cream*. When a standard of identity is established by the government, it is a legal description of what the item is. The buyer and seller can negotiate more easily when such terms are defined in this manner. Many of these standards are also used to establish truth-in-menu requirements. (See

**Appendix A.**) Web searches of "crop explore," "organic certification" and "good agricultural practices" are useful for finding the latest standards.

Other standards that assist in promoting buying and selling in the food product market are those of *fill* or *weight.* Canned items must be filled to a specific level in the can, and cereals must settle only to specific levels in the packages. Standard sizes for barrels, hampers, bushels, crates, and other package units have been established, along with their fraction units.

There are also laws to control market interaction and to establish procedures for manufacturing and handling. Some laws, such as the Pure Food, Drug and Cosmetic Act, provide standards for food sanitation. Others, such as the Federal Trade Commission Act, the Agricultural Marketing Act, and the Perishable Commodities Act, deal with how the market is to function and what constitutes legal actions in the market. These acts protect both buyers and sellers. They also give order, reliability, and stability to the marketplace.

## Purchasing Controls

Good purchasing is not simply deciding what, when, and how much; it requires further efforts in accounting, controlling, and monitoring the flow of materials within the operation from receipt until they are used. All of these functions must be well coordinated, organized, and simplified as much as possible. Management should scrutinize the purchasing system to see if any procedure can be streamlined or eliminated. The system should provide procedures for an adequate flow of information between departments and for maintaining good records to provide accountability.

Much of the paperwork for purchasing procedures can be handled by a computer, which helps simplify, speed, and improve the purchasing process.

Purchasing needs arise from different areas in an operation. Various units within the food and beverage department may prepare requisitions. Storeroom personnel or a check of the computer file may alert the buyer to the need to bring stores up to a par or maximum level. All this information is consolidated by the buyer and usually put on a purchase order, call sheet, or uploaded into a PDA and the order placed. Note that each purchase order carries its own number, and this must usually appear on all related invoices, packages, and shipping papers so they can be readily identified with the order. **Exhibit 10.6** shows a daily order and receiving record. These applications help ensure accurate order placement and the delivery date. All order lists and purchase orders should be properly authorized and executed. Many purchase orders include a copy that is signed by the seller and returned to the buyer, indicating that an exchange of goods will occur. **Exhibit 10.7** shows a Computerized Detailed Purchase Record with Vendor, PO number and extension.

### *Receiving*

Receiving is an important step in the sequence of obtaining menu supplies. If done efficiently it allows management to see that the goods delivered meet the terms in every respect as described in the specifications, the purchase order, and other purchase documents, so menu items can be the be both the proper cost and the proper quality. Deliveries should be accompanied by proper invoices and package lists. After a thorough check is made of quantity, quality, temperature, and weight, any discrepancies are noted both on the receiving sheet and on the *invoice* acknowledging receipt that is returned to the delivery person. The invoice copy is sent to accounting with the receiving report. If items are returned, a form such as the one shown in **Exhibit 10.8** might be

**Exhibit 10.6**
Daily Order and Receiving Record

**DAILY ORDER AND RECEIVING RECORD**

FOR THE ______________ PERIOD

(OPERATION STAMP)

| QUOTE FROM AND PRICE | | | | ITEM DESCRIPTION | SUPPLIER ITEM CODE | DATE ORDERED | QUANTITY ON HAND | QUANTITY ORDERED | PERIOD REFERENCE NUMBER | RECEIVED | | DATE INVOICE PROC. | REMARKS |
|---|---|---|---|---|---|---|---|---|---|---|---|---|---|
| FROM | UNIT PRICE | FROM | UNIT PRICE | | | | | | | DATE | QUAN. | | |
| | | | | | | | | | | | | | |

used. Such a form might also be used to note any credit that might be given or any discrepancy in the delivery.

### *Storage and Inventory*

It is the responsibility of receiving personnel to move the goods to storage as quickly as possible. To ensure the safety of perishable incoming food items, it is important that foods be placed in proper storage immediately after they are received and checked. Food safety and quality can be compromised if products are left out too long. Of course, it is important to try to prescribe delivery hours so that delivery does not occur during peak times when employees are unable to give attention to proper receiving procedures. If an operation purchases in large volume, it is in a prime position to do this. Smaller accounts are more likely subject to their geography on established routes. These customers still need to make their desires known.

The storage step should include *inventory control* procedures. Goods should be either removed from packing cases and stored on shelves or stored in their packing cases on shelving. The order of arrangement of items should follow some logical method of use, such as *first in, first out* (FIFO), in which items are stored and used in the same order as they are received. Minimum and maximum quantities should be established for each item. When supplies reach the minimum on the ROP (reorder point), a notice is given to the buyer taking the stock up to maximum or par.

Sometimes inventory must be taken daily—for example, of quantities of foods on hand in the kitchen or available bar supplies. This inventory form may also be used as a requisition to bring

## Exhibit 10.7 Computer Generated Detailed Purchases

Date: 5/15/03
Time: 02:03 PM

*ChefTec Software*
**Detailed Purchases**
Culinary Software Services

Categories: All selected
Start Date: 5/07/03
End Date: 5/14/03
Select By: All stocked, Decreases, Increases

| Item | Date | Vendor | Invoice | Quantity | Units | Cost |
|---|---|---|---|---|---|---|
| alfalfa sprouts | 5/7/2003 | Sysco | 8882265 | 10 | lb | $18.50 |
| almonds, slivered | 5/7/2003 | Sysco | 8882265 | 5 | lb | $27.00 |
| apple, Granny Smith | 5/7/2003 | Sysco | 8882265 | 1 | case | $30.00 |
| avocado | 5/7/2003 | Sysco | 8882265 | 1 | case | $20.00 |
| bacon, sliced | 5/7/2003 | Sysco | 8882265 | 15 | lb | $12.00 |
| baguette | 5/7/2003 | Sysco | 8882265 | 30 | ea | $9.50 |
| basil, fresh | 5/7/2003 | Sysco | 8882265 | 5 | bunch | $6.50 |
| beans, black | 5/7/2003 | Sysco | 8882265 | 1 | case | $35.00 |
| beans, great northern | 5/9/2003 | Shamrock Foods Co. | No Invoice | 1 | case | $20.00 |
| beans, green | 5/7/2003 | Sysco | 8882265 | 25 | lb | $40.00 |
| blueberries | 5/7/2003 | Sysco | 8882265 | 1 | case | $48.00 |
| bread, pumpernickel | 5/7/2003 | Sysco | 8882265 | 1 | case | $20.00 |
| broccoli | 5/7/2003 | Sysco | 8882265 | 1 | case | $40.00 |
| butter | 5/7/2003 | Sysco | 8882265 | 50 | lb | $52.00 |
| cabbage, green | 5/7/2003 | Sysco | 8882265 | 1 | case | $12.00 |
| cabbage, red | 5/7/2003 | Sysco | 8882265 | 1 | case | $15.00 |
| cajun spices | 5/7/2003 | Sysco | 8882265 | 5 | bottle | $150.00 |
| cantaloupe | 5/7/2003 | Sysco | 8882265 | 1 | case | $30.00 |
| caraway seed, whole | 5/7/2003 | Sysco | 8882265 | 5 | bottle | $15.00 |
| carrot | 5/7/2003 | Sysco | 8882265 | 1 | case | $11.50 |
| celery | 5/7/2003 | Sysco | 8882265 | 1 | case | $8.00 |
| cheese, Bel Paese | 5/13/2003 | Willow River Cheese Co. | 65765 | 10 | lb | $36.50 |
| cheese, Bel Paese | 5/13/2003 | Willow River Cheese Co. | 65765 | 10 | lb | $36.50 |
| cheese, Edam | 5/13/2003 | Willow River Cheese Co. | 65765 | 10 | lb | $24.00 |
| cheese, Edam | 5/13/2003 | Willow River Cheese Co. | 65765 | 5 | lb | $15.00 |
| cheese, Emmanthal | 5/13/2003 | Willow River Cheese Co. | 65765 | 10 | lb | $34.95 |
| cheese, Emmanthal | 5/13/2003 | Willow River Cheese Co. | 65765 | 15 | lb | $34.50 |
| cheese, Jarlsberg | 5/13/2003 | Willow River Cheese Co. | 65765 | 10 | lb | $36.95 |
| cheese, Jarlsberg | 5/13/2003 | Willow River Cheese Co. | 65765 | 10 | lb | $35.95 |
| cheese, Jarlsberg | 5/7/2003 | Sysco | 8882265 | 10 | lb | $42.00 |

**Exhibit 10.8** Returned Merchandise Report Form

**RETURNED MERCHANDISE OR SHORTAGE REPORT**

Name of Supplier ______________________ Invoice # __________

Amount of Correction ______________________

Reason ______________________

______________________

______________________

______________________

______________________

Date __________ Deliveryman's Signature ______________________

By __________ Company ______________________

items up to par level. It is important that management establish good policies and procedures for taking inventories.

Purveyors will work with their customers to provide customized inventory/requisition forms that echo storage order, simplifying inventory and ordering. Using palm-held PDAs with barcode software further streamlines these processes.

There are two main kinds of inventories: perpetual and physical. A *perpetual inventory* is derived solely from records, such as delivery reports indicating what items have been received and put into storage and from requisitions indicating what items have been withdrawn from storage. *Present stock* plus *deliveries* minus *outgoing stock* equals *perpetual inventory.* In most operations today, the perpetual inventory is maintained by computer. As food items are received, the kind, amount, and cost are fed into the computer. As the items are withdrawn, the computer updates the inventory. Thus, one can quickly find out from the computer how much of an item is on hand, plus what the maximum/minimum level should be, the average usage, and other valuable information. A perpetual inventory sheet is shown in **Exhibit 10.9.**

In a physical inventory each item is counted visually and recorded on a form or swiped into a PDA. (See **Exhibit 10.10**.) Often, two individuals are assigned to participate in a physical inventory, one of whom comes from accounting or a department other than receiving. The inventory sheets must be signed by whomever does the counting. PDA devices sometimes have surface recording areas for signatures. Counting devices, such as bar code scanners, that record items by name, size, value, and other data needed for inventorying, are frequently used.

Some experience and knowledge are required to take a good inventory. Those taking the inventory must, for example, know the various can sizes and package quantities to obtain accurate data. Bar codes simplify the process.

It is usually considered normal to have a 1 to 2 percent difference between the perpetual inventory and the physical inventory, but if it varies much from that range, management should investigate. The food and beverage department or the accounting division may also maintain perpetual inventories of items on hand.

**Exhibit 10.9** Computer Software Generated Inventory-On-Hand Report

Date: 7/15/1999

**Time: 4:24 PM**

Inventory Date: 12/15/1998

*ChefTec Software*

**Inventory On-hand**

Grouse Mountain Grill

Beverage Room

| Item | Cost/ Unit | Units | Open | Purchases | Sales | Theoretical in Stock | Actual End | Theoretical Usage | Shrinkage | Start Date |
|---|---|---|---|---|---|---|---|---|---|---|
| drink, "Gatorade" | $0.41 | can(s) | $7.41 | $28.00 | $28.83 | $6.59 | $6.59 | $28.83 | $0.00 | 11/30/1998 |
| drink, "Snapple" | $0.44 | bottle(s) | $7.07 | $53.00 | $55.21 | $4.86 | $4.86 | $55.21 | $0.00 | 11/30/1998 |
| juice, "Kerns" | $0.30 | each | $1.20 | $18.00 | $11.40 | $7.80 | $7.50 | $11.70 | $0.30 | 11/30/1998 |
| juice, "Knudsen's" | $0.50 | bottle(s) | $16.50 | $42.00 | $48.00 | $10.50 | $10.00 | $48.50 | $0.50 | 11/30/1998 |
| juice, "Ocean | $0.37 | each | $2.20 | $33.00 | $20.54 | $14.67 | $14.67 | $20.54 | $0.00 | 11/30/1998 |
| juice, "Tropicana" | $0.56 | bottle(s) | $8.17 | $37.00 | $35.88 | $7.29 | $7.29 | $35.88 | $0.00 | 11/30/1998 |
| juice, "V-8" | $0.68 | bottle(s) | $8.77 | $92.00 | $80.50 | $18.27 | $18.27 | $80.50 | $0.00 | 11/30/1998 |
| juice, "V-8" 46 oz | $1.75 | can(s) | $17.50 | $42.00 | $29.75 | $29.75 | $28.00 | $31.50 | $1.75 | 11/30/1998 |
| juice, apple | $2.54 | bottle(s) | $7.62 | $330.01 | $317.31 | $20.31 | $20.31 | $317.31 | $0.00 | 11/30/1998 |
| juice, pineapple (lg) | $1.25 | can(s) | $11.25 | $30.00 | $31.25 | $10.00 | $10.00 | $31.25 | $0.00 | 11/30/1998 |
| juice, pineapple | $0.28 | can(s) | $1.69 | $9.00 | $5.62 | $5.06 | $5.06 | $5.62 | $0.00 | 11/30/1998 |
| juice, tomato (sm) | $0.30 | bottle(s) | $4.50 | $24.00 | $25.20 | $3.30 | $3.30 | $25.20 | $0.00 | 11/30/1998 |
| pop, assorted | $7.25 | case | $3.63 | $68.00 | $54.38 | $7.25 | $7.25 | $54.38 | $0.00 | 11/30/1998 |
| pop, ginger ale | $8.83 | case | $8.83 | $53.00 | $44.17 | $17.67 | $17.67 | $44.17 | $0.00 | 11/30/1998 |
| syrup, assorted | $26.00 | box(es) | $32.50 | $130.00 | $136.50 | $26.00 | $26.00 | $136.50 | $0.00 | 11/30/1998 |
| water, "Evian" | $8.00 | case | $8.00 | $160.00 | $24.00 | $144.00 | $144.00 | $24.00 | $0.00 | 11/30/1998 |
| water, "Geyser" | $0.63 | each | $8.13 | $70.00 | $76.88 | $1.25 | $1.25 | $76.88 | $0.00 | 11/30/1998 |
| water, "Pellegrino" | $13.50 | case | $27.00 | $270.00 | $13.50 | $283.50 | $270.00 | $27.00 | $13.50 | 11/30/1998 |
| water, club soda | $1.27 | liter(s) | $25.33 | $190.01 | $173.54 | $41.80 | $41.80 | $173.54 | $0.00 | 11/30/1998 |
| water, tonic | $0.80 | liter(s) | $7.20 | $40.00 | $45.60 | $1.60 | $1.60 | $45.60 | $0.00 | 11/30/1998 |
| | | | $210.48 | $1,709. | $1,258. | $861.46 | $645.41 | $1,274.1 | $16.05 | |



Some operations use a slightly different inventory system for meats. When a meat order is received, it is tagged with the proper receiving information, indicating the weight, the value, and the name of the item. A copy of this record is used to indicate on the perpetual inventory that this item is on hand. When the item is withdrawn from storage for use, the tag is taken off and sent through proper channels, and the item is deducted from the perpetual inventory.

### *Issuing*

*Issuing* is another critical step in the series of activities related to purchasing. Requisition forms are usually completed to indicate what items various departments want to withdraw from stores.

**Exhibit 10.10** Physical Inventory Worksheet

Date: 7/17/98

Time: 2:22 PM

*ChefTec Software*

**Physical Inventory Worksheet**

The Green House

Walk-in

| Inventory Item | Quantity | Unit |
|---|---|---|
| SHELF 1 | | |
| passion fruit | ____________ | lb |
| lemon | ____________ | lb |
| lime | ____________ | lb |
| apple, Granny Smith | ____________ | lb |
| apple, lady | ____________ | lb |
| banana | ____________ | lb |
| SHELF 2 | | |
| cabbage, napa | ____________ | case |
| broccoli | ____________ | case |
| brocoflower | ____________ | lb |
| cabbage, green | ____________ | case |
| SHELF 3 | | |
| cheese, fresh mozzarella | ____________ | lb |
| cheese, bruder basil | ____________ | lb |
| cheese, carambola | ____________ | lb |
| cheese, gorgonzola | ____________ | lb |
| cheese, cambozola | ____________ | lb |
| cheese, blue Castello | ____________ | lb |
| cheese, havarti | ____________ | lb |
| cheese, mexican (queso anejo) | ____________ | lb |
| cheese, fontina | ____________ | lb |
| cheese, lowfat cheddar | ____________ | lb |
| cheese, herb brie | ____________ | lb |
| cheese, monterey jack | ____________ | lb |
| SHELF 4 | | |
| egg, extra large | ____________ | dozen |
| egg, small | ____________ | dozen |
| grapes, green seedless | ____________ | lb |
| grapes, Concord | ____________ | lb |
| grapes, red seedless | ____________ | lb |
| lettuce, baby mixed | ____________ | case |
| lettuce, iceberg | ____________ | case |
| lettuce, romaine | ____________ | case |
| lettuce, bibb | ____________ | case |
| lettuce, baby red oak leaf | ____________ | case |
| lettuce, Boston | ____________ | case |
| cauliflower | ____________ | case |
| chard, green | ____________ | lb |
| END CURRENT LOCATION | | |

These are signed by the person needing the items and approved by the manager or the person's supervisor. The requisition is then presented to the storeroom, and the goods are withdrawn. When tagged meats and other tagged items are issued, the tag is removed and sent with the requisition to the accounting department to indicate issue. Specialized computer software may be used for these processes.

Issues from the storeroom plus *direct deliveries,* which go directly from the receiving dock to production, should be totaled to obtain the amount of food used for the day, week, or month.

## Value Analysis

A *value analysis* is a review of the purchasing process to evaluate whether the best possible purchasing job has been done and determine whether it can be improved. After purchase, the items themselves are analyzed to see if they meet the needs for which they were purchased. If they prove inadequate, an alternative should be investigated. An operation purchasing No. 189 tenderloins may decide to eliminate all the trimming and labor necessary to bring them to a usable condition by purchasing No. 190 tenderloins, which come well trimmed. Also, prepared grapefruit sections may be purchased instead of fresh whole grapefruit, which require labor to section. Many operations obtain most of their beef packaged and deboned, with most fat removed, and divided into quantities based on the cooking methods required to cook each item. This removes the necessity of cutting the meat after it is received, which, if done improperly, can result in high waste. In addition, the shipment of bones and fat, which are usually waste products, is eliminated. Working out solutions such as these helps operations simplify procedures and reduce costs.

Value analysis also includes observing items as they are processed to see whether preparation procedures can be improved and to ascertain the actual cost of preparation.

The supplier must also be appraised. A supplier should not be judged by the price of items alone but by the quality of service performance and sometimes sanitation. Some purveyors perform more or better services and, in turn, charge higher prices.

Some value analysts say that value is *quality divided by price* ($Q + P = V$). If quality can be increased and price held stable, value is increased. If quality drops and price remains stable, value is lost. If price drops but quality remains the same, value increases. If price rises but quality remains the same, value is lost. Buyers should constantly attempt to increase value.

There are numerous variables in the quality or price of items purchased on the market, and buyers must seek constantly to equate one with the other. One of the prerequisites for doing this is to know the yield of items purchased. Schuler's restaurant in Michigan weighs every rib before roasting and keeps accurate track of the number of servings that are obtained from each rib. Schuler's can look over a year's record of rib yields and obtain accurate portions costs. It can also evaluate the method it uses to purchase ribs and see whether it can be improved to get a better product at a lower cost.

There are many things that can be done by buyers, if value analysis is practiced, to bring down costs and improve procedures. It is usual for buyers and purveyors to work together to improve purchasing methods and reduce costs. It may be possible for a group of foodservices to consolidate their orders with one purveyor so that the supplier can make more deliveries in a given area, which reduces costs and results in a more favorable price for each operation. This is known as *cooperative,* or *co-op,* buying. Since some foodservices are slow in paying bills, those who pay promptly can help to defray the cost burden this places on the purveyor. By guaranteeing payment within a specified period, some enterprises may obtain lower charges from purveyors. It may be possible to work out arrangements, such as cost-plus buying or providing guarantees to a purveyor for a specific amount of business during a period, in return for the purveyor's seeking out the best possible products for the enterprise, adding only a specific amount above costs for this service. Orders may be placed farther in advance so that purveyors can do a better job of searching the market and making the most favorable buys.

Simplifying purchasing procedures and reducing paperwork can do much to assist in reducing costs, as can improving the purchasing task to reduce inventories kept on hand.

Good value analysis is a constant search for ways to improve purchasing performance and, if consistently practiced, can lead to worthwhile savings. It involves the search for facts, the analysis of those facts, and then taking action on the basis of the information gained.

# Preparing the Food: Production

Menu writing and food production are mutually dependent. The success of the menu depends very much on how well the kitchen can prepare what is listed. Conversely, successful production depends on how thoughtfully the menu itself is planned. No menu writer should place items on a menu when workers lack the skills necessary to produce them or when equipment is lacking. If the menu properly reflects the assets and limitations of the kitchen in layout, equipment, and personnel, the food can be prepared so that it meets the expectations of the guests.

Good forecasts, recipe preparation, and portioning information must be provided so that the kitchen knows the quantities, methods, and times required to prepare items. Coordination of service, management, purchasing, and other functions of the operation is essential. It is necessary for management to study the production system used and develop the flow of information and the controls that are required. Patron forecasts must be coordinated with accurate purchase quantities, and the production department needs to know what quantities to expect and when foods are to arrive. The production department should also be advised of menu forecast requirements so that personnel can be properly scheduled. This information should be entered to on a production sheet, as shown in **Exhibit 10.11.** The kitchen must also establish procedures to ensure that items are produced in the proper quantity and quality at the time when they are required.

Management should ensure that the kitchen knows exactly what each menu item is and what it consists of. A menu item and its recipe should be explained to production in precise detail as to ingredients required, their amounts, and the procedures used to produce the item. *Standardized recipes* should be used and controls established to produce exactly the same item each time it is offered to a patron. Costs should be accurately known. The most successful enterprises are those that establish good information flow so the production department knows what is to be done and how to do it. Failure of management to see that the production department is informed and is able to perform as expected can result in an unprofitable menu because the work may be haphazard and the product frequently disappointing. It is easy to put items on a menu. It is not so easy to put them on the menu *with adequate planning* for the expected result, the item cost, and customer enthusiasm.

It is sometimes said that a menu promises and the kitchen reneges. There should be a complete understanding by the menu planner, the patrons, and the kitchen of what is promised in the menu. What is promised must be delivered. Grilled fish should be prepared on a grill over

**Exhibit 10.11** Production Sheet

**PRODUCTION SHEET**

DAY ______ DATE ______

| AMOUNT | • ITEM • | LOCATION | TEMP. | TIME | REMARKS |
|---|---|---|---|---|---|
| | | | | | |
| | | | | | |
| | | | | | |
| | | | | | |
| | | | | | |
| | | | | | |
| | | | | | |
| | • REHEATS • | | | | |
| | | | | | EMPLOYEES' CAFETERIA |
| | | | | | |
| | | | | | |
| | | | | | |
| | | | | | |
| | • VEGETABLES • | | | | |
| | CASES OF SPINACH | | | | |
| | PANS OF BAKING POTATOES | | | | |
| | | | | | |

AMOUNT OF RIBS LEFT: ______ • OUT AT • ______

BAKED POTATOES LEFT: ______

AMOUNT OF GREENS LEFT: ______

AMOUNT OF SPINACH LEFT: ______

GREENS IN WALK-IN ICED: ______

FISH PROPERLY ICED: ______

STOCKPOTS CHECKED (SIMMER): ______

TOTAL NUMBER OF DINNERS: ______

**GENERAL REMARKS:** ______

*briquettes,* sometimes in combination with wood—not on a flat-top unit. Items prepared on a flat-top "grill" are actually *griddled*. A consommé is not a bouillon or rich broth but a special soup produced by specific methods from specific ingredients.

Each item on the menu should be thoroughly tested and standardized so the food production department can produce the correct quantity and quality. Recipes must be complete, precise, and carefully followed. **Exhibit 10.12** shows a standardized recipe using an application that enables you to store recipes. A wide variance in quality, portion size, or appearance of menu items can result in dissatisfied customers. Service personnel should check foods they receive, and if they are not correct, let the food production department know. **Exhibit 10.13** is a nutritional analysis of the recipe, completed with a related software function.

Foods should be prepared at the proper time. Some can be prepared ahead, and are often better if they are. Examples are soups, marinated salads and casseroles. Most should be prepared as close to service as possible. Foods that sit too long after preparation can lose culinary and nutritional quality and may become contaminated and unsafe. Production schedules will help indicate the amount of food required and the times it should be ready for service. Checks should be made to see how much is carried over after the service period. Times that an item becomes *86* or *sells out* should be noted.

Good organization is required of supporting units. Thus, the butcher shop should have items on hand as they are needed. The pantry cooks should be informed as to the amounts of pantry items required and have them ready at the proper times. Purchasing must take place sufficiently in advance to allow proper preparation. Guests dislike ordering an item from the menu and then being told the operation has sold out of it. (From a marketing perspective, "running out" implies incompetence; "selling out" implies popularity. Neither is desirable.)

There is a certain amount of prep work to be done in the kitchen. Soup stocks need to be made in advance, as well as sauces, dressings, breads, and other preparations. Long before service begins, a check should be made to see that supplies and foods required are on hand. Good organization and coordination will help make the mealtime rush run smoothly and be up to quality standards.

Spending some time with kitchen staff to train them regarding what is intended by the menu as written can help them provide tasty, satisfying meals that will make patrons want to return.

## Kitchen Staff

The people working in the kitchen must be trained to produce what the menu lists. In a quick-service, limited-menu operation, the skill, abilities, and knowledge of food production employees may be limited to preparing only a few simple items. If the menu is more extensive, more production skills, judgment, experience, and training, will be required.

Often the reason employees produce foods of poor quality is that they have little knowledge of what is expected. It is essential that food production people know enough about production principles to understand what is happening when they prepare foods. Workers may know from experience that a steak or hamburger changes color when subjected to heat and that it becomes slightly more firm and less moist. However, if they understand the reasons for these phenomena—how some of the more desirable changes can be encouraged and some of the less desirable ones limited in the food production task—higher quality items will be produced. Teaching employees why procedures are important is frequently required to ensure adequate performance. A manager who tries to discourage an employee from cooking hamburgers too far in advance of service may not be able to prevent this practice until the employee learns that heat and extended holding increase the toughness and shrinkage in meat. Employees must be shown what their finished products should look and taste like. This type of thorough training usually results in higher productivity.

Personnel organization should be functional. In many operations, a chef may be in charge and

**Exhibit 10.12** Software Generated Recipe

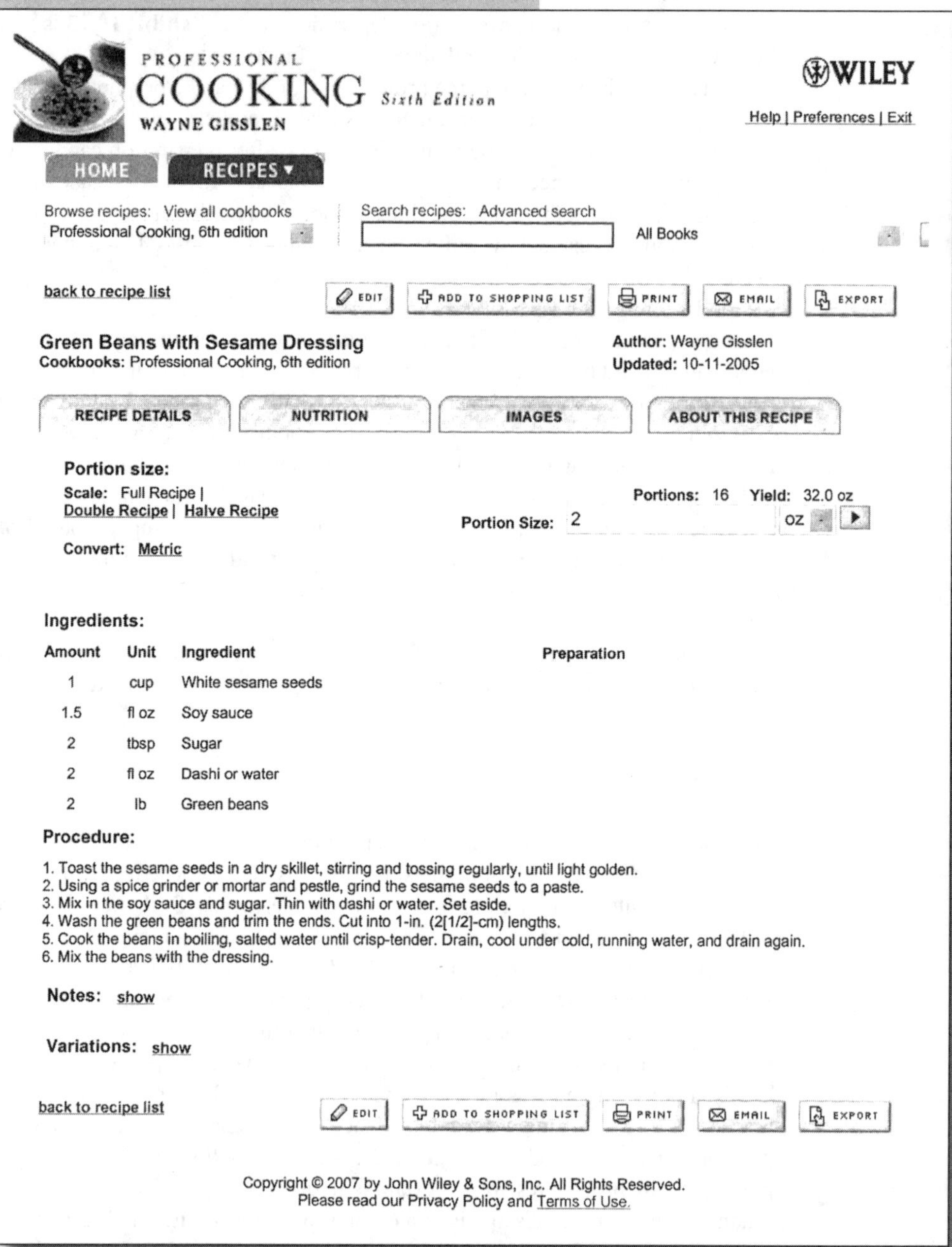


Professional Cooking, *6th edition CD-ROM; Gisslen, Wayne; Copyright©2007; Reprinted with permission of John Wiley & Sons, Inc.*

**Exhibit 10.13** Software Generated Nutritional Analysis

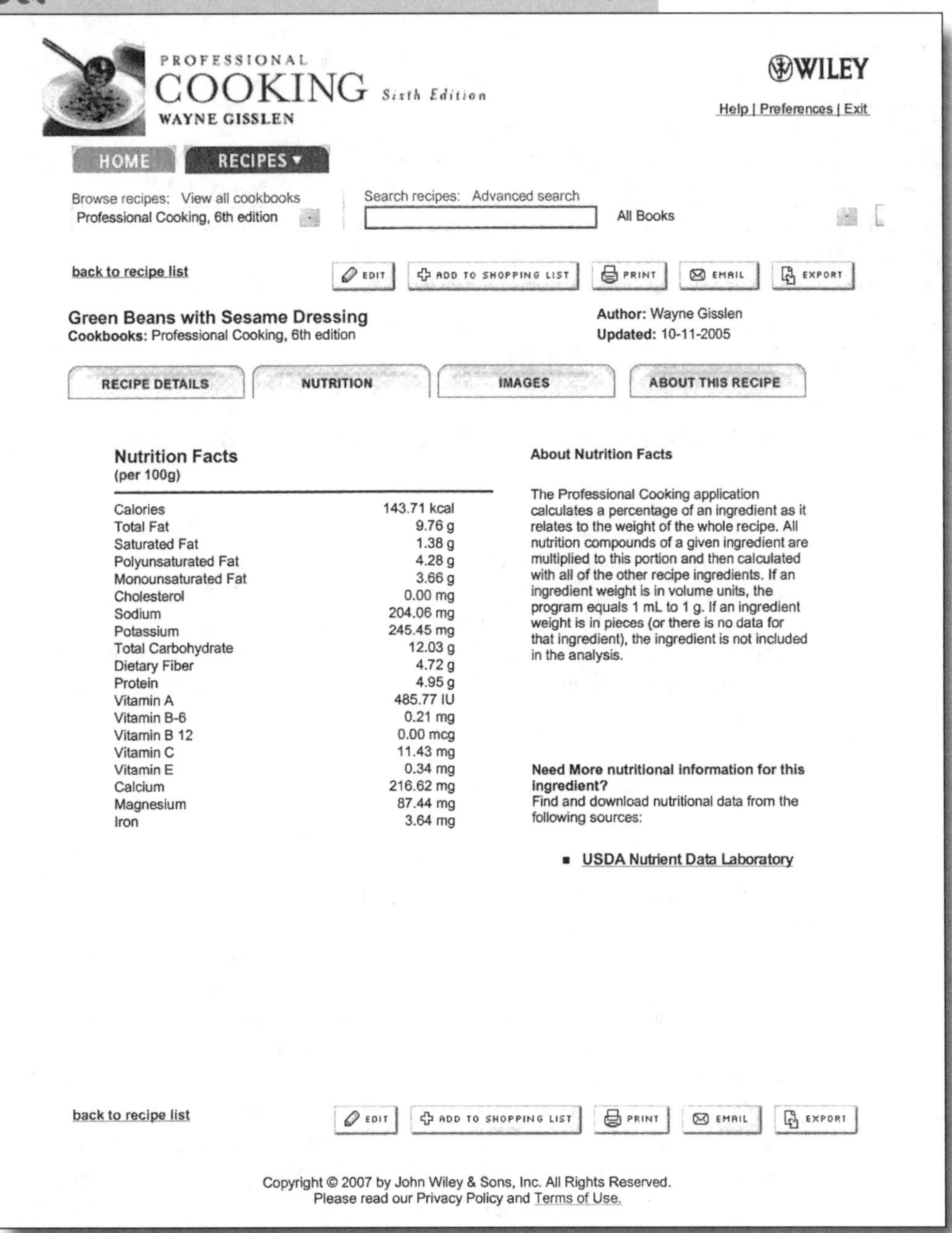


Professional Cooking, *6th edition CD-ROM; Gisslen, Wayne; Copyright©2007; Reprinted with permission of John Wiley & Sons, Inc.*

**Exhibit 10.14** Food Production Plan

| Day: Tuesday | | Date: 2/3<br>Volume Forecast: 412 | | | Meal: Dinner | |
|---|---|---|---|---|---|---|
| **Menu Item** | **Forecast** | **Portion Size** | **Production Method** | **Portions on Hand (Prepped)** | **Needed Production** | **Left Over** |
| Grilled chicken salad | 84 | 6 oz | Recipe #46 | 0 | 72 | — |
| Curried eggplant | 76 | 6 oz | Recipe #42 | 12 | 76 | 7 |
| Crab rolls | 115 | 3 rolls | Recipe #57 | 16 | 99 | 13 |
| Pod thai | 84 | 8 oz | Recipe #39 | 0 | 84 | — |
| Wasabi tuna steak | 53 | 8 oz | Recipe #41 | 0 | 53 | — |
| TOTAL | 412 | | | | | |

may have assistants, called *chefs du parti* in formal, continental kitchen organizations. In other operations, a kitchen manager may be in charge of several head cooks. Whatever the organization, it should lead to efficient, smooth production of high-quality menu items. Those in charge should constantly seek to develop high standards of performance that employees must meet. The organization should ensure that workers are adequately supervised and are assisted by supervisors or co-workers when they need help.

## Control Procedures

Labor can be controlled through the use of procedures, such as production schedules, time clocks, payroll records, training records, and health records. Written and communicated *production schedules* are vital to a smooth production process. (See **Exhibit 10.14.**) Employees must be properly scheduled to see that work is done. Improper scheduling is often the cause of high labor costs, poor-quality food, and waste. Sales histories are a big help in predicting quantities to prepare and reducing excess production and cost. Comparing the amounts prepared with amounts sold can give valuable information and reduce waste. As much as possible, one application should do the work of many, to reduce the amount of duplication and to eliminate errors. Some applications in use have been presented here. However, each operation should design a system to meet its needs. There is less uniformity here than in purchasing, and one will find that the number and kind of specifics used may not be consistent. Using a computer to do much of this has reduced time and costs, and generated more accurate information in this area.

As with purchasing, management should study the kitchen production system to see that it is efficient. Procedures should be coordinated to bring the efforts of the production team together so that menu items will be produced as planned at the right time. This coordination is a complex, sometimes difficult task, but is a critical part of implementing a menu.

## SUMMARY

The purchasing and kitchen production systems are critical factors in ensuring a menu's—and therefore the operation's—success. When purchasing and production are performed well, the menu can be expected to live up to its potential.

Purchasing entails supplying an operation with the right products at the right price at the right time. Thus price, while important, is not the only criterion. How a product *performs* is the essential measure of successful purchasing.

Purchasing includes a number of steps that need to be coordinated to ensure good results. The first step is to determine need. This may be done by checking inventory levels, the weekly menu, or discussing needs with production, or by the production staff, who may use a *purchase requisition* to request items. Searching the market is the next step. This may include accepting bids from suppliers and choosing from among the offers. The third step is to negotiate services from the chosen purveyor. Fourth, deliveries are inspected and received, and the next step involves storage and issuing. The last step involves evaluating the entire process.

Controls must be established in all phases of purchasing to ensure the proper handling and security of products until issue.

*Value analysis,* an important, often neglected step in purchasing, is a method used to evaluate the purchasing process itself, the product, and the supplier.

Kitchen production can make or break a menu. If food products are not what guests expect, the results will show up on the bottom line. An operation must have the right staff and implement procedures to ensure that foods are prepared according to standardized recipes and as close to service times as possible.

There are a number of procedures that can assist in production and purchasing to ensure proper coordination and communication among departments, tight controls, and good record-keeping. These include the production plan, standardized recipe, production schedule, and carry-over food report.

## QUESTIONS

1. Set up a flow chart that shows the steps in procuring items, starting at the definition of need and ending at production. Indicate some of the procedures used in the various steps.
2. How can using a standardized recipe control quality and quantity?
3. List the factors a good specification should include.
4. What should be included in a thorough value analysis? What are the benefits of implementing this process?
5. How do production plans, standardized recipes, and production schedules work together to ensure efficient production?

## Mini-Cases

10.1 a. Critique the menu from the Assignment Café at the end of this chapter from the standpoint of purchasing and production challenges. Attempt to provide a thorough list of each specific problem related to the concepts discussed in this chapter.

b. Search the market for each item listed on the menu. Report the results of your search.

c. Keeping in the spirit of each menu offering, rewrite the selections so that each is more practical in terms of purchasing and production.

### Assignment Cafe

**Appetizer**

Escargot in Garlic Oil—Fresh French Snails are marinated in Virgin Olive Oil and Organic Elephant Garlic and served with Crusty Baguette.

**Salad**

Chopped Salad—Iceberg, Endive, Hearts of Romaine and Organic Baby Carrots are chopped and served with Bulgarian Sheep's Milk Cheese dressing.

**Entrees**

Flying Fish—Flown in fresh from Barbados, marinated in Passion Fruit Nectar and Charcoal Grilled. Served with Fiddlehead Ferns, Fava Beans, and Thai Jasmine Rice.

Pit-roasted Pork—Pulled Pork from Suckling Pig, marinated with Adobo Spices, wrapped in Banana leaves and slow-roasted for 24 hours. Served with Breadfruit and Taro.

**Dessert**

Tropical Organic Fruit Plate—Finger Bananas, Parcha, Guava, and Pineapple drizzled with Mango and Cuban Rum syrup.

**Beverage**

Free Trade Coffee—Rich Blue Mountain roast.

**10.2** Four Girls Griddle is a corporate franchise that specializes in whole-grain waffles served with preserved fruit topping.

The corporation purchases container ship loads of grain and has patented recipes for the fruit toppings, which are produced at the peak of strawberry, blueberry, peach, and cranberry seasons.

a. What are specific purchasing opportunities and challenges faced by Four Girls?

b. What are specific kitchen opportunities and challenges?

c. What factors in regards to purchasing and production would need to be considered if Four Girls goes international?

**10.3** Alumni Bistro is an entreprenuerial fine-dining establishment opened by you and a classmate.

a. Starting small, and "on a shoestring," what specific purchasing and production challenges might you face?

b. What procedures should you use to minimize these concerns?

# 11

# SERVICE AND THE MENU

## Outline



## Objectives

After reading this chapter, you should be able to:

1. Describe the importance of the role of service in fulfilling the objectives of the menu.
2. Explain the concept of service as it relates to the hospitality industry.
3. Identify the essential elements of good service.
4. Differentiate among types of table service, and identify what sort of service might best be matched with various types of menus.
5. Explain the methods in which payment is secured from guests

# Introduction

Service is the point at which all of the work previously done to create a menu comes to fruition. The patron has been introduced to the enterprise in a few important steps: The patron has arrived, taken in the atmosphere, been seated, and received the menu. The menu's objectives are to communicate and sell the items the menu offers and to satisfy patrons. How patrons are served can be crucial in fulfilling a menu's goals.

The manner in which food is served is a critical step in the menu's success. Poor service can ruin a dining experience. Timing is always important. In situations where guests prefer a leisurely meal orders must be placed or *fired* appropriately but *picked up* without delay once prepared. Hot food should be hot; cold food should be cold. This is often stated, but violated almost as frequently. Water, butter, bread, and other foods must be replenished as needed, condiments must appear at correct times, beverages should be replenished, and all other service details handled in an efficient manner. Serving should be suited to the dining pace of each party. Service that proceeds smoothly and efficiently—without patrons noticing that it is very hard work—should always be the goal. The meal event is about the guest in successful operations.

# Serving Guests

When a menu does not measure up to expected profit potential, and purchasing and production procedures seem to be efficient, managers must then evaluate whether service is consistent with the goals of the operation and the menu.

In the hospitality industry, when we serve guests, we strive to satisfy their needs. Our customers seldom take a tangible product out of our establishment to keep. The service is an intangible product.

As an industry, we render very personal service. We not only take care of sustenance and nutritional needs, but also seek to please people in the manner in which we do so. Patrons in a foodservice are guests, and while they are, our managers and employees have the responsibility to see to it that they are courteously treated and are made happy and satisfied. In that way, we prosper.

Being responsible for a guest's comfort and pleasure is an ancient tradition in many cultures. The principle of *sanctuary* arose from the tradition that people must be cared for and not injured while in an establishment. Under an ancient Irish custom, guests were offered wine and a pinch of salt as they came through the door. Today, when a maitre d' or host greets guests as they come into the dining area to be seated, the same spirit of hospitality should prevail. We should be proud to maintain these traditions and see that guests are properly greeted. There is a great satisfaction and reward in knowing we can please others with our food and service. When this pleasure is lost, employees lose something dear that should come with their jobs.

When the foodservice industry first began, the feeling was that customers were guests coming into the home of the proprietor, and in many areas of the world customers are still treated as if they were coming into a home rather than into a business establishment. The Middle Eastern merchant goes through a considerable ceremony to convey pleasure that you are the patron. Guests should be made to feel that they are friends whom we wish to serve and make happy. Doing this is not an act of servility but the pride and dignity of the host. Developing a sense of ownership drives the concept of empowerment. In much of today's service industries, courtesy is lacking. This loss in civility can be like the loss of rain to a fertile plain. If rain fails to come, the plain dies and becomes a desert.

Good service is not always simple to provide. It takes polished skill, concentration, knowledge, and diplomacy. Employees who give outstanding service work at it. They study their jobs to find out how they can make serving efficient and improved. They are constantly alert to details that please their guests. This pays off in personal satisfaction—and gratuities for the server—and in happier clientele and higher profits for the establishment.

Employees who serve guests directly must have pleasant personalities and enjoy interacting with patrons and fellow workers. They should desire to perform excellently. Some guests try a server's patience and are difficult to serve, but every customer must be provided with polite, equitable attention. Servers must not hesitate to inform management of any guest issue that they are unable to settle. The server who can successfully solve problems with enthusiasm and grace is a true professional.

Servers must have stamina. They often work under pressure. Their appearance should be appropriate: neat and clean. Hands and fingernails should be immaculate. The hair should be washed and restrained. The uniform should be crisp and freshly laundered.

The type of service provided will depend on the type of operation, the food, and what patrons expect. Patrons of a quick-service operation will not expect—nor will they have time for—fine seated service. Some establishments will have different sorts of service within one operation. The major types of service are discussed here.

## *Mise en Place* in Service

*Good service does not begin with the appearance of guests*. It depends very much on the preparations made for patrons before they arrive. *Mise en place* means getting ready for the job to be performed. Service personnel should see that the necessary tasks are performed so that service can proceed smoothly and efficiently when guests are at hand. Without this, the job may be poorly done, and both the server and the guest will be dissatisfied with the results. An atmosphere of chaos does not create an experience of hospitality.

Before service begins, butter or olive oil must be prepared for bread. Ice, silverware, napkins, water, condiments, juices, salads, dressings, and other items needed for service should be assembled and put where they can be obtained quickly.

It is essential for service personnel to be briefed on how foods are prepared and how they are to be served. Sometimes this is the job of the dining room supervisor, and a short meeting just before service may be held to cover the menu and its items for the day.

Dining room employees also should see that all dining areas are clean and in good order. Chairs, tables, and other items in the area should be wiped, polished, or vacuumed and be in good condition. In a full-service operation, linen should be crisp and neat, and tables should be properly set with sparkling glassware and bright silverware. Menus should be clean and in good condition.

Managers and employees should continually ask themselves, "Would I dine here?" Supervisory personnel should submit survey reports to management on the condition of the dining room. Management should see to it that repairs, maintenance, and other needs are taken care of. Outside areas should also be checked. Patrons are likely to think that a poorly maintained operation will offer poor food and service. Having an eye for detail is imperative to the professionally run dining room.

## Greeting Patrons

A maitre d'hôtel, headwaiter, host, or hostess may greet patrons as they enter a foodservice establishment, take them to their places, seat them, and give them menus. If some individual is not immediately available to do this, the first employee encountered should greet patrons and make them feel welcome. Benjamin Franklin said, "The taste of the roast is often determined by the handshake of the host." Greetings should be warm but not so overly friendly that guests are embarrassed. Knowing exactly how to give a greeting takes training and practice. All patrons should be made to feel welcome.

It is not easy to greet 200 guests during a meal and escort them to their tables, but this fact should not be conveyed to guests. Every greeting should be distinctive and special so that guests feel that they have been especially identified and welcomed. A mechanical greeting from someone who simply grabs menus and recites, "This way, please," will do little to make guests feel welcome. Making eye contact is an essential element of the greeting.

## Serving Food

A good serving system is needed so that orders will be accurately taken, properly given to the production department, correctly prepared, and, finally, appropriately served. Just as standard recipes are taught and followed in the kitchen, so must staff be trained in standards of service as proscribed in a service manual developed for the operation. This should cover rules of conduct and procedures for doing the job, from greeting guests to resetting tables. Service personnel should be informed about how to place orders for different kinds of foods. They should also be told the times required for preparation and manner of firing each course.

Different foods require different kinds of service, and servers should know these requirements. There is also a proper sequence to be observed in giving good service. Water should be brought immediately. The server must time the service so that groups of foods, such as a main dish and accompaniments, come to the table together. Personnel should not serve food that is not up to standards. A good motto is: "If you are not proud of it, don't serve it." Thus, food must not only taste good but be served at the right temperature, be appealing, and be of sufficient quantity to satisfy patrons. Servers should be aware of the operations standards when serving menu items.

Servers need to have good organizational abilities and be competent to follow the systems used. They should be trained in how to present the menu, how to pour beverages, how to take orders, how to communicate with guests, and how to convey orders in detail so that production personnel can produce them as guests desire them. Servers should also follow a system designed to help them remember which guest has ordered which menu item. This usually requires a set procedure for writing down the orders and numbering the guests, starting always in one specific place and going around the table clockwise.

A good system for placing orders in the kitchen is also needed. Historically, orders were given in written form. A circular rack was used so that orders came in proper rotation as the rack was turned. Now a computerized point-of-sale system is the norm, establishing a record that the order was placed. Whatever the method, it should be simple and workable. Service personnel should develop a system for firing orders so that they are ready on time but are not prepared so far in advance that they lose quality. A good communication system is needed to inform service personnel when their orders are ready. This may be done electronically or by using back waiters or runners.

A method needs to be worked out for keeping hot foods hot and cold foods cold after they are prepared and while they are awaiting pickup by service personnel. Infrared lights over a table on which hot foods are placed may be sufficient for hot foods, though items should spend as little time under lamps as possible. A refrigerated shelf may be used for cold foods. The method should include storage areas where dishes can be properly warmed or chilled as appropriate to the food item.

Some method for checking foods as service personnel pick them up can be devised, so that the right foods are taken and foods for each course are presented in order. An expediter can serve these functions.

Tableware must be handled properly. Glassware should be handled so that the fingers do not touch the tops or insides. Silverware should be held by the handles. A dish should be held with four fingers under it and the thumb steadying it at the edge. The use of trays is more sanitary and professional than balancing dishes. Alternatively, carts may be used, although often the space required is too dear.

Serving should proceed from a selected spot at the table. It is traditional to serve women first, but wherever serving starts, service goes to the server's left in a clockwise fashion. As the server moves left from the next guest to be served, he or she turns to the right, using the right hand to put down the main dishes and items from the guest's right side. With the left hand, the server puts down items to be served at the guest's left, such as a salad. While this procedure is common, there is no reason why the service *has* to move in this fashion, as long as *good service occurs.*

It is proper, after guests have been served and sampled their food, to ask whether everything is satisfactory. Prompt attention should be given to complaints. It is not necessary to interrupt. Observation and eye contact with guests often suffice to perform what is sometimes referred to as *check-back*. Servers should never argue with patrons. If necessary, supervisory personnel should be called to the table to correct any problems or smooth over difficulties. It is also wise after service has ended to have some means for addressing problems that have arisen in the dining area during a shift. A problem can be reviewed and discussed, and perhaps a recurrence can be avoided. It is a grave error to allow the same mistakes to occur over and over again without making an attempt to correct them.

# Types of Service

The type of service an operation uses depends on the type of establishment and its clientele. Putting good service into effect is a matter of training staff and having the right combination of equipment and atmosphere to make it work. At minimum, a standard for a particular type of service should be achieved so that patrons will be served properly and consistently.

## Counter Service

Counter service is fast and usually has a high rate of customer turnover, since patrons can be seated, can order, and can eat in a short period of time.

Contrary to some thinking, counter service does not necessarily save space. The space needed behind the counter, the service area, counter and seats, plus aisle space for traffic, is rather large. In some cases, seated service requires less space.

A server may have from 8 to 20 seats to serve at the counter. Everything possible should be within easy reach so that the server has little need to go to other parts of the operation and leave the guests. The counter should also be close to the production area to permit orders to be placed, picked up, and served quickly. Work simplification is desirable, and work center arrangements should be studied. Using a communication system that gets orders into the kitchen without service personnel having to take them there speeds service and allows servers more time to help guests.

In some counter-service operations, places may be preset for guests: a placemat, knife, and spoon to the guest's right, and forks and a napkin to the left. Only water and a menu need to be handed to guests when they sit down. Menus and napkins may also be available on the counter for self-service, or the placemat may double as a menu. Water should be placed to the guest's right. If other silverware is needed, it is put in proper position after the guest has ordered.

Good counter service is not easy; it is fast-paced work. The individual doing it must have exceptional organization skills and work quickly. While their minds are constantly on orders and details of service, service persons must also manage to be pleasant and make guests feel welcome. Good teamwork is required with other counter servers and with production personnel. When not serving, it is usually necessary to keep busy cleaning up behind the counter, replenishing supplies, and otherwise organizing the station. Salt and pepper shakers, sugar bowls, and napkin dispensers always need filling; failure to perform these tasks during down periods can mean that service goes slowly at peak times. Getting behind can make counter work hard and frustrating. Learning how to keep ahead is essential.

After a guest has been given a check—with a *thank you*—and has paid and left, the soiled dishes should be removed quickly. Some sorting of items is usually desirable. Counters should be arranged so that containers used for dishware are handy and workers do not have to walk long distances with used tableware. Buspersons should remove bus trays and tubs filled with dirty dishware and replace them with new ones.

## Cafeteria Service

In cafeteria service, guests go to a counter and select foods that are plated up by service personnel and then take these foods to a table. There may be modified types, such as in some cafeteria operations where service personnel stand at the end of the line, pick up the tray, and take it to a table for the guest. Water, silverware, and beverages may also be served by service personnel. Good cafeteria service should put about six people per minute through the cafeteria line. A cafeteria line over 50 feet long is usually not as efficient as a shorter one. Cafeteria service is often associated with institutional foodservice or senior citizen markets.

A *scramble* cafeteria system is one in which guests go to specific locations to get certain types of foods. For instance, beverages may be obtainable at an island in the center, salads and entrees at one counter, and desserts at another counter. Hot entrees may be obtainable at a central location. None of these locations is connected to another; guests must pick up their trays and move to the next location without sliding trays along in a line. This prevents the formation of long lines and allows the individual who wants only a few items to get them quickly and leave the service area. Thus, patrons do not wait in line while someone else wanting more items takes time to make the selections. Patrons pay as they leave the service area before proceeding to a table. Some cafeteria lines have a checker who adds up the cost of the food on the tray and hands the guest a slip with the total on it. The guest then moves to the cashier and pays the check. In some cases, guests may pay at the door after finishing a meal. Plastic cards are swiped at each station so that an accurate total is paid prior to the guest departing. The scramble system has been used very successfully in college residential feeding.

Some cafeteria lines can move more quickly by having alternative lines where only certain foods are obtainable. Thus, some cafeterias have cold-food or snack-food lines that relieve the hot-food line of patrons who want only a limited amount of food.

Self-bussing of trays may be used. If this is done, the station where guests carry their used dishes should be located on the way out. If this is not done, most guests will walk out and leave their dirty dishes on the table. Many cafeterias find it more efficient to have floor personnel who bus dishes; clean tables; replenish salt, pepper, sugar, and condiments; see that napkins and glasses are available; and do other general dining area work. It is important that they work diligently and establish good procedures for clearing tables.

## Buffet Service

Buffet service has the advantage of reducing service personnel, but the disadvantage of wasting food. In times of high food cost, the cost of the food may outweigh the advantage of a lower labor cost, with a result that final costs are higher. Normally, a buffet is served at a set price with guests taking whatever and as much as they want. Sometimes food portions may be controlled, as when service personnel serve guests or a service person carves the meat. However, second helpings are usually allowed and are often a good part of the appeal to guests. Methods such as using scramble stations, using 9-inch plates rather than 12-inch plates, and serving filling, appealing bread or starch products help with costs.

Buffets make it possible to display and merchandise food attractively. They are being used more and more to give simple, fast service for continental or regular breakfasts, light lunches, or for meals where service must be fast and guests wish to come and go quickly. They are also good for handling large numbers of people who may wish to eat at different times. When seated service cannot handle the numbers to be served, a buffet may work best. For instance, the Hawaiian Regent Hotel of Honolulu found that with large tours in the hotel, the seated-service restaurants could not handle the demand. For a set price to tour groups, guests were invited to go to the banquet room of the hotel, in which a buffet was set up. This enabled the hotel to accommodate them adequately with good guest satisfaction. Likewise, special dinners, receptions, luncheons, and other special occasions lend themselves to buffet service.

A *smorgasbord* is a Swedish buffet that includes a large assortment of cold foods after which hot foods and then a dessert and beverage are offered. A true smorgasbord must include pickled herring, rye bread, and Swedish *mysost* or *gjetost* cheese. Similarly, a Russian buffet must include caviar in a beautiful glass bowl (or in a carved-ice bowl), rye bread, and sweet butter.

Some health food markets and delis have buffets where guests plates or to-go containers are weighed and a per-ounce price is charged.

It is possible for a buffet to be combined with other kinds of service. For instance, guests may pick up only cold foods and order the remainder of the meal. Salad bars often include soup and are included with entrees or added on. The price is highest for the guest who is not ordering additional food. Many salad bars also include dessert items. Or they may eat the cold and hot foods and then be served a dessert and beverage, or just a beverage. Water is usually poured by service personnel.

When buffet service is used, dishware, silverware, and a napkin are usually preset on the table for each cover. Plates and other dishes needed for the buffet are set at the front of the line where guests can pick them up. If silverware, napkins, and water are to be picked up by guests, they are usually placed at the end of the line or in a separate area. Desserts may be set on another table. Using separate tables can speed service. It is also faster to have personnel behind the buffet to serve guests than to have guests serve themselves. It is best to have meat carved at a separate station. Similarly, waffle, omelet, and pasta stations may be separate.

Foods served at buffets should be colorful and attractively arranged. It is aesthetically desirable to have different shapes and sizes of dishes. *Réchauds, bains-marie,* and other holding units can be used to keep foods hot. The heights of foods and dishes should be varied to give an interesting pattern. Buffet tables can be arrayed with candles, flowers, or more elaborate decorative pieces to give an attractive display. Health departments regulate the temperature and manner of displayer items.

The buffet table should be neatly spread and have a cover on the front that completely hides the underside and table legs from view. Table linen or snap-drapes in various colors can be used to achieve an interesting effect. The table shape can vary, but consideration should be given to flow, since some shapes will not speed service and may even hinder it.

A good system for replenishing the buffet should be developed. Some buffets have service from both sides, and these are difficult to replenish. The table should be as close as possible to the food preparation point, but this convenience factor will be influenced by service considerations. Some operations may have a rule that when a dish is about two-thirds empty, it will be removed from the table and another, full one put in its place. The one removed is returned to the preparation area to be cleaned and refilled later. Vessels containing food should be about two to three inches from the table edge. It is important to not have one food in back of another—making guests reach over the food in front—unless the food in front is on tiers so that it is easy to reach. Sanitation should be observed to assure food safety.

Food served at a buffet should be of a kind that lends itself to self-service. Items that require a thin sauce or are not easy to handle should be avoided. It is usual to put both cold and less impressive foods first in line. However, the proper sequence may be dictated by the type of food and when it should be eaten during the meal.

## TABLE SERVICE

The foodservice operation that serves its guests at their tables in the familiar way is classified as seated-service establishment. This is the basic, traditional method of serving food and drink. The term is used to distinguish it from other service methods, such as drive-through, carry-out, or delivery.

There are three distinct types of table service used in this country—American, Russian, and French—and English service is seen occasionally.

### *American Service*

The simplest and least expensive type of seated service is American service. It is also fast and does not require a great amount of labor. The table setting for this service places the knife (blade side in) at the right, the soup spoon next, and the tea spoon next on the outside. Normally settings are made so that the first utensil on the outside, on either the left or right, is used first, then the next utensil to it, moving in as the courses progress. Placing the soup spoon inward from the teaspoon violates this rule, but it is a more common way of positioning the soup spoon. However, it is not improper to put the soup spoon on the outside, where it is used first with the first course, the teaspoon next, and the knife on the inside. The dinner fork and then the salad fork are placed on the left. In the center, a service or hors-d'oeuvre plate is set. The water glass is set at the tip of the knife and about an inch away. The coffee cup and saucer may be placed on the table next to the teaspoon to speed service. The ashtray, in restaurants that have smoking sections, and salt and pepper should be in the center. (See **Exhibit 11.1.**)

If a wine glass is used, it is placed to the right of the water glass. The bread and butter plate should be above the forks and perhaps a bit to the left of them. The butter knife should be at a right angle to the forks on the butter plate with the blade turned toward the forks. For a normal meal the table may be covered first with a *silence cloth* and then a table cloth. However placements alone may be used for an informal meal. In recent years, some establishments have taped fresh white butcher paper to the table for each new party. Silverware and dishes should not be any closer than a half-inch from the edge of the table. Chairs should be out from the table, away from the tablecloth.

In American service, food is dished onto plates in the kitchen. The server takes plates to the dining area. Coffee is often served with the meal. The coffee cup and saucer may also be on the table to speed service.

**Exhibit 11.1**
American Service Setting

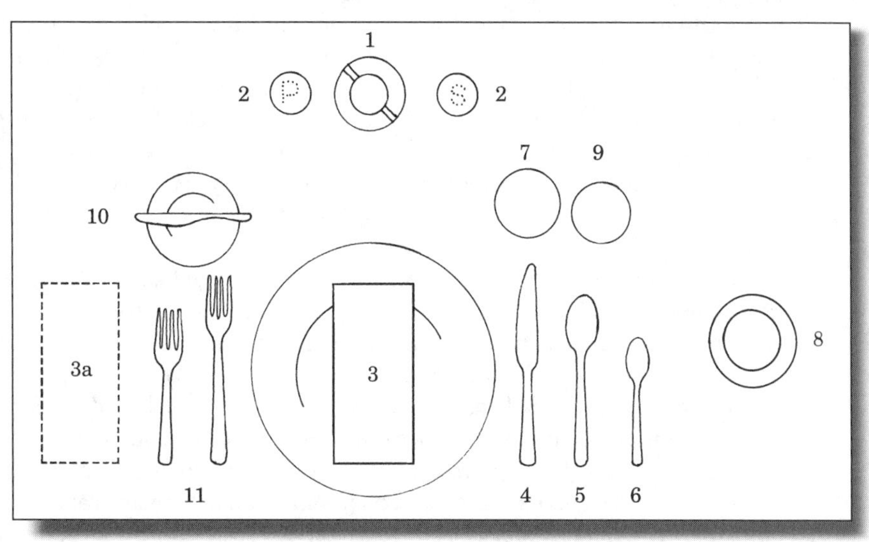

American service, or a modified version of it, is used both in operations requiring fast turnover, such as coffee shops, and in fine-dining restaurants. When it is used in a fine-dining operation, the coffee cup and saucer are not on the table, since hot beverages generally are served later in the meal. Procedures vary with different operations.

It is common to serve foods from the left and beverages from the right, moving around the table from right to left. However, much variation occurs. All clearing may be done from the right using the right hand. The general rule in all service is that when serving at the guest's right, use the right hand; when serving at the guest's left, use the left hand. Whatever method is used, it should be based on what is easiest to do most efficiently and quickly for the guest. It is not proper to remove dishes or to start a new course in American service, or in any other type of service, until everyone at the table has finished eating the present course. The server must take cues from the guest, however, and clear if patrons are plainly uncomfortable with dirty dishes set before them. Some guests who have not finished their food do not intend to, or prefer to save the remaining food to accompany the next course. When in doubt, the server need only ask politely.

Soiled dishes are removed from the table as follows; items to the left, from the left-hand side using the left hand; central items and items to the right, from the right-hand side using the right hand.

American service is frequently used at banquets because a large number of guests can be handled quickly by a limited number of servers. The banquet head server is responsible for all dining room service. This person directs set-up of tables, arrangements for the head table, setting of silverware, decorations, and, in general, all service. If wine is served, the head server will plan its service. On the average, between 15 and 25 guests will be assigned per server. Buspersons may assist.

### *French Service*

The most elegant service is French service, but it is also slower and more expensive. At one time it was even more elaborate than it is today. It consisted of three or four table settings, with each setting accommodating a number of courses. Guests would sit down and eat all the courses for a given setting. Then they would leave the table while it was being cleaned and reset. They would then sit down again and eat another series of courses. As many as 48 or more dishes might be served at one meal. The goal was for guests to taste many foods rather than eat large quantities. To a large extent, the purpose of all the food was to make for a lavish display. French service reached its peak in the court of Louis XIV. At that time it was called the grand cover (*le grand couvert*). A good part of the service was the ceremony that accompanied each setting and the serving of the courses. The French service of today is descended from this service but has been considerably simplified.

Modern French service is performed from a cart called a *guéridon,* which has a *réchaud* or heating unit on it. The réchaud is usually heated by an alcohol lamp, but some guéridons may be equipped with a small bottle of butane or propane, which fires a gas burner. Food is brought raw, or partially finished, to the cart, where it is prepared next to the table. Meats, poultry, and fish may be cooked in the kitchen, but they will be carved or deboned on the cart. For some food, preparation must occur in the kitchen, and only service is done from the cart. Salads, desserts, and other foods may be prepared completely at the guéridon from raw foods.

Employees who perform French service will be more numerous and more skilled than in any other service. A dining room using a French service has a *maitre d'hôtel* in charge. In Europe, this person exercises much more authority than in the United States. Maitres d'hôtel are in charge of making reservations and assigning tables, as well as determining service procedures. They may greet guests, and some may even take guests to the table and give them menus. In some operations, the menu may be presented by a captain, waiter, or *chef du rang.* Captains usually supervise about four servers. The *chef du rang* has charge of a table and is assisted by an apprentice, called the *commis du rang.* The dress is usually formal, with servers wearing white gloves. Large napkins, called *serviettes,* are carried on the arm (not under it). Apprentices wear white server jackets with a white shirt, black bow tie, and dark pants.

The chef du rang usually takes the orders and gives them to the commis, who takes the orders to the kitchen. The commis gives the orders to the *aboyeur* (announcer) who, in a loud, clear voice, calls them out. (The aboyeur is said to have been introduced by Escoffier, who wanted to reduce loud talking in the kitchen between cooks and servers.) When the foods are ready, the commis takes them to the chef du rang, who takes them to the guéridon. The commis must also bring plates, dishes, and other items required for service.

As the chef du rang prepares the food on the réchaud or in other equipment and then dishes it onto the service dishes for guests, the commis serves the guests. Service is from the right, with the right hand, except for items to be placed on the left. (If the server is left-handed, the service may differ.) Removal of plates and dishes is usually from the right with the right hand, but this too can vary. Plates and items should not be removed until all individuals have finished eating. Second servings are not given in French service.

If guests want alcoholic beverages before the meal, the chef du rang will take the orders and serve them. However, today some dining rooms prefer a cocktail server to do this. A wine steward, or *sommelier,* should bring the wine list, make recommendations, take the order, and then later serve the wine. Glasses should be on the table so that the first wine to be consumed is poured into the glass on the right. The progression is inward, as it is with the silverware. However, some variation in wine glass placement may occur. It is not proper to have wine glasses inverted on the table when guests come in, since this indicates that the room is not ready for them.

All the silverware needed may not be on the table, but will be put down as required. It is not considered proper to have more than three or four pieces of silver on either side at one time. A *service* or *show plate* is usually in the center when the guest is seated, and an hors-d'oeuvre plate may be on top of this, if required. The napkin can be put on the hors-d'oeuvre plate with the fold to the left, making it possible to pick it up with the right hand so that it opens as it is lifted. It is proper for the server to pick the napkin up, open it, and hand it to the guest. (See **Exhibit 11.2.**)

The dinner fork and other forks should be to the left of the service plate. The cocktail fork can be on the plate on which the item is served, or placed in a position on the service plate so that the diner's right hand can pick it up. The dinner knife is to the right of the service plate with the cutting edge toward the plate. The soup spoon is usually to the right of the knife. The dessert fork and spoon are placed lengthwise across the top of the service plate. The handles should point to the side of the plate where that utensil belongs. (Note dotted lines on **Exhibit 11.2.**) If a bread and butter plate is used, it is above and to the left of the dinner fork, with the butter knife at a right angle to the dinner fork and the blade facing the forks.

The number of courses in a French meal is limited today. A dinner can consist of a soup, main course, salad, and dessert, served in that order. The salad is served as a separate course after the main course. If there are more courses, an hors d'oeuvre may start the meal, a fish course may then follow, with a poultry course next. The main course should then be served. The salad may be followed by a cheese course. Coffee in a demitasse is proper after the meal.

**Exhibit 11.2**
French Service Setting

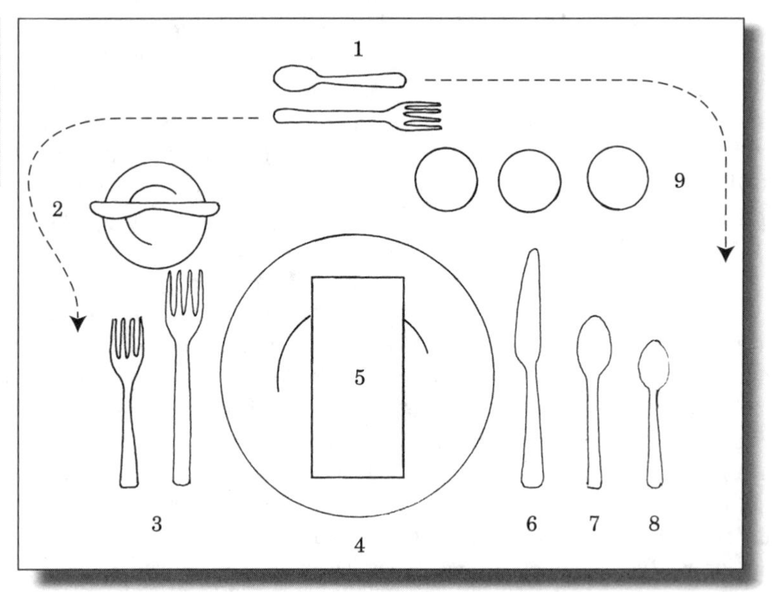


The service plate is left on the table until the main course is served. It is replaced when the salad course is served and remains until the end of the meal. Soup is brought from the kitchen in a tureen or other container, placed on the réchaud, ladled from there into cups or bowls, and served to the guests.

Different foods require different eating utensils, dishes, and methods of service. The knowledge of what is correct and proper takes considerable study and training. The skill required of the chef du rang is considerable and must be coupled with showmanship.

Fingerbowls are proper after each course, at the end of the meal, and at any other time required. Fresh napkins also may be given at any time during the meal. In very formal French service, rolls and butter are not served, and salt and pepper are not on the table. Water is not served—only wine—and, at the most, only three wine glasses are in place at a setting. More are placed on the table as needed. Ashtrays are not on the table, since guests are expected not to smoke until the meal is completed. However, if guests request them, they are brought to the table. It is never proper to instruct the guest on etiquette. The customers' desires take place over convention unless regulations prohibit guests from wish fulfillment, be it smoking or drinking more than is legally allowed, for example.

The chef du rang usually presents the check and collects the money. The commis clears. In some large dining rooms, the person who comes around with the cart of attractive pastries or other foods may be called the *chef du trancheur. Curry persons* are those who serve condiments and other food accompaniments.

French service usually requires more equipment and space than American or Russian service. French service is designed for the operation that specializes in emphasizing the finest dining, decor, service, food, conversation, and wine. It connotes leisurely dining. French service has many traditions, but modern practice may vary these. Purity for purity's sake is not necessarily the most desirable procedure to bring about good service; some license may be permitted, providing it leads to better service. However, tradition and some of the finer points of French service ought to be preserved because they help maintain those aspects that give much grace and elegance to dining.

## *Russian Service*

Russian service has great elegance and showmanship. It is also efficient and relatively fast. It requires less labor and skill than French and is also suitable for elaborate banquets. A version of Russian service is sometimes found on cruise lines.

The table setting for Russian service follows what is common for French service, but a water glass may be on the table just above the tip of the knife, and an ashtray may be on the table though a smoke-free atmosphere is often dictated by guest preferences or legislation. A bread and butter plate and butter knife are also used. Since Russian service is considered slightly less formal than French, the rules for setting the cover are not quite as strict.

In Russian service, the plates for the course to follow are put down in their proper places before the guests. If the food is to be served hot, the plates should be hot; cold foods should be served on chilled plates. Sometimes, salads are brought to the table already dished onto cold plates. To put the plates down for a course, a server uses the right hand to place the item on the right side of the guest. The movement is then to the next guest on the left, and proceeds clockwise. When all plates are in position, the serving dish is picked up by the waiter and held in the palm of the left hand, or by the left arm and palm for a tray or large platter. The right hand serves the food from the serving dish to the guest's plate at the right. Soup may be ladled into soup dishes from a tureen or it may be brought to the table in small, individual serving dishes on a tray. Servers take these, one by one, and pour the hot soup from them into the guest's soup bowl or cup.

Considerable dexterity is required to perform good Russian service. The silver tray on which the portions of food have been dished can be heavy and very hot. Some servers wrap a towel lightly around their left arm before putting on the jacket to protect the arm from being burned by the hot dish. The serving dish must be held securely and balanced, while the right hand manipulates a large spoon and fork to grasp the portion and move it without spilling it. The right hand holds the serving spoon with its bowl facing up. Directly over this, with the tongs up, is the fork. The spoon is used to scoop up the item while the fork, with some pressure on the top, holds the item on the spoon as it is being transferred to the guest's plate. The spoon can be used to pick up a bit of sauce from the platter and pour it over the plate. The spoon is also used to serve some vegetables and sauces.

Coffee is served after the meal. Coffee cups, saucers, cream, and sugar are not on the guests' table; coffee may be poured from a buffet and served.

## *English Service*

English service, sometimes called formal family service, is most often used in family-run inns and on special occasions in other foodservices. Foods are brought from the kitchen on platters and in serving dishes. The host, who remains at the table during the meal, carves the meat while the hostess serves the vegetables, salad, dessert, and beverage. The host places the meat portion on a hot plate and then passes it to the hostess, who puts on the other foods and then passes it to a guest. The host is served last, or next to last before the hostess. The meat should be in front of the host, with plates used for service immediately in front of him. It is desirable to work out a passing pattern that requires the least handling of plates. Sometimes servers may take plates and carry them to the guests instead of having them passed. It is proper to have the first course placed on the table when guests come into the dining room. Water can be poured and butter placed on the butter plates. Coffee cups and saucers will not be placed on the table but brought at the conclusion of the meal or when dessert is served. Small tables may be placed to the right and left of the host and hostess where service dishes can be placed when the service is ended. However, servers may remove these.

Place settings may be similar to those used in American service, but they can vary. Normally, knives and spoons are to the right and forks are to the left. The order of placement is from the outside in as the courses occur. Wine glasses are placed to the right of the water glass, which is placed just above the tip of the knife.

### *Organization for Table Service*

The organization of the dining area may vary considerably with different operations. Some foodservices may have no head of services but will have the task handled by the assistant manager. Some may use a host or hostess who, in addition to supervising the dining room, greets and seats guests. Sometimes the individual in charge is a head server. In the most formal establishments, a maitre d'hôtel performs this task. Whatever form of organization is established should be effective and well managed. The number of hours usually clocked in a foodservice is substantial, and unless the operation is managed properly, costs can be unnecessarily high. Usually, the ratio of hours used for service, or front-of-the-house, to the hours used for food production, or back-of-the-house, is 10 to 7.

The individual in charge of service should see that service proceeds properly, servers are neat and clean, and they follow established policies. This person should handle any personnel problems and, perhaps, may be responsible for hiring and firing, although management may reserve this right. This person should schedule servers and assign work stations and days off. Work assignments should be made on an impartial and balanced basis. Schedules should be posted sufficiently in advance so that members of the service staff can make personal plans. A rotation system may be used in allocating days off and the stations at which servers will work. In this manner, the same day off will not always fall to the same person and, from time to time, the days off will fall on a weekend. Of course, this can be varied if personnel wish to have the same days off regularly. This is often desired by students, parents, and other care givers who must juggle a variety of considerations. Any system devised should be fair to all. Some stations in the dining area are better than others because they are easier to work or provide more tips. Unless there is a policy that establishes stations based on seniority or clearly defined merit, the work stations should be rotated.

A policy should be established for taking breaks and eating meals. Often, meals are taken before a shift begins in conjunction with a crew meeting. Communication about service issues, specials, VIP guests, or parties are shared. Other methods include rotating employees during down times or at the end of a shift. A policy should exist on what can be eaten by employees for free or for minimal payment. Some type of records must be maintained on employee meals so that costs can be calculated.

## Room Service

Room service may represent an important source of revenue for the food and beverage departments of hotels, as well as a significant factor in the satisfaction of guests. Special organization may be required to give good room service. Some hotels even have special room-service production kitchens.

Considerable space must be available so that small mobile tables can be stored and ready to be moved with the items ordered to guests' rooms. Mise en place is extremely important since distances are usually great, and hot foods can quickly cool down and cool foods can warm up. A special service elevator is desirable if the amount of room service is large.

Most hotels and motels with room service have a special telephone number for guests to call for room service. The order is taken and transferred to the proper personnel so beverages and foods can be prepared. Timing is important, since items must be ready at the same time. Service personnel must be trained to assemble all of the tableware required, as well as butter, bread, and other items.

Many operations follow the practice of having the check time-stamped when the order is received, when it leaves for the guest's room, and when the signed check is brought back.

**Exhibit 11.3** represents a room-service menu for breakfast. The illustrated menu offers a full range of products. Other properties greatly abbreviate the choices and request that guests leave their doorknob order form hanging outside of their rooms by a specified hour.

## Delivery, Drive-through, and Self-Service

More Americans are working outside of the home than in the past. In addition, fewer people know how to cook or are inclined to cook on a daily basis. People are eating out more frequently at many different types of establishments. Delivery serves the purpose of allowing the meal to be brought to the home, hotel/motel room, or office of the diner, who may be weary from work or travel or simply does not wish to bother cooking or going out. Pizzas have been at the forefront of delivered foods, but Chinese food, steaks prepared to doneness, and even fine-dining options all may be available for delivery. Often, a single service will pick up food from any of a number of cooperating fine-dining establishments that do not employ a full-time delivery person. Foodservice operations that are near or in office buildings will often fax or e-mail menus to businesses and deliver lunches directly to customers' desks or workstations.

Drive-through options are generally associated with the quick-service industry. This service is very popular and often provides the bulk of business for quick-serve restaurant (QSR) operations. Some have opted to close their lobby dining areas during late hours, and a few exist exclusively with drive-through business. Many fine-dining and family operations are countering with the curbside option, whereby diners may pick up regular sized or family-portioned dishes at the host stand or special door, with some having parking spaces designated for this purpose. Quick-service restaurants have devoted much effort to studying the most efficient systems to move patrons rapidly through the process and on their way.

Whether the food is picked up or delivered, it is critical that the order be accurate, as guests often do not discover an error until there is distance between themselves and the establishment or delivery person. Rather than complain, many customers will simply not return. They will instead try one of the myriad of competing options available. Suitability of packaging and user friendliness of plasticware and condiments are also important factors to consider. Condiments, napkins, and straws supplied should be of sufficient quantity to satisfy the customer, but only provided if desired so as not to create needless cost for the establishment. Drivers wish to be able to eat and drive and do not wish to have to make undue effort or create a mess in the process. Similarly,

**Exhibit 11.3**
Room-Service Menu

# BREAKFAST

Served from 6:00 a.m. until 11:00 a.m.

## NATURALLY HEALTHY

**GRAPEFRUIT AND ORANGE SEGMENTS** 6.50

**FRESH SEASONAL BERRIES** 8.00

**PLAIN YOGURT** 6.00

**FRESH SLICED SEASONAL FRUIT PLATE** 17.50

**GRANOLA YOGURT PARFAIT** 11.00
*Sun-dried cranberries, shredded coconut, low fat yogurt and berries*

**MANGO TANGO BREAKFAST SHAKE** 10.00
*Blended fresh pineapple, orange, mango and low fat yogurt*
*Fat 1g, Saturated Fat .5g, Cholesterol 5mg, Carbohydrates 48g, Protein 5g, Calories 210*

## HOT AND COLD CEREALS

**HOT OATMEAL** 7.00
*Served with brown sugar, raisins, milk or cream.*

**DRY CEREALS** 6.50
*Rice Krispies, Corn Flakes, Special K, Raisin Bran, Cheerios, Fruit Loop or Granola*

## AMERICAN BREAKFAST

*Fresh eggs prepared any style with choice of crisp bacon, sausage or chicken-apple sausage, breakfast potatoes, toast or English muffin, chilled juice and coffee, tea, decaffeinated coffee or milk* 24.00

## THE CONTINENTAL

*Choice of chilled juice, coffee, tea, decaffeinated coffee or milk. Choice of three miniature pastries: croissants, muffins, Danish, pecan sticky bun, apple-cinnamon strudel or buttermilk scones* 16.00

## FROM THE BAKERY

*Served with butter and preserves*

**TOAST OR ENGLISH MUFFIN** 5.00

**DANISH PASTRY, MUFFIN OR CROISSANT** 5.50

**TOASTED BAGEL** *includes cream cheese* 7.50

- Chef's signature item

- Hilton Eat Right nutritional values are determined through database analysis with the Food Processor SQL from ESHA Research, Inc., Salem, OR 97302 and available ingredient product data. This data is based on average serving size and standard portion guidelines. However, slight variations in nutritional values may occur due to seasonality use of alternate suppliers and menu item preparation. Cholesterol is indicated in milligrams

A $4.00 Delivery Fee. 15% Service Charge and Applicable Taxes will be Added to Your Account.

Jan-07 2

*Courtesy of Palmer House® Hilton*

**Exhibit 11.3**
*Continued*

## OFF THE GRIDDLE

*All items from the grill are served with Vermont maple syrup and butter*

**CRÈME BRULEE FRENCH TOAST** 14.00
*Vanilla cream dipped Brioche, topped with powered sugar and berries. Served with bacon or sausage*

**BELGIAN WAFFLE** 15.00
*Plain, blueberry or orange waffle, with Grand Marnier vanilla sweet mascarpone. Served with bacon or sausage*

**BUTTERMILK PANCAKES** 15.00
*Plain, blueberry, banana or orange. Served with bacon or sausage*

## BREAKFAST SPECIALTIES

*Served with breakfast potatoes, toast, butter and preserves*
*(Egg beaters or egg whites may be substituted for fresh eggs)*

**EGGS BENEDICT** 16.00
*Canadian bacon*

**TWO EGGS ANY STYLE** 15.00
*Choice crisp bacon, ham, sausage, Canadian bacon or chicken-apple sausage*

**FLUFFY THREE EGG OMELETTE** 17.00
*With choice of three of ingredients: tomatoes, peppers, green onion, mushrooms, ham, sausage, bacon, spinach, salsa, jalapeno pepper, Swiss Cheese, American Cheese or Cheddar Cheese*
*Add proscuitto, asparagus, crab, smoked salmon* 1.50

**SMOKED SALMON** 20.00
*Cream cheese, sliced tomato, red onion, hard boiled egg and capers, with your choice of a toasted bagel or toast points*

**GRILLED 6oz. PRIME ANGUS FILET MIGNON** 38.00
*with two eggs any style*

## SIDE ORDERS

**BANANAS, STRAWBERRIES OR BLUEBERRIES** 5.00

**ONE EGG ANY STYLE** 5.50

**BREAKFAST POTATOES** 5.50

**HAM, SMOKED BACON or MAPLE SAUSAGE** 5.50

**CANADIAN BACON** 6.00

**CHICKEN APPLE SAUSAGE** 6.00

- Chef's signature item

- Hilton Eat Right nutritional values are determined through database analysis with the Food Processor SQL from ESHA Research, Inc., Salem, OR 97302 and available ingredient product data. This data is based on average serving size and standard portion guidelines. However, slight variations in nutritional values may occur due to seasonality use of alternate suppliers and menu item preparation. Cholesterol is indicated in milligrams

A $4.00 Delivery Fee. 15% Service Charge and Applicable Taxes will be Added to Your Account.

Jan-07 3

office workers or lone diners in a motel room need to have everything necessary at hand to enjoy their meal and tidy themselves afterward. Containers that are usable for storage of leftovers may be preferred by some in a position to save what remains.

Self-service is another way of reducing both labor cost for the establishment and product cost for the consumer. Some consumers are willing and may prefer to get their own ice, beverage, and refills, pick up their food from the counter when ready, or bus their own table when finished. The degree to which self-service is acceptable to the guest is something that must be carefully studied. Guests must perceive that their efforts translate into greater value, not merely greater profit for the establishment. Certainly one motivation for dining out may be to be waited on and self-service will not appeal to everyone.

# Handling the Guest Check

After guests have finished their food and after-meal drinks, the server should bring them their check. If the server knows who is to get the check, it can be placed upside down to that person's right. If not, it should be placed in the middle of the table. If there are to be separate checks, this is preferably known when first taking orders. Today's point-of-sale (POS) systems make calculating separate or split checks simple. The check should be totaled and placed on a tray. The server should always say thank you. It is proper for guests to check the accuracy of the check. Prompt attention should be given to questions about charges. The menu prices should be indicated in such a manner that guests do not receive an unpleasant surprise when the bill arrives.

**Exhibit 11.4**
Full-Service Sales Check

*Carlisle's*

| Date 4-13 | Server LPG | Table No. 11 | No. Persons 4 | No. 2370 |
|---|---|---|---|---|
| 1 | | | | |
| 2 | | | | |
| 3 | 1 mush | | | 7.00 |
| 4 | 1 shr | | | 9.00 |
| 5 | | | | |
| 6 | 1—SALM rice bl. ch | | | 22.00 |
| 7 | 2—SNAP B-BSc bl. ch | | | 24.00 |
| 8 | 3—PASTA veg v.o. | | | 19.00 |
| 9 | 4—LAS fr | | | 21.00 |
| 10 | | | | |
| 11 | | | | |
| 12 | | | | |
| 13 | | | SUBTOTAL | 102.00 |
| 14 | | | TAX | 7.14 |
| 15 | | | TOTAL | 109.14 |
| | | | | No. 2370 |

| Date 4-13 | GUEST RECEIPT | Persons 4 | AMOUNT OF CHECK | 109.14 |
|---|---|---|---|---|

*Carlisle's*

Some system for maintaining a record of service checks should be established. Historically, each service person, at the start of a shift, was assigned a set of consecutively numbered checks. Later, the accounting department verified the count of checks turned in by guests to cashiers. Missing checks were investigated. Some operations charged servers a fixed sum for each missing check. Today, the POS systems have mostly made these methods obsolete, although small operations may still employ these techniques. Rarely will guests walk out without paying, but this does occur, and it may be necessary to make allowances for it. Some operations say that to anticipate walkouts numbering one-tenth of a percent of sales is not unrealistic.

**Exhibit 11.5**
Quick-Service Sales Check

**Frank-n-Burger**

Go Stay

| No. of Items | 1 | 2 | 3 | 4 | 5 | W | WO |
|---|---|---|---|---|---|---|---|
| Hamburger | .70 | 1.40 | 2.10 | 2.80 | 3.50 | | |
| 1/4 lb. Burger | .80 | 1.60 | 2.40 | 3.20 | 4.00 | | |
| Cheeseburger | .80 | 1.60 | 2.40 | 3.20 | 4.00 | | |
| Dbl. Chsburger | 1.40 | 2.80 | 4.20 | 5.60 | 7.00 | | |
| Combo Meal | 2.10 | 4.20 | 6.30 | 8.40 | 9.50 | | |
| Salad | .90 | 1.80 | 2.70 | 3.60 | 4.50 | | |
| Fries | .60 | 1.20 | 1.80 | 2.40 | 3.00 | | |
| Lg. Order Fries | .80 | 1.60 | 2.40 | 3.20 | 4.00 | | |
| | | | | | | | |
| Cola-Lg. | .80 | 1.60 | 2.40 | 3.20 | 4.00 | | |
| Cola-Reg. | .70 | 1.40 | 2.10 | 2.80 | 3.50 | | |
| Cola-Sm. | .60 | 1.20 | 1.80 | 2.40 | 3.00 | | |
| Orange-Lg. | .80 | 1.60 | 2.40 | 3.20 | 4.00 | | |
| Orange-Reg. | .70 | 1.40 | 2.10 | 2.80 | 3.50 | | |
| Orange-Sm. | .60 | 1.20 | 1.80 | 2.40 | 3.00 | | |
| Van. Shake | .80 | 1.60 | 2.40 | 3.20 | 4.00 | | |
| Choc. Shake | .80 | 1.60 | 2.40 | 3.20 | 4.00 | | |
| Straw. Shake | .80 | 1.60 | 2.40 | 3.20 | 4.00 | | |
| | | | | | | | |
| Coffee | .40 | .80 | 1.20 | 1.60 | 2.00 | | |
| Milk | .40 | .80 | 1.20 | 1.60 | 2.00 | | |
| Hot Choc. | .40 | .80 | 1.20 | 1.60 | 2.00 | | |

Subtotal ______
Tax ______
Total ______

In addition to its own serial number, a service check should have a space for identifying the server who used it and for showing the number of persons served, the date, and, perhaps, the table number. Only one item should be put on a line on the check. Some checks provide a ticket that the server can tear off and keep as a record. Many different kinds of checks are used. Each is designed to fit some special need of the operation. (See **Exhibit 11.4** and **11.5.**)

When point-of-sale (POS) electronic cash registers or computers are used, servers no longer have to go to bartenders or to the kitchen to place orders. These devices transmit orders by sending a coded message through to a printer at the bar or kitchen. Even more sophisticated is the use by the server of a small hand-carried POS register. The server can code in the number of the item ordered, and this will send out a signal that will activate the electronic register or main computer so it prints out the order and sends it to the bar or kitchen. Some handhelds allow the server to write in words using a stylus.

Some operations use expediters, who review orders coming to the kitchen to ensure that only the foods ordered on the check are taken out. Portions are checked by this individual to see that they are adequate but not too large. The appearance and garnishing of the food is also inspected at the same time. For instance, if fish is on the plate, the expediter will see that a slice of lemon goes with it.

There are many occasions for mishandling of checks, especially when service personnel collect the money. If a check goes out at 6:00 P.M. and is still out at 10:00 P.M., one might suspect that the check is being held by the service person as a *floater*—being reused for orders and presented to different guests. To avoid this and other abuses, an operation should establish control procedures. POS systems employ a management function key whereby the supervisor can view open checks.

Computers and electronic cash registers automatically time-stamp checks and price items as the order is placed. They make it difficult to use a check again, as a floater, since the machine reads the number and will reject a check the second time around, since the number has already been placed in the machine. (Collusion between the bar person and the server, however, can still make it possible to misdirect funds. For example, the bar person fills the order without putting the check through the machine, the server collects for it and doesn't bring in payment, and the two share the proceeds.)

In some cases, such as hotel room service, checks require special handling. Checks are usually issued to room service personnel, and when an order is placed, they are filled out and placed with the order. A record must be maintained of which room service checks are out and who has

**Exhibit 11.6**
Room Service Control Sheet

**ROOM SERVICE CONTROL SHEET**
**Coffee Shop**

| PERSON RECEIVING ORDER | ROOM NUMBER | NO. OF GUESTS | WAITER'S NAME | TIME ORDER PLACED | TIME ORDER LEAVES | TIME TRAY PICKED UP | CHECK NUMBER |
|---|---|---|---|---|---|---|---|
| | | | | | | | |

DAY OF WEEK ______ DATE ______ HOTEL ______

them. **Exhibit 11.6** shows one form historically used to keep track of room-service checks. (Chapter 9 describes other methods to control checks and to maintain records.)

When guests have charges put on their bills, as in a hotel or club, it is important that these charges be forwarded promptly to the accounting department. Some hotels give a guest a code number that identifies his or her account. When a charge is given to a cashier by a guest, the total of the charge, the department where the charge is being made, and the guest's code number are put into the cash register. Electronic registers or computers are connected with the accounting department, so charges are automatically put on the guest's bill as they are made. This prevents hotel guests from checking out before all charges are on their bills. Generally, only guests with a credit card number on file are allowed to charge. The guest invoice carries fine print stating that late charges, if made, may be posted to the credit card on file following check-out.

## Gratuities

Some guests indicate on a sales check that a gratuity is to be given to an employee. The amount of the *tip* is noted by the accounting department and added to the employee's wage check. However, some operations pay the employee the amount of these tips each day. For this, it is necessary to maintain a record for internal payroll and IRS tip-reporting purposes. Usually, this information includes the check number, gratuity, and total charge. If the employee is paid out on a per-shift basis, the employee getting the gratuities should sign for them to substantiate the paid-out tips reading on the register or to reconcile the cash. This is also proof that the service person received the tip. The sheet is turned in to the food and beverage department daily. Servers are taxed on their cash and charge gratuities.

Tipping varies from 10 to 20 percent of the total bill, depending on the style of service; 15 to 18 percent is most common. The term originated as an abbreviation of "to ensure promptness." It was historically given as a token of appreciation for service above the ordinary. There is no expectation for patrons to tip if the service is poor. In many states, server hourly wages have been held to just over $2.00 per hour. As this is far from a living wage, the guest is paying the bulk of the server income. . Some states do mandate an hourly rate as high as $6.00 or $7.00 for servers. Often establishments now add a *service charge* automatically for parties over a specified size or for occasions where the diner might not be aware of the custom. An example would be prom night or diners from locales where service is included in the check. Although diners should not be forced to tip for bad service or pay for food not up to standard, neither should the server be put in the position of being cheated out of wages earned through good service by a guest who simply is too tight or mean-spirited to pay for services rendered. Customers in other retail establishment do not specify the salaries of those providing services. It would be curious indeed to reduce payment to an attorney or physician who kept clients or patients waiting. It is common for servers to give about 15 percent of their tips received to buspersons, bartenders, or others who have assisted in service.

**SUMMARY**

Good service is crucial for a menu to fulfill its purpose. Good service requires many things: proper preparation, courteous and helpful communication, and timely and efficient delivery of product. Servers should follow established procedures for the kind of service being given. Major kinds of service are counter, cafeteria, buffet, and seated service. Seated service includes American, French, Russian, and English service. Each requires different table settings and number and tasks of service personnel. Dining rooms can be organized differently according to kind of service and need. Service should be concluded with a courteous presentation of the bill. Controls can be established to ensure that checks will be handled carefully.

**QUESTIONS**

1. What are the most important considerations involved in getting ready for service (*mise en place*)? In greeting guests?
2. What are examples of menus (and their markets) for which French service might be the most appropriate choice? Russian? English? American?

3. What are examples of buffet menus (and their markets) for which "all you can eat" would be most suitable? Plates weighed by the ounce? Scramble stations? What are some of the forms of service suitable to the various forms of buffet service?
4. What are the primary responsibilities of the person supervising the dining room?
5. Describe some common methods used for controlling collection of funds.

## MINI-CASE

*11.1* Using the mini-case illustration menus (Menus #1, #2, and #3 on pages 310, 311, & 312), describe the sort of service that would be best suited to the needs of the guests you envision as patronizing the operations. Describe the guests' characteristics and explain how your service system will best meet their needs.

## Menu #1

# Dig it Drive Through

## Start Me Up

**Groovy Granola Snack Mix**
Rolled oats baked with maple syrup, tossed with dried blueberries, cranberries, and assorted mixed nuts

**Cheese It**
Cubes of certified organic Feta, white cheddar and pepper jack cheese for munching

## Fruits and Veggies

**Fruit Cup**
Seasonal fruit drizzled with local honey and topped with a glop of vanilla yogurt

**Green machine**
Mixed field greens with edame, carrot curls and green onions served with a packet of dressing on the side. Choice of: Goddess, Vinaigrette or Tahini

## Big Chow Down

**All Chow Downs served with choice of Blue Corn Tortilla Chips or Dried Apple Crisps**

**Veggie Burger**
Our own patty of grains with portobello mushroom, black beans and peppers topped with greens and a slice of organic tomato

**Wild Caught Salmon Salad Sandwich**
On whole grain bread- toasted or not

**ABJ**
Our own almond butter with locally made fruit spread on whole grain bread

## Sweet Treats

**Carob Brownie Bliss**
Sweetened with stevia and packed with almonds- this concoction is pure heaven

**Frozen Tofutti**
Dairy free with a fresh flavor daily. Today is: Blueberry

## Good Hydration

**Mineral Water**
bottle of still or sparkling

**Fresh Squeezed Juices**
Your choice of combination of: Apple, Orange, Carrot, Kale, Wheat Grass and Celery with a dash of lemon, ginger or parsley

## Menu #2

# *Pompado Luxe*

## *Appetizers*

*Oysters on the half shell*
Todays oysters shucked table side, with a squeeze of lemon and serving of freshly grated horseradish

*Roasted Wild Mushroom Caps*
Served in their own juices with wild garlic and fresh herbs

## *Soup*

*Cold Vegetable Puree*
Yukon gold potatoes, celery, leeks and fresh chives served table side

## *Fish Course*

*Chilled Maine Lobster Claw*
With stone ground mustard sauce and Meyer lemon removed from the shell for your dining pleasure

## *Entree Course*

*Ostrich Briand*
Ostrich filet for two served with roasted squash, wild rice pilaf, and caramelized pearl onions carved table side

*Duck Breast*
Pan sauteed before you with cranberries, shallots and Burgundy

## *Salad*

*Chopped Salad*
Iceberg, Romaine, Tomatoes, Carrots, Beets, Cucumbers, Green Beans and Red Onion are tossed table side with champagne lemon dressing

## *Cheese Course*

*Trio of Cheeses*
Artisan cheeses are selected by our chef and served with toasted baguette for your pleasure

## *Dessert*

*Cherries Jubilee*
Fresh Michigan Cherries are swirled in butter, brown sugar and brandy and flambe'd at your table. Served warm over vanilla bean ice cream

## Menu #3

Bingsfordworth House Bed and Breakfast Sunday Brunch

### Beverages

Freshly Squeezed Juices
Choice of Orange Carrot or Garden Tomato

Pot of Tea
English or spiced herbal with lemon, cream or soy milk

House ground Coffee
Robust rich roast with cream and sugar cubes

### Breads and Cereals

Porridge
Local farm grain cereal served hot with dried fruit and cream

Muffins
Oversized Corn, Berry or Whole Grain served with jam and creamery butter

Everything Bagel
Toasted and served with cream cheese

### Side Dishes

Potatoes Anna
Delicate layers of wafer thin potatoes layered with mild onion and seasoned with fresh ground pepper

Steamed Asparagus
Served with butter or lemon

### Main Course Selections

Smoked Salmon Side
Carved to your liking and served with Capers, Chopped Hard Boiled Egg and Spring Onion

Omelet
Egg White Omelet served with your choice of mushrooms, country ham, tomatoes or cheddar cheese

### Desserts

Seasonal Fruit
Berries with a dusting of sugar and fresh whipped cream

Truffles
Sinfully rich dark chocolate hand rolled confections

# 12

# THE MENU AND THE FINANCIAL PLAN

## Outline



## Objectives

After reading this chapter, you should be able to:

1. Explain the need for a menu to cover capital costs and (for commercial operations) contribute to profit.
2. Identify some of the basic costs of going into the foodservice business and indicate how to calculate them.
3. Review some of the methods used to calculate whether a menu is providing an adequate return on investment.
4. Discuss some of the factors that contribute to the success or failure of a restaurant and describe how the positives could be accentuated and negatives avoided.

# Introduction

This book has dealt with the menu as a basic document to be used in planning, operating, and controlling a foodservice operation. In the process of our exploration, we moved from a simple view of the menu as a mere listing of what is offered to a more complex definition. We established the menu as a merchandising tool, and elaborated on that to say that the menu is *the* management tool of *primary importance* for initiating and controlling all work systems in a foodservice operation.

This chapter discusses financial aspects of establishing a foodservice operation and how the menu can be evaluated in terms of satisfying the required return on capital investment. When a commercial menu does not produce enough revenue, or when the institutional menu does not stay within reasonable budgetary limits, financial failure is certain to result.

Operators can analyze the operating results generated by a menu and determine whether the menu has met the financial requirements placed on it. The menu itself creates expenses, and it must create income (or cover costs) to be effective. Several financial tests are used to determine whether the menu performs monetarily. A variety of strategies help managers adjust menus to meet financial needs.

# The Menu and Financial Planning

Any menu requires an initial capital investment. The menu will dictate the facilities, equipment, decor, inventory, and space needs. The menu reflects the essence of the operation that is planned and will have a strong influence on atmosphere, staffing, and a host of other factors.

## Capital Investment

A well-planned menu can help eliminate or alleviate many capital costs. Many institutions, for instance, can operate quite well with only ovens and steam-jacketed kettles for heavy cooking equipment. It takes a very well-planned menu in order to scale down capital needs to this level. Other operations might need specialty equipment, such as smokers or rotisseries. These may be essential for some menus, but many operations do not have the money or floor space to satisfy only a fraction of their market.

Some equipment is used so infrequently that it is preferable to do certain operations by hand. The capital needed to start an operation, be it a not-for-profit hospital kitchen or a for-profit large restaurant, can easily reach hundreds of thousands or even millions of dollars. Regardless of the intended nature of the operation, there will be some payback of capital required, either in the form of moneys owed to lenders and investors, or in terms of community benefit. The savvy menu planner tries to keep initial capital investments at a minimum so as to maximize the payback ratio.

## Foodservice: A High-Risk Business—Or Is It?

Banks frequently label the industry as "easy to enter without business skill, high-risk, and low-profit." The number of business failures in the foodservice industry is very high-among the highest of any industry. A study published in the Cornell Quarter on *Why Restaurants Fail* counters the premise that up to 90 percent of restaurants fail with figures between 25 to 30 percent. Although preliminary, this may provide assistance to counter the image of restaurants as extremely risky. At any rate, controlling business costs is one of the primary management tasks in any foodservice operation. Since cost control is so complex, and the cost of perishable supplies and the amount customers are willing to pay change so frequently, a business can quickly get into financial trouble. Bankers have seen many losses when making loans to the foodservice industry, making them quite wary of making new loans, particularly to entrepreneurships.

It usually takes several investors going in together to spread out the capital requirements and risk of starting a new foodservice business. Sometimes limited partnerships are formed with investors who act as silent partners, or small corporations are founded with numerous shareholders investing relatively small amounts of capital. Often these investors are passive, meaning they contribute only capital, while in other situations several people with good professional knowledge of the foodservice business may get together and contribute their expertise, as well as money. Regardless of the type of business formation-corporation, partnership, limited partnership, or individual owner-the investor(s) seldom supply all of the capital needed for start-up. Some form of borrowing must take place.

Most foodservice operations require from $500,000 to well over $1 million in capital investment. Lenders frequently require the investors to put up between 30 percent or more of the total capital cost for an operation. This can mean that the investors must invest between $150,000 and $500,000 of their own money. Franchises may be purchased for as little as $10,000, but the more financially worthwhile ones require the franchisee to invest considerably more. Given these large dollar amounts, one can easily see why lenders and investors are so careful when making a commitment, and why the wise use of the capital investment is so critical. It is theorized that many restaurants have failed due to underfinancing.

Running a foodservice operation is a complex task, requiring not only good merchandising skills, but also a knowledge of food purchasing, production, and service, as well as high competency in finances. A large business organization can afford to hire managers who specialize in one of these areas, but most foodservice operations are small, and the investor or a very limited group of partner-managers usually try to perform these tasks themselves. The requirement that one person be experienced and successful at such diverse activities means that only a few, unusually talented people are successful running a foodservice operation. This situation may explain why there is such a high failure rate of foodservice business. Review of studies and discussion with industry professionals suggest that many who enter the industry and do not succeed had the impression that if they enjoyed hosting dinner parties or had a great recipe for something, they could run a restaurant. Too many people simply had no idea as to the hours, complexities and superb skills of diplomacy that are required to run a foodservice operation. In addition, the foodservice industry is highly competitive. Even hospital, airline, and other institutional foodservice operations are in a competitive environment. Competition means that only the best survive.

## The Feasibility Study

Any planning for the opening of a new foodservice operation should include a *feasibility study* designed to answer the basic question: "Will the commercial business turn a profit?" If it is an institution, a better question would be, "Can it be operated within budgetary cost constraints?"

The feasibility study is the detailed report produced to show exactly what the capital investment is to be used for and precisely how the operators plan on generating the revenues to pay back that investment. It must be realistic and constructed from the best available facts and projections. The study must take the broad estimates from the initial phase of considering the project and show a detailed accounting of what really will be designed, built, and purchased. While preliminary estimates, such as using the historical rule-of-thumb of $65 per square foot for equipping a kitchen, were fine for determining the general cost of a facility, such estimates are not good enough for a formal feasibility study. Vendors estimate that the rule-of-thumb expenses have more than doubled and are rising. Investors and lenders want to know exactly what their money will be used for, and how their investment will pay off.

Independent contractors will work with an operator to write a feasibility study. These companies usually have expertise in writing such studies, and some lenders give more credence to studies written by well-known accounting firms specializing in the hospitality industry. Feasibility studies can be expensive (another possible pre-opening capital cost), but if they can convince a banker to lend an operator money, they will be worth the price.

A typical feasibility study looks like the one shown in **Exhibit 12.1.** The study essentially shows the prospective lenders or investors exactly what the foodservice operation is all about—its concept, the potential market, how it will satisfy the market, what it will cost to accomplish these objectives, and what kind of return one may expect.

A complete feasibility study can easily be hundreds of pages in length. Accounting firms generally offer services concerning the creation of pro-forma financial statements. Vendors often can help in writing what equipment will be needed. Marketing consultants can help analyze the market and propose advertising budgets. But with all these parties participating, the operator must not become a passive observer. The manager should understand fully and approve all aspects of the study.

## Restaurant Business Plan Software Packages

An alternative to a professionally done feasibility study is the use of restaurant business plan software packages. This software is available through vendors and can often be downloaded onto the computer of the entrepreneur. Software varies in quality and usefulness, so it is wise to shop around and get references. The user must do a good deal of the information-gathering legwork. This has the advantage of forcing the operator-to-be to realize the extent of planning and detail that must be worked through to hope to achieve success. A Master Check List & Project Guide from one software program is illustrated in **Exhibit 12.3.** In this package, each worksheet form has a detailed explanation with suggestions as to how to go about obtaining required information. **Exhibit 12.4** shows data entry for the Vivando Restaurant, **Exhibit 12.5** shows sources of funding for the same, **Exhibit 12.6** shows the statement of income with percentages matched up to national averages, **Exhibit 12.7** is a balance sheet, and **Exhibit 12.8** shows a break-even analysis and scenario worksheet.

**Exhibit 12.1** Feasibility Study

| Feasibility Study | | |
|---|---|---|
| **Section** | **Subsection** | **Description** |
| **Introduction** | | Brief summary of the proposed foodservice operation |
| | Mission | Brief statement of the type of operation planned |
| | Location | Description of the facility's site |
| | Capital | Summary of how much it will cost and where the money will come from |
| **Target Market** | | Purpose of the operation |
| | Demographics | Who the operation is meant to serve |
| | Competition | Which other operations directly compete for the same market |
| | Economy | Overall economy of the market and economic trends |
| **Site Selection** | | Describes location in more detail |
| | Community | Description of the community, its history, future, etc. |
| | Site Location | Specific description of the site |
| | Traffic Flows | How people will gain access to the site, how many potential customers come by the site, how they will know the operation is there |
| | Rent vs. Buy Analysis | Analysis of the cost of purchasing the site and building a facility versus renting the site |
| | License Costs | Cost of all permits, licenses, and fees |
| | Utility Costs | How accessible utilities are, cost to provide on site |
| **Operation's Concept** | | What the operation plans on providing the market, how it plans on satisfying that market |
| | Theme/Style | Description of the theme, ambiance of operation |
| | Menu | What food and beverages will be offered and when |
| | Marketing | Public Relations Community involvement and customer relations |
| **Capital Budget** | | Statements concerning what money is required for opening the operation and what that money will be used for |
| | Design | The design of all facilities |
| | Equipment | Equipment needed to operate the facility |
| | Supplies | Inventories and operating supplies required |
| | Pre-opening | Detailed list of all money used prior to opening |
| | Working Capital | List of funds necessary to operate on a continuous basis |
| **Financial Statements** | | Pro forma statements of projected income and expenses and projected cash flows for a five-year period |
| | Income and Expense Statements | In a not-for-profit operation, this may be called a *statement of revenue* and expenditures or an *operating budget* |
| | Balance Sheet | Shows projected assets, liabilities, and owner equity |
| | Cash Flow | Shows how cash will come into the operation, how it will go out, and how much will be left for debt and as a return on the investment |
| **Staffing Plan** | | Description of how the operation will be staffed |
| | Personnel Policies | Policies and procedures concerning hiring, training, compensation, benefits, discipline, and advancement |
| | Service Standards | Descriptions of what each job requires of an employee; also describes education, experience, and other requirements |
| | Organization Chart | Shows lines of authority and relationships among employees |
| **Advertising Plan** | | Description of planned advertising, promotions, and other marketing to be done in the first year of operation |

**Exhibit 12.2** Short-Form Feasibility Study

## SHORT-FORM FEASIBILITY STUDY

**Operating Data**

| | | |
|---|---|---|
| **Lunch:** turnover 2.5 times | Occupancy average: 75% | Check average: $12.00 |
| **Dinner:** turnover 2 times | Occupancy average: 80% | Check average: $24.00 |

Bar and special catering income per day (average): $600
Number of seats in restaurant: 100
Closed Mondays, Christmas, and New Year's Day; average days of operation per year: 312
Return on investment desired: 12%
Net profit before taxes projected: 5%

**Feasibility Data**

| | |
|---|---|
| **Amount invested** | $1,000,000 |
| **Return on investment** | ($1,000,000 × 12%) = $120,000 |
| **Sales needed @ 5% profit to yield ROI** | ($120,000 ;÷ 5%) = $2,400,000 |
| **Average check required to yield sales*** | $19.80 |
| **Number of checks needed to produce sales** | ($2,400,000 ÷ 19.80) = 121,212 checks needed |
| **Checks per day needed** | ($121,212 ÷ 312 days) = 388.5 checks per day |
| **Income per day needed** | ($19.80 check average × 388.5 checks) = $7,692 |

**Estimated Income per Day**

| | |
|---|---|
| **Lunch:** 100 seats × 75% occupancy × 2.5 turnovers × $12.00 check average | $2250 |
| **Dinner:** 100 seats × 80% occupancy × 2 turnovers × $24.00 check average | $3840 |
| Bar and catering business per day | $600 |
| Estimated income per day | $6,690 |
| Income needed per day | $7,692 |
| Income surplus per day | $–1,002 |

Detail of how to calculate average check:

| | | CALCULATION | LUNCH | DINNER |
|---|---|---|---|---|
| 1 | Seats | from part 1 | 100 | 100 |
| 2 | Occupancy percent | from part 1 | 0.75 | 0.80 |
| 3 | Turns expected | from part 1 | 2.5 | 2 |
| 4 | Checks required | Line 1 × 2 × 3 | 187.5 | 160 |
| 5 | Check | from part 1 | $12.00 | $24.00 |
| 6 | Total | Line 4 × 5 | $2,250 | $3840 |
| 7 | Overall average check | Total in line 6 ÷ total in line 4 | | $19.80 |

**Exhibit 12.3** Software-Generated Feasibility Study

**Master Check List & Project Guide**

**Please follow the instructions below in the order they are written.**
**When you have finished a task, return to this page, check it off and move onto the next step.**
**Use the "Help" link above open the help file Manual3.doc in the same directory as this workbook.**

Status Task

☐ **1. Enter general information about your project.**

| | | | |
|---|---|---|---|
| Company name | ____________ | Projected opening date | ____________ |
| Business name (DBA) | ____________ | Hours of operation | ____________ |
| Type of business | ____________ | Breakfast | ____________ |
| Form of business D | ____________ | Lunch | ____________ |
| Company address | ____________ | Dinner | ____________ |
| Contact person | ____________ | Days of the week open | ____________ |
| Phone | ____________ | Holidays closed | ____________ |
| Fax | ____________ | Length of Lease | ____________ |
| Email address | ____________ | Square footage | ____________ |
| Website address | ____________ | | |

☐ **2. Enter food & beverage cost assumptions.**

Estimated food cost (%)
Estimated wine & beer cost (%)
Estimated liquor cost (%)
Estimated Costs—Other (%)

Since we need a history of costs & sales to accurately determine food and beverage costs we make an educated guess here. Sources for your data might come from a chef or manager with experience in these areas.

☐ **3. Record your sources and amounts of funding for your project.**

Go To Funds Worksheet

☐ **4. Record loans or notes data (if needed).**

Go To Loan Calc 1 Worksheet
Go To Loan Calc 2 Worksheet
Go To Loan Calc 3 Worksheet

These forms will create a loan amortization schedule that will automatically create monthly and yearly totals for interest and principal and automatically link them where needed.

☐ **5. Record your startup expenses.**

Go To Startup Worksheet

Note: these are only expenses that you will incur before you actually open for business. If you will be writing a check for something before opening day, enter it here. Otherwise it will be expensed during year 1.

☐ **6. Record your breakfast, lunch, and dinner menus.**

Go To Breakfast Worksheet
Go To Lunch Worksheet
Go To Dinner Worksheet

☐ **7. Record your yearly sales information.**

Go To Sales Worksheet

In the "Covers" cells, enter the average amount of customers you would expect on that day for that shift.

*virtualrestaurant.com*

**Exhibit 12.3** *(Continued)*

- ☐ **8. Record your payroll information.**
  Go To Payroll Worksheet — Under "Position" fill in the job title (e.g. waiter). Under the days of the week, fill in the hour total for that employee, for that shift. Under "Rate" fill in the hourly rate for that job.
- ☐ **9. Record amortization and depreciation information.**
  Go To Amort Worksheet — Please read the help file for this page for detail instructions.
- ☐ **10. Record your yearly expenses.**
  Go To Detail Worksheet — This form will be partially filled in from the data you have already entered. Proceed to record all of the expense information that is applicable using yearly totals.
- ☐ **11. Record cash flow information.**
  Go To Cashflow Worksheet — Please read the help file for this page for detail instructions.
- ☐ **12. Record Industry Average information (Income IS).**
  Go To Income IS Worksheet — Please read the help file for this page for detail instructions.
- ☐ **13. Review Year 1 Income Statement (Income A)**
  Go To Income A Worksheet — Please read the help file for this page for detail instructions.
- ☐ **14. Record your 5 year projection data.**
  Go To Income 5 Worksheet — This form projects income and performance for years 2–5.
- ☐ **15. Record break-even information.**
  Go To Break-even Worksheet — Please read the help file for this page for detail instructions.
- ☐ **16. Review balance sheet information.**
  Go To Balance Sheet Worksheet — Please read the help file for this page for detail instructions.
- ☐ **17. Check Loan Calculation sheets**
  Go To Loan Calculation Worksheets — Please read the help file for this page for detail instructions.
- ☐ **18. Check Comparison Graph**
  Go To Comparison Graph — Please read the help file for this page for detail instructions.
- ☐ **19. Check The Cover sheet**
  Go To Cover Sheet — Please read the help file for this page for detail instructions.
- ☐ **20. Review the Entire Document**
  Check for accuracy and errors before printing.
- ☐ **21. Print Your Reports**
  Go To Utilities Sheet — Please read the help file for this page for detail instructions.

**Exhibit 12.4** Software Project Guide Sheet; Data Entry Page

**The Master Checklist and Project Guide Sheet: Vivando Restaurant**

This sheet is your "Homepage" for the workbook. It is where you will enter basic information that will be stored and transferred to various other sheets and it provides a step-by-step guide for data entry and report building.

☑ Sheet Protected　　Go To Utilities Worksheet　　Help

**Master Check List & Project Guide**

**Please follow the instructions below in the order they are written.**
**When you have finished a task, return to this page, check it off and move onto the next step.**
**Use the "Help" link above open the Microsoft Word Help File "Manual3.doc".**

**Status　Task**

☑ Entered　**1. Enter general information about your project.**

| | | | |
|---|---|---|---|
| Company name | Vivando, Incorporated | Opening date | 7/15/2005 |
| Business name (DBA) | Vivando Restaurant | Hours of operation | |
| Type of business | Full Service Restaurant | Breakfast | 7:30–10:30 |
| Form of business D | LLC | Lunch | 11:30–3:30 |
| Company address | 65 Deacon Street<br>Cambridge, MA 02138 | Dinner | 5:00–11:00 |
| Contact person | Elliot Davies | Days of the week open | 7 days per week |
| Phone | 509-943-5109 | Closed on Holidays | Christmas, Thanksgiving |
| Fax | 509-888-9898 | Length of Lease | 10 years |
| Email address | elliot@vivandorestaurant.com | Square footage | 2600 |
| Website address | http://vivandorestaurant.com | | |

☑ Entered　**2.F Enter food & beverage cost assumptions.**

| | | |
|---|---|---|
| Estimated food cost (%) | 32% | Since we need a history of costs & sales to accurately determine food and beverage costs we make an educated guess here. Sources for your data might come from a chef or manager with experience in these areas. |
| Estimated wine & beer cost (%) | 38% | |
| Estimated liquor cost (%) | 25% | |
| Estimated Costs—Other (%) | 0% | |

☐ To Do　**3. Record your sources and amounts of funding for your project.**
Go To Funds Worksheet

**Step #1—General Information**

Enter all this information about your company. If you don't have the information or it does not apply, just leave it blank.

When you have finished with this step (or any other step), Click into the little square (checkbox) next to the red "To Do" cell. When the box is checked the "To Do" cell will change color to green and now say "Entered".

*virtualrestaurant.com*

**Exhibit 12.5**
Software Project Guide Sheet; Data Entry Page

## Sources of Funding

Vivando, Incorporated
June 29, 2005

| | | |
|---|---|---|
| **Total start-up capital** | **$541,250** | |
| **Loans & Notes** | **325,000** | |
| **Stock Sales** | **206,250** | Cash from Shareholder (C-Corp, S-Corp or LLC) |
| **Cash** | **10,000** | Cash Investments by Owner or Partners |

**Loans & notes**

| | Amount | Rate | Term | Payments/Year | Source |
|---|---|---|---|---|---|
| **#1** | 250,000 | 8% | 10 | 12 | BankNorth |
| **#2** | 25,000 | 8% | 5.00 | 12 | Elliot Davies |
| **#3** | 50,000 | 8% | 5.00 | 12 | Rex McClean |
| **Total** | **$325,000** | | | | |

**Stock Sales**

| | |
|---|---|
| Total shares | 101,000 |
| Outstanding shares | 25,000 |
| Price per Share | 7.50 |

**Shareholders**

| | Amount paid in | Shares | % |
|---|---|---|---|
| **#1** | 37,500 | 5000 | 4.95 |
| **#2** | 75,000 | 10000 | 9.90 |
| **#3** | 56,250 | 7500 | 7.43 |
| **#4** | 37,500 | 5000 | 4.95 |
| **#5** | | | |
| **#6** | | | |
| **#7** | | | |
| **#8** | | | |
| **#9** | | | |
| **#10** | | | |
| **Total** | **$206,250** | **27500** | **27.23** |

**Cash Investment by Owner, Partners and Limited Partners**

Number of limited partners
Cash paid in by each partner
**Cash Investment by Limited Partners**

**Investment by Owners and General Partners**

| Source | Amount |
|---|---|
| Marsha Davies | $10,000 |

*virtualrestaurant.com*

**Exhibit 12.6** Software-Generated Statement of Income

| Statement of Income | | | |
|---|---|---|---|
| **Year 1 (showing industry average expenses)** | | | |
| SALES | Yearly | | % |
| Food | 1,010,888.54 | | 73.24% |
| Wine | 312,546.65 | | 22.64% |
| Liquor | 56,823.00 | | 4.12% |
| Other | | | |
| TOTAL SALES | $1,380,258 | | 100.00% |
| COST OF SALES* | | Cost | |
| Food | 323,484.33 | 32.00% | 23.44% |
| Beverage | 106,265.86 | 34.00% | 7.70% |
| Other | | | |
| TOTAL COST OF SALES | $429,750 | | 31.14% |
| GROSS PROFIT | 950,508 | | 68.86% |
| OTHER INCOME | | | |
| INTEREST INCOME | | | |
| TOTAL INCOME | 950,508 | | 68.86% |
| CONTROLLABLE EXPENSES* | Yearly | % | |
| Salaries & Wages | 455,485.20 | | 33.00% |
| Employee Benefits | 55,210.33 | | 4.00% |
| Operating Expenses | 69,012.91 | | 5.00% |
| Marketing | 69,012.91 | | 5.00% |
| Energy & Utility Expenses | 41,407.75 | | 3.00% |
| Administrative Expenses | 55,210.33 | | 4.00% |
| Repair & Maintenance | 41,407.75 | | 3.00% |
| Other | | | |
| TOTAL CONTROLLABLE EXPENSES | 786,747.17 | | 57.00% |
| OCCUPATION COSTS | 69,012.91 | 5.00% | |
| INCOME BEFORE INT. & DEP. | 94,748 | | 6.86% |
| INTEREST | 24,926 | | 1.81% |
| DEPRECIATION | 53,803 | | 3.90% |
| RESTAURANT PROFIT | 16,019 | | 1.16% |
| INCOME TAXES | 4,005 | | 0.29% |
| NET INCOME | $12,015 | | 0.87% |

*virtualrestaurant.com*

**Exhibit 12.7** Software-Generated Balance Sheet

| Balance Sheet | | | |
|---|---|---|---|
| Vivando, Incorporated<br>For the Period Ending July 31, 2006 | | | |
| **ASSETS** | | | |
| Current Assets | | | |
| Cash | $160,421 | | |
| Accounts Receivable | $6,000 | | |
| Deposits and Prepays | $41,000 | | |
| Inventory | $18,000 | | |
| Total Current Assets | | $225,421 | |
| Fixed Assets | | | |
| Leasehold Improvements | $150,901 | | |
| Building | $— | | |
| Vehicles | $20,000 | | |
| Start Up Costs | $110,750 | | |
| Furniture, Fixtures & Equipment | $125,000 | | |
| Less Accumulated Depreciation and Amortization | $(53,803) | | |
| Total Fixed Assets | | $352,849 | |
| Other Assets | | $— | |
| TOTAL ASSETS | | | $578,269 |
| **LIABILITIES AND CAPITAL** | | | |
| Current Liabilities | | | |
| Accounts Payable | $9,600 | | |
| Short-Term Notes Payable | $— | | |
| Total Current Liabilities | | $9,600 | |
| Long-Term Liabilities | | | |
| Loans and Notes | $295,279 | | |
| Total Long-Term Liabilities | | $295,279 | |
| Stockholders' Equity | | | |
| Capital stock | $216,250 | | |
| Retained Earnings—Net Profit or (Loss) | $57,140 | | |
| Total Stockholders' Equity | | $273,390 | |
| TOTAL LIABILITIES & NET WORTH | | | $578,269 |

*virtualrestaurant.com*

**Exhibit 12.8** Software-Generated Break-Even Analysis

Break-Even Analysis and Scenario Worksheet

Vivando, Incorporated
June 27, 2005

| | Income Statement | Option #1 | Option #2 | Option #3 | Option #4 |
|---|---|---|---|---|---|
| Weekly Sales | $26,543.43 | $32,000.00 | $40,000.00 | $20,000.00 | $15,000.00 |
| Gross Sales | $1,380,258 | 1,664,000 | 2,080,000 | 1,040,000 | 780,000 |
| Less: Cost of Sales | 456,458 | 550,293 | 687,866 | 343,933 | 257,950 |
| Gross Profit | 923,800 | 1,113,707 | 1,392,134 | 696,067 | 522,050 |
| Profit Margin | 67% | 67% | 67% | 67% | 67% |
| Controllable Costs | $686,995 | $686,995 | $686,995 | $686,995 | $686,995 |
| Other Costs | | | | | |
| Occupation Cost | $71,500 | $71,500 | $71,500 | $71,500 | $71,500 |
| Interest | $24,926 | $24,926 | $24,926 | $24,926 | $24,926 |
| Depreciation | $53,803 | $53,803 | $53,803 | $53,803 | $53,803 |
| Total Fixed Costs | $837,224 | $837,224 | $837,224 | $837,224 | $837,224 |
| Restaurant Profit | $86,576 | $276,484 | $554,910 | $(141,157) | $(315,174) |

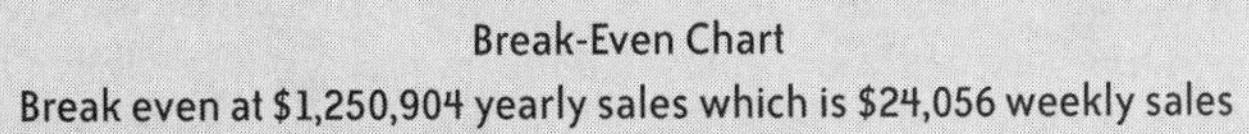

Break-Even Chart
Break even at $1,250,904 yearly sales which is $24,056 weekly sales

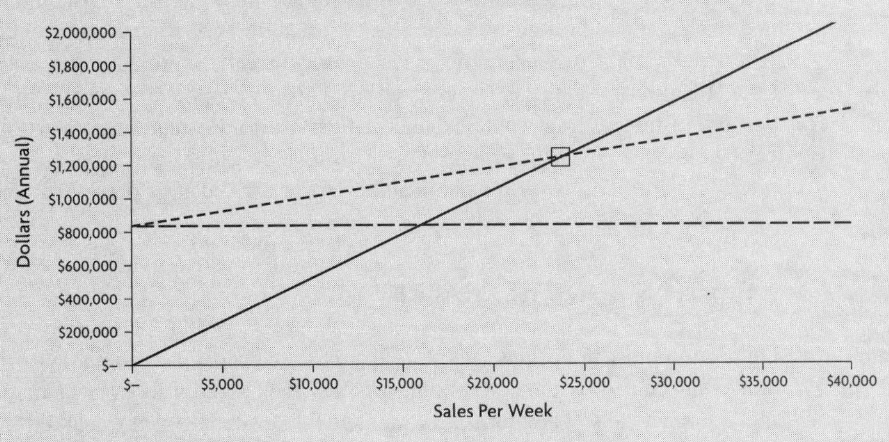


*virtualrestaurant.com*

## Analyzing the Feasibility Study

Data gathering is the first and most time-consuming stage of the process. Learning what residential and commercial demand is likely to be, area demographics, traffic patterns, and future growth generators such as planned development is the purpose of this all-important step. A 2005 *Cornell Quarterly* study on restaurant failure stated that having a defined target market was not critical to success, nor was a marketing strategy in terms of advertising and promotions. Much more important to the successful restaurant operators were public relations, community involvement, and customer relations.

The concept development step involves the use of the data about the population to form a restaurant concept that meets needs and preferences. Elements of the concept include theme, menu, service, style, hours of operation and atmosphere. The Cornell study cited a well defined business concept as the first essential component of success. Losing focus and trying to be too many things to too many people were cited as reasons for failure and closure. Failed owners were unable to describe their concept beyond food production.

### *The Short-Form Feasibility Study*

A *short-form feasibility study* is no substitute for an in-depth one, but it can help managers and investors see whether a project is worth the time and investment that a detailed feasibility study will entail. Suppose a steak house is planned in a downtown area that will be about average in profitability. A short-form feasibility study for this operation is shown in **Exhibit 12.2** appearing on p. 318.

The example shown states that the project is not feasible. The investor would need to go back to the drawing board in terms of concept. Could adding breakfast, expanding the bar, or raising the check average produce desired numbers? What difference would lowering expectations regarding return make? If the original numbers were based on realistic projections, this project may not be feasible. The turnover data can be obtained from observing potential competing restaurants in the target area, or from averages obtainable from the National Restaurant Association. (The check average can be calculated from the proposed menu using the various items' popularity indexes times the sale price per item). The estimated bar business can be arrived at by observing potential competitors.

## Location and the Market

One of the most important factors affecting the success of a foodservice is *location.* Location alone may be decisive in determining whether the operation even has a market. Some have put it more emphatically. The pioneering hotelier E. M. Statler once remarked that the three things a hotel had to have for success were "location, location, location."

Site analysis is the stage of the process whereby it is determined if the selected site could support a new restaurant concept.

Some operations get 95 percent of their patrons from within a relatively small area, while others may have to try to lure them from a considerable distance away. It is important to know how many potential customers go by the door, either on foot or in a car. It is also important to know whether or not they *might* come in. For instance, an operation might be successful on one side of the street and unsuccessful on the other side. For example, on the Ohio State Turnpike, patronage at a reststop restaurant was good from westbound traffic but poor from eastbound traffic. It was found that people traveling west had set a goal—to reach the state border—and eastbound travelers had not.

The speed of traffic also may be important. Signs must be strategically set to coincide with speed so that people can see them ahead and make plans to stop, but not so far ahead that they forget they wanted to stop. The orange roof of Howard Johnson's and the golden arches of McDonald's were designed purposely to show up some distance away so that motorists could identify the unit before they come to it.

An operation also must be suited to its traffic. A quick-service or carryout operation may do well near a public transit center, but a sit-down unit may not. People are in a hurry to catch their trains and want something quick or packaged to carry out.

Parking can be important for some locations, though not for others. A foodservice in an area to which people come in cars should have parking spaces for a fourth of the house seats. An establishment on a highway should have one parking space for every two seats.

The location should be investigated for environmental factors. The location may be undesirable because it is in a deteriorating neighborhood, or it may be desirable because it is in a growing or trendy area. The Cornell study found that a poor location can be overcome by a great product and operation but a good location cannot overcome bad product operation.

Competitor analysis is the survey of area restaurants that might compete with a new operation. It should also be realized that delivery, food market take-out, and quality prepared groceries are all potential competition. The effect of it can vary. It may be harmful if the new operation will be in direct competition with existing services. However, it may not be harmful if it is a complementary type of operation. Many foodservices may be in close proximity to each other but each, by emphasizing a different product, do well. Thus, within a block one may find a McDonald's, a Chik-Fil-A, a Waffle House, and a Taco Bell—all doing well. However, if two operations in the same area offer identical or similar products, one or the other might fade away. That said, successful owners viewed competition to be a tool for self-measurement but not necessarily something to use to develop strategic defenses. Countering this was the 2005 *NRA Table Service Trends Report* citing competition as the biggest challenge their business will face. All sectors reported that the intense competitive environment continues to gain momentum.

A market may already exist for a certain location, or there may be a market potential that must be developed. Even with an existing market, it should be well researched because it may not be durable. A very popular owner may sell, and the buyer may think the operation itself has inherited the clientele, only to find that the market has followed the former owner who has set up in a new location. Unless a market is well established, well defined, and can be held, it is best to interpret market projections conservatively. Studies show there is a tendency to overestimate, rather than underestimate, markets.

It is important to get an estimate of the number of potential customers. A close examination of the market is also necessary. Income levels, social habits, eating patterns, ethnic background, age level, sex, and many other factors about potential customers should be investigated as completely as possible when designing an operation and a product for the market.

Thorough market studies ask questions such as the following:

- Who are the customers or potential customers? Can they be well defined?
- What do they want? Can it be produced for them at a price they are willing to pay?

- Why will they come to this establishment?
- What are their incomes? When do they get paid?
- What is their age level?
- What is their background?
- How many potential customers are there in the area?
- What percentage of these may come in on a single day?
- Where does the market eat now? What do they eat?
- What is the competition for this market?
- Is the market steady and reliable?

## Own, Rent, or Lease?

To determine initial capital costs, it is important to look at how the operation is to be established. There are a number of ways in which a foodservice can be established. How an operation is established will identify the costs that must be paid by the foodservice owner/operator. A new building can be erected on a site with land, equipment, and furnishing belonging to the owner. Or the land can be leased, with the landowner putting up the building and equipping and furnishing it. Or the site and building can be built by someone else and leased; the owner of the building then furnishes and equips it. Sometimes furnishings and equipment are leased. Lease contract forms are available on some of the same Websites that sell business plan software.

There are various ways to make arrangements to rent or lease a building and its furnishings. A straight rental of so much per year, per month, or for a certain period, such as five years, may be negotiated. This rent may be 2 to 20 percent of gross sales, depending on what is included. The highest rent would be for a *turn-key operation,* called a grade 4 lease, with everything ready to go. Even some utilities may be paid by the landlord. A *net, net, net* lease is one in which the business owner pays rent, taxes, and insurance. In a *net, net* lease, the operator pays only the rent and taxes. A *net* lease means that only rent is paid, and the owner of the building pays rent and taxes and insurance on the property. The owner of the business, and not the landlord, usually must carry insurance on equipment and furnishings he or she owns. When the landlord owns them, the procedure will vary.

The length of the lease is also variable. If the operator has a good track record in business, the lease may be written for ten years. Because of the instability of many foodservices, leases often run for five years or less. The lessee must usually guarantee the lease. This guarantee may require that two to four months rent be applied on the final months of the lease. Options to renew the lease may be included. Transfer of the lease may be allowed only with the property owner's approval. The seller of the lease may be able to get the lease guarantee back from the new buyer.

A grade 3 lease is one covering an improved building shell with food preparation equipment in it. New furnishings and some decor and other changes may be required. Rent is lower than for a grade 4 (*turn-key*) operation.

A grade 2 lease is for a slightly improved shell. It should include partitions, finished floor (carpet, tile, and hardwood), finished coiling, heating, ventilation, electrical wiring and outlets, air-conditioning runouts for wiring, finished plumbing with fixtures, and drains. The rental cost for this may run 6 to 8 percent of gross sales.

The lowest-level rental arrangement is a grade 1 lease, which is an unimproved shell. It has completed exterior and interior walls and has basic floors. Almost any business could go into it. The rent per square foot per year for this type of lease will vary, depending on the area.

A rental agreement can also be set up so that a variable rent is paid, usually based on a percentage of gross sales. Another method is to pay a minimum rent plus a percentage of gross sales. Rent can also be set at a sliding percentage of gross sales with the percent figure declining as sales increase. Usually, there is not much difference in the final amount. A return of 3 to 7 percent on the capital investment by the operator is normal, and the profit for the individual leasing the property is around 2 to 5 percent of sales.

It is important in negotiating a lease to establish exactly when it is to start. If rent is paid during remodeling or building for a period before the business opens, the property may have to be operated for some time to erase the rental deficit and begin to show a profit.

## Other Capital Costs

*Preopening expenses* are funds spent prior to opening for nonassets. These include salaries paid to management or key workers before opening, preopening marketing expenses, licenses and other prepaid expenses, utility deposits, and other costs incurred before the first patron walks in the door. *Working capital* is the money necessary to keep the operation afloat once it is in business. This includes sufficient cash reserves to meet emergencies, funds needed to buy supplies, money needed to buy inventories at quantity discounts, and money tied up in cashier's banks.

Preopening expenses often overwhelm a new operator. **Exhibit 12.9** is a use of funding and startup costs summary from business plan software. It shows the range of costs and the sobering total of over half a million dollars. It can seem like everybody in town has his or her hand out. A lawyer must set up the business entity. An accountant must set up the books. Both are expensive; costs for these professional services often amount to several thousand dollars prior to opening. Next are the various permits, licenses, and fees collected by federal, state, and local government. Liquor licenses in some locales are very costly, often as high as $100,000. Moreover, liquor licenses usually take several months to obtain, and hearings and bonds are usually required at a minimum. Insurance for the business is expensive, liability insurance being the principal expense, especially for those serving alcoholic beverages. Insurance companies want payment in advance, and many localities will not issue permits to do business without proof of adequate insurance. The utility companies usually want deposits before hooking up their services. Newspapers and other advertising services usually want payment up front. The menu printer wants payment before delivery of the menus. All of these expenses, while often individually quite small in amount, add up. The conservative entrepreneur will not underestimate these costs.

Stocking the business with inventories can be quite expensive. Suppliers often require payment of purchases either C.O.D. or weekly. With exceptionally good credit, an operator may secure 30-day accounts, but the new operator should not plan on this. The average food inventory turns over about once every two weeks. This means that if opening inventory is $5,000 (a very small inventory), about $130,000 will be spent per year on food ($5,000 × 26 turns = $130,000). One way of looking at this is that the food inventory is like a checking account, where funds move in and out rather quickly. However, some operators look at food inventory as a savings account,

**Exhibit 12.9** Use of Funding and Start-Up Cost Summary

## Use of Funding & Startup Costs Summary

Once you have determined how much money you will be getting from outside sources, you have to determine what you will do with it. This is a very important part of planning and not as simple as it seems. The first thing to remember is that **all of the items within this summary worksheet will be payable BEFORE you open the restaurant.** So you will have either paid or owe the money for these items before opening day. Because of this, these items are not treated as expenses, but as *startup costs*. This means that they have to be treated differently for legal and accounting reasons. Instead of being deducted at 100% as they occur (as a normal expense would be) they must be amortized or depreciated over a specified time period. Each item must be extended to the proper category for this purpose. This worksheet is intended to make this process easier.

Total start-up capital $541,250

| | Amount | Furniture & Fixtures | Start Up Costs | Leasehold Improvements | Deposits and Prepaid | Inventory |
|---|---|---|---|---|---|---|
| Security Deposit | $3,500 | | | | $3,500 | |
| Construction rent | $6,500 | | $6,500 | | | |
| Construction utilities | $1,000 | | $1,000 | | | |
| Liquor License | $25,000 | | | | $25,000 | |
| Yearly License Fees | $3,800 | | $3,800 | | | |
| Utility Deposits | $2,400 | | | | $2,400 | |
| Insurance—Prepaid | $5,600 | | | | $5,600 | |
| Advertising & Promotion | $15,000 | | | | | |
| Graphics & Printing | $4,600 | | | | | |
| Architects/Decorators | $14,000 | | | | | |
| Bank Note Payment | $4,500 | | | | | |
| Other Note | | | | | | |
| Legal & Accounting | $6,000 | | $6,000 | | | |
| Inventory | $10,000 | | | | | $10,000 |
| Uniforms | $2,600 | | $2,600 | | | |
| Contractor Fees | $40,000 | | $40,000 | | | |
| Cash Reserves | $90,500 | | | | | |
| Furniture, Fixtures, Equipment | $125,000 | $125,000 | | | | |
| Leasehold Improvements | $144,000 | | | $144,000 | | |
| Payroll training | $5,000 | | $5,000 | | | |
| Consultants | $5,500 | | $5,500 | | | |
| Research & Development | $1,500 | | $1,500 | | | |
| Landscaping | $2,000 | | $2,000 | | | |
| Debt consolidation | | | | | | |
| Building | | | | | | |
| Vehicle 1 | $20,000 | | | | | |
| Vehicle 2 | | | | | | |
| Employment Agency Fees | | | | | | |
| Dues & Subscriptions | $450 | | $450 | | | |
| Other | $2,000 | | $2,000 | | | |

Using spreadsheet software, such as Microsoft Excel, when you enter data for each item the program will automatically extend that number to the proper column/category. If you add more items to list, **you will have to extend that amount manually to the proper extension category** (Depreciation, Balance Sheet, etc.). **Be sure to verify this and the entire sheet with your accountant.**

The *cash reserves* item is a protected cell and is automatically generated. It is the difference between the Total Startup Capital at the top of the page (in this case, $541,250) and all the expense items you have entered. It represents the funds you have leftover after all the start-up money has been spent. Note: if this is a negative number, you may want to consider raising more money! **The total of the Amount column should always equal the Total Start-Up Capital.**

*Adapted from the Virtual Restaurant Website at* www.virtualrestaurant.com

where funds sit and gain interest. The "interest" is gained through volume purchasing. When commodities are at their lowest, savings of 20 percent or more can be obtained. With freezing and other long-term storage techniques, products can last six months or longer. Not everything can be stored, but some operations can save money this way. Product quality, energy costs, and loss to theft must be taken into considerate.

Liquor inventories are said to turn about eight times a year. Many operators keep two inventories, one for wines and one for beers and spirits. Beers and spirits should turn over about every two weeks. Liquor can usually be bought at a large discount when buying cases at a time. Here, again, operators who stockpile goods will see the cost per unit drop, but the total investment cost rise. If you borrowed your capital at 10 percent interest, you must factor in the cost of borrowing against any savings in purchase price. You break even if the liquor price drops by 10 percent while you have to pay the bank 10 percent for the cash to buy the goods. Liquor may be the most theft-prone supply item in an establishment.

One final preopening expense is the hiring of staff. Skilled employees should already be trained and on the payroll when the business opens.

Once the doors of a new business are opened, working capital, in the form of cash for cash registers and money to buy more food, liquor, and incidentals, is needed. There should be a positive cash balance in the operation's checking account. As a rule of thumb, many operators maintain a checking account balance equal to one-half their normal monthly expenses. Thus, an operation with a projected $100,000 a month in expenses would keep a balance of $50,000. In addition, the operation must keep a *contingency fund* for emergencies. This is usually kept in the form of a savings account or certificate of deposit. This contingency amount could be one additional month's expenses. This contingency amount may be replaced by a short-term borrowing agreement with your banker. For a fee, most banks will negotiate a short-term loan agreement that allows you to borrow on an as-needed basis.

# Financial Analysis and the Menu

A foodservice operation's financial success is measured by profitability and return on investment. Both will be discussed here in detail.

## Measuring Profitability

The most common measure of how well a business is performing is to calculate the ratio of net profit to total assets. This ratio is called a *return on assets,* and it shows how well the assets entrusted to the management have been used. A return on assets of from 3 to 6 percent is considered a fair return.

For example, if assets are $500,000 and $25,000 is the profit, the ratio of profit to assets is 5 percent ($25,000 ÷ $500,000 = 0.05).

Another way is to measure net profit. Net profit in this instance is the profit left over after operations, interest, other non-operating expenses, and business taxes have been paid. It is the amount of money that is left to the owners and investors after everyone else has been paid.

## Return on Investment

Another way of measuring the financial health of an enterprise is to calculate the *return on investment* (ROI). The ROI is the amount of money obtained as a profit in relation to the actual amount invested by the owners of the firm. ROI is compared across investments; that is, investors calculate the ROI of the foodservice enterprise and compare the results to what would have happened had they invested their capital in an interest-bearing bank account, or government bonds, or the stock market. Given the higher risk of investing in foodservice, investors usually demand that the ROI be higher than that on relatively safer investments such as government bonds.

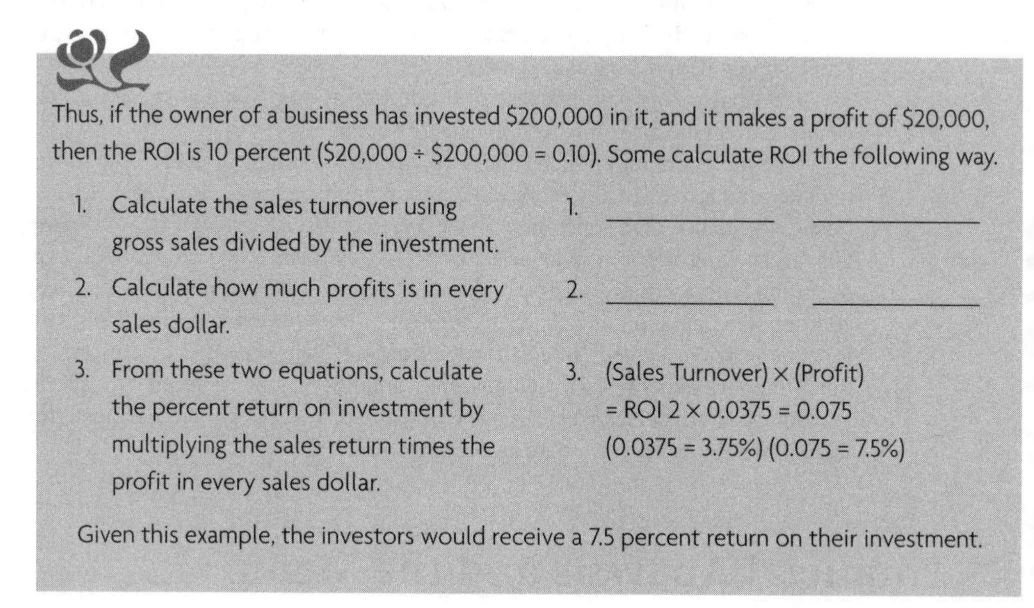

Thus, if the owner of a business has invested $200,000 in it, and it makes a profit of $20,000, then the ROI is 10 percent ($20,000 ÷ $200,000 = 0.10). Some calculate ROI the following way.

1. Calculate the sales turnover using gross sales divided by the investment. — 1. ____________ ____________
2. Calculate how much profits is in every sales dollar. — 2. ____________ ____________
3. From these two equations, calculate the percent return on investment by multiplying the sales return times the profit in every sales dollar. — 3. (Sales Turnover) × (Profit) = ROI 2 × 0.0375 = 0.075 (0.0375 = 3.75%) (0.075 = 7.5%)

Given this example, the investors would receive a 7.5 percent return on their investment.

One advantage of investing in a foodservice is that often it may be highly leveraged. *Leverage* is the amount invested versus the asset value. If investors can get very high amounts of financing, they can often buy an operation that generates high profits on relatively little investment. An operation where 90 percent of the assets were bought using loans and 10 percent of the assets were purchased using investor dollars could easily yield an ROI of 15 percent or more. Naturally, finding lenders willing to take the risk of loaning a high percentage of the asset value of the business is difficult.

## Cash Flow

Simply stated, *cash flow* is the dollars generated in the operation during an operating period—dollars that can be used to pay off obligations plus other costs. In other words, cash flow is the excess of dollars available after cash paid out for operating costs and represents money coming from all sources in the operation. The sources of cash include sales revenue, but may also include additional cash put into the operation by investors, or the net proceeds from additional loans secured during the period. Cash flow is calculated by deducting payments made during the period to banks to repay principal on loans and other cash uses not ordinarily recorded on profit-and-loss statements. Cash flow represents the actual money left over at the end of the period without the artificial contrivances of general accounting practice. As such, it is a good management tool usable primarily to determine what strategies to use to continue operations.

A foodservice needs cash to operate. Due to accounting conventions such as the expensing of depreciation on assets, and accruing certain expenses not yet payable, the usual financial statements do not completely show what cash is really available to the operators. Operations may do well and be profitable from an accounting viewpoint, but poor managerial decisions, such as allowing inventory levels to rise or spending too much money redecorating, may leave little money left to pay creditors when their bills are due. Dollars, not book profits, are what is needed to actually pay the bills. Cash-flow calculations can indicate the number of dollars available to do this.

Management should know the cash flow, and the impact on future cash flows, before making capital investment decisions such as remodeling or expansion. Of course, such decisions can be funded from cash reserves, but management must be assured through a *cash flow analysis* that such investments will pay off. No investment is sound if the cash flows generated after the investment are insufficient to repay the loans. As a minimum, an investment must generate the cash flow to pay off the amortized loan payments, but ideally an investment should return additional cash to the owners as well.

Thus, if an operation has $1,320,000 in gross sales per year, the working capital should be between $33,000 and $26,400 ($1,320,000 ÷ $40 and $1,320,000 ÷ $50). This shows a low ratio of required working capital. However, foodservices can operate fairly well at lower ratios.

# Liquidity Factors in Foodservice

Most businesses must function in the black, and most standards of evaluation are often based on dollar revenue in relation to debt and can be obtained either from the balance sheet or the profit-and-loss statement. *Liquidity* means that *current assets,* such as accounts receivable, inventory, cash, and other convertible assets, exceed *liabilities,* or debts, such as capital payments due within a current operating period and accounts payable.

To determine whether a business is liquid or not, a factor called the *current ratio* may be used. This is a comparison of current assets to current liabilities. Assets should be greater than liabilities. A ratio of 1:1 is barely satisfactory; a 2:1 or 3:1 ratio is better. Foodservices can often function well on a low ratio because they have little or no accounts receivable and thus can have a good cash flow. If a business has no historical records, then it is necessary to make the best possible estimate of potential liabilities. This is always assumed

Thus, if one has $250,000 in a business but owes $500,000, the solvency ratio is 1:2 ($250,000 ÷ $500,000).

to be the case when new operations are set up and some projections for the future have to be made.

Another standard method of evaluating liquidity is to calculate the ratio between dollar sales and working capital. Working capital is current assets less current liabilities. The ratio between dollar sales and working capital can indicate good or poor financial management. A high ratio of sales to working capital, such as $40:$1 or $50:$1 is desirable.

Another ratio called the *solvency ratio* may be used to indicate worth with respect to debt. This is a ratio of the business owner's equity to the amount owed to creditors.

A 1:1 ratio is considered adequate, but the more dollars an owner has in the business, as compared to money owed, the stronger the business will be. A ratio of 1:4 indicates the foodservice is operating more on the money of creditors than on that of ownership, a condition that might initially seem desirable but could be quite dangerous if creditors press for payment. Creditors generally prefer to see the solvency ratio at around 2:1.

# Noncommercial Operations

Measuring the operating efficiency of a noncommercial operation is often simpler than for a commercial operation. The most common check is to compare operating costs and sales to budgeted figures. Many institutional foodservices are operated for reasons other than dollars and it may be important to compare other factors as well. However, a satisfactory performance must be achieved in income, or an operation cannot last very long or achieve other goals.

# Menu Strategies and Financial Success

How might a change on menu affect the feasibility of an operation? The menu prices may be changed to increase the check average (usually this requires a change in the menu items to increase the perceived value). Prices could be lowered and anticipated volume could be increased. A restaurant in a downtown area with a large business luncheon market can perhaps increase volume with a less elaborate menu and longer serving times or office delivery business. Clearly, the feasibility study should not be thought of as a static or fixed document. It is highly dynamic, and changes in operating assumptions can have a significant effect on the feasibility of a project. Studies suggest that the approach must not be so rigid as to disallow for adjustment for opportunities. The restaurateur must remain alert to external as well as internal business factors.

However, while a foodservice must operate at a satisfactory financial level, the success of an operation is not measured in dollars alone. The experience of eating out, whether born of necessity or for pleasure, is important to patrons. A creative and profitable menu can help fulfill the expectations of the owner and customer alike.

**Exhibit 12.10** is a summary of findings from *Why Restaurants Fail,* titled "Elements of Restaurant Success and Failure." The conclusions are well worth considering.

**Exhibit 12.10** Why Restaurants Fail, by Parsa, et al.

## Elements of Restaurant Success and Failure

*Elements of Success:*

1. Have a distinctive concept that has been well researched.
2. Ensure that all decisions make long-term economic sense.
3. Adapt desirable technologies, especially for record keeping and tracking customers.
4. Educate managers through continuing education at trade shows and workshops. An environment that fosters professional growth has better productivity.
5. Effectively and regularly communicate values and objectives to employees. In one instance, new owners credited communication of their values and objectives to their employees as a major element in the successful repositioning of their restaurant to better meet the needs of the growing neighborhood businesses by adding lunch to their dinner-only concept.
6. Maintain a clear vision, mission, and operation strategies, but be willing to amend strategies as the situation changes.
7. Create a cost-conscious culture, which includes stringent record keeping.
8. Focus on one concentrated theme and develop it well.
9. Be willing to make a substantial time commitment both to the restaurant and to family. One successful owner refused to expand his business into lunch periods because he believed that his full-service dinner house was demanding enough from his family.
10. Create and build a positive organization culture through consistent management.
11. Maintain managerial flexibility.
12. Choose the location carefully, although having a good location seems to be more a moderating variable than a mediating (causal) variable in restaurant viability.

*Elements of Failure:*

1. Lack of documented strategy; only informal or oral communication of mission and vision; lack of organizational culture fostering success characteristics.
2. Inability or unwillingness to establish and formalize operational standards; seat-of-the-pants management.
3. Frequent critical incidents; managing operations by "putting out fires" appears to be a common practice.
4. Focusing on one aspect of the business at the expense of the others.
5. Poor choice of location.
6. Lack of match between restaurant concept and location. A night club failed, for instance, because it opened across the street from a police station. The owners thought that the police station would be a deterrent for potential criminal elements and bar fights, but unfortunately it was also a deterrent for customers, who were afraid of police scrutiny and potential DUI tickets. The club was closed within eighteen months.
7. Lack of sufficient start-up capital or operational capital.
8. Lack of business experience or knowledge of restaurant operations. The owners of a successful night club expanded their business by investing more than $1.5 million in renovating an old bank building for a fine-dining restaurant. With no knowledge of restaurant operations, they opened the restaurant with zero marketing budget as they relied primarily on free publicity and word of mouth. In less than one year, the restaurant was closed, with more than $5 million in debt. The owners tried to salvage the business by converting it into a night club, but with no success.

*Parsa, et al,* Cornell Hotel and Restaurant Administration Quarterly *(Volume 46, Issue 3) pp. 317–318, copyright 2005 reprinted by Permission of Sage Publications*

## Exhibit 12.10 *(Continued)*

*Elements of Failure: (continued)*

9. Poor communication with consumers. One restaurant failed to take off after a major renovation because the owners did not communicate to their clientele their reason for closing or their timetable for reopening. Their customers were long gone by the time they reopened.
10. Negative consumer perception of value; price and product must match.
11. Inability to maintain operational standards, leading to too many service gaps. Poor sanitary standards are almost guaranteed to kill a restaurant. One operation was exposed by a local television station for poor sanitary practices. Though the sanitary conditions subsequently improved—as reported by the same television station—the damage was done and the restaurant was closed. It later reopened as a successful full-service restaurant.
12. For ethnic restaurants, loss of authenticity; for all restaurants, loss of conceptual integrity.
13. Becoming everything to everyone; failure of differentiation or distinctiveness.
14. Underestimating the competition. A contemporary restaurant located near an established restaurant adjacent to a golf club failed when it could not draw the golfers from their traditional haunts. Owners thought that their new restaurant would have no problem attracting the golfers.
15. Lack of owner commitment due to family demands, such as illness or emotional problems. In an extreme example, a child with a long-term illness prevented an owner from devoting necessary time to the restaurant, which soon closed.
16. Lack of operational performance evaluation systems. In one instance, new owners did not know how to calculate food cost and relied on employees to maintain proper inventory controls.
17. Frequent changes in management and diverse views of the mission, vision, and objectives. In an example that is common in partnerships, the owners of a failed restaurant could not agree on its direction after just one year of operation.
18. Tardy establishment of vision and mission statements of the business; failure to integrate vision and mission into the operation; lack of commitment in management or employee ranks.
19. Failure to maintain management flexibility and innovation.
20. Noncontrollable, external factors, such as fires, changing demographic trends, legislation, economy, and social and cultural changes.
21. Entrepreneurial incompetence; inability to operate as or recruit professional managers.

*Parsa, et al,* Cornell Hotel and Restaurant Administration Quarterly *(Volume 46, Issue 3) pp. 317–318, copyright 2005 reprinted by Permission of Sage Publications*

## Summary

The foodservice business is a risky one, and many enterprises fail. One factor in making the foodservice business a high-risk one is the high capital required to open and run an operation. Factors within the operation also make for risk. The products are mainly perishable and must be produced to order; the industry is very labor intensive, and labor costs are high in comparison with material costs. Management must be strong in many areas, including financial management, accounting, and balancing the demands of the business with a healthy personal life.

Anyone wanting to start a foodservice should use extreme caution in going into the venture. Many costs must be analyzed in the initial planning to assure sufficient cash balance to carry through the early stages of operation. Hidden costs, such as licenses, bonds, advertising, and insurance, should be considered.

A detailed *feasibility study* should be made before opening. The potential market, competition, and finances should be carefully and realistically assessed.

A number of tests should be made to see whether there will be sufficient *return of investment* (ROI). These tests include comparing current assets to current liabilities, calculating the ratio of dollar sales to working capital, using the *solvency ratio,* and calculating the ratio of net profit to total assets. Non-profit operations usually make a test based on budgetary considerations. If all tests turn out satisfactorily, management can then turn with greater confidence to bringing the menu to life. If everything has been properly done, the result should be a menu that performs successfully for the operation.

**QUESTIONS** Use the following summaries of a foodservice balance sheet and a profit-and-loss statement to answer questions 1 and 2.

| BALANCE SHEET | | | |
|---|---|---|---|
| Assets | | | |
| Cash | $ 86,000 | | |
| Payroll account | 51,000 | | |
| Accounts receivable | 6,500 | | |
| Inventory | 12,400 | | |
| Total current assets | | $155,900 | |
| Equipment and furnishings | $ 62,000 | | |
| Building and grounds | 112,000 | | |
| Total assets | | | $329,900 |
| Liabilities | | | |
| Accounts payable | $ 37,400 | | |
| Note payable in 30 days | 8,500 | | |
| Vacation payable | 1,800 | | |
| Taxes payable | 3,200 | | |
| Total current liabilities | | $ 50,900 | |
| Mortgage payable | $108,000 | | |
| Ownership | 171,000 | | |
| Total liabilities and ownership | | | $329,900 |
| *Profit-and-Loss Statement* | | | |
| Gross sales | $878,000 | | |
| Cost of goods sold | 270,000 | | |
| Gross profit | | $608,000 | |
| *Operating Expenses* | | | |
| Labor | $265,000 | | |
| Other operating expenses | 181,000 | | |
| Capital and Occupation Costs. | 86,000 | | |
| Total operating expenses | | $532,000 | |
| Profit before taxes | | | |
| Reserve for taxes | | $ 23,000 | |
| Net profit | | | |

1. Calculate the following:
   a. Profit before taxes
   b. Net profit
   c. Percent of costs of goods sold, gross profit, labor cost percentage, total labor and operating costs, and capital and occupational costs
   d. Net profit percentage
   e. Current ratio (assets to liabilities)
   f. Ratio of dollar sales to working capital
   g. Solvency ratio
   h. Return on initial investment
   i. Percent of net profit to total assets
2. Would you say this business is in good financial shape and that it is well operated? Explain your answer.
3. Set up a feasibility study as was done in the text, based on the following facts.

   A downtown cafeteria serves three meals a day. Breakfast occupancy is 62 percent, 83 percent for dinner and lunch combined. There are 240 seats. Turnover rates are: breakfast, 1½ times; lunch, 3 times; and dinner, 2 times. The average check for breakfast is $5.60, lunch $6.20, and dinner $13.60. Additional income per day of nonrush-hour income is $372.00. The cafeteria is closed on Sundays and holidays, so use 312 days as the number of operating days of the year. The business wants an 8 percent profit on sales and 12 percent ROI. It has $810,000 in capital invested.

   Project the expected income per day and the income per day needed to make the operation a successful one.

# MINI-CASE

*12.1* You have always dreamed of owning your own restaurant someday. Your first job was washing dishes as a teenager. The long hours and hard work did not discourage you. You started taking career exploration classes in high school that led to a firm commitment to your goal. You continued working in a variety of types of establishments, learning much by observing the successes and failures of management. You chose to continue your education in college and worked your way up the managerial ladder following graduation. Supportive family members and loyal customers believe in you and are willing to invest in you. One of your potential investors is a Realtor who has a location in a mixed residential and commercial neighborhood that is currently in the process of gentrification. The building is on the edge of this development, close to a major road. Residents of the area are a mix of elderly, long-time owners, young professional couples, and singles and a segment of impoverished public-housing projects with a population that is unemployed or making minimum wage. The ethnic mix is a mirror of what the United States is predicted to be in 30 to 40 years. Incomes vary from minimal to upper middle class. Competition includes a quick-service Mexican restaurant and an old-fashioned 24-hour Diner, as well as a white-tablecloth fine-dining Asian-Italian fusion restaurant that is open for dinner only Tuesday to Saturday. A farmers market thrives on the weekends, drawing in a variety of visitors. One office building rents to a variety of up-and-coming professionals and a social service agency.

Design a menu for your market. Include hours of operation, meal periods, prices, and physical layout. Explain the assumptions that your concept is based on.

# 13

# ETHICAL LEADERSHIP IN RESTAURANT MANAGEMENT

## Outline



## Objectives:

After reading this chapter, you should be able to:

1. Define the qualities of leadership that contribute to the successful operation of a foodservice establishment.
2. Explain the elements of effective communication in the workplace.
3. Outline methods by which various sources of workforce motivation can be discovered, nurtured, and maintained.
4. Identify ways a dynamic approach to problem solving helps to avoid crises and the need for crisis management.
5. Define the concept of business ethics and elaborate on the relationship between leadership and ethics.
6. Discuss the ethical responsibility of the foodservice industry with regard to providing nutritional value to its patrons.
7. Discuss the ethical responsibility that foodservice operators have in terms of their membership in the foodservice industry and community.

# Introduction

A well-defined concept crafted to meet the needs of a ready market, executed through a carefully planned menu, will not result in a successfully operated foodservice enterprise without *leadership*.

A leader provides a motivating climate within an operation by developing effective procedures to accomplish agreed upon goals and focusing on achieving desired results. The goals will be established to meet the organization's shared vision or mission. The operation's mission, goals, and methods are to be communicated and effectively coached to ensure they are constantly being upheld. Every operation has a manager, but not every manager is a *leader*. Managers occupy the position of being in charge but may not have the respect of the crew. Leaders are followed by staff that looks up to them. If *motivation* is inwardly directed, *leadership* focuses on the external elements. Leadership helps to organize human and financial resources in order to produce quality products and service. *Management* often plans and organizes mechanically, reacting to events (putting out fires), pushing product*s*, and controlling people. In contrast, those with *leadership* skills, envision the future, oversee development of quality (pulling guests toward our offerings), and provide a motivating climate for people, while controlling *things*. Consequentially, a manager might suffer from burnout from the job, while a leader looks forward to going to work.

# Leadership in the Foodservice Industry

In today's world, change occurs rapidly. The management styles used in the slower paced environment of the past are no longer effective in responding to the competitive, challenging environment of the hospitality industry today. To be a successful leader today, you must have the ability to keep both the tasks required to carry out the business at hand and the needs of your employees in an effective blend. Businesses do not compete simply for guests but for quality staff as well. For this reason, it is essential to ensure a good quality of life for employees so workers stay with the operation and don't look elsewhere for better opportunities. A leader is a person who establishes the standards that help to create and maintain a quality group of employees.

The same characteristics a manager uses to secure employee performance are used in the manifestation of leadership—but with some differences. Employees work for managers primarily because they are paid to do so. However, if true leadership exists, staff still work for the pay, but will also have a real desire to work to accomplish the common goals of the organization. A manager has to work to build high morale among employees. With leadership in place, employees build morale among themselves; it comes naturally. Employees will do what the manager says or they may no longer have a job. People will follow a leader because they believe in the leader's vision. Respect for a manager develops among staff and helps to make that manager a leader. Employees develop respect for a manager based on many reasons. It may be because of the manager's fine character, or innate respect for staff, or the manager's personal abilities, knowledge, and professional accomplishments. When this respect develops, the leader and employees can achieve great success together. A leader's vision becomes indoctrinated into the staff, and this gives everyone common purposes and goals at the workplace.

## Leadership and Self-Awareness

A successful and effective leader typically embodies integrity, enthusiasm, and a sense of self-awareness. A leader sets an example for employees both professionally and personally. The ability to see the big picture, prioritize, delegate, manage time, communicate effectively, engage in team building, and problem solve are all crucial skills. A healthy sense of humor sets a positive tone at the work place, which in turn helps to provide sustenance and pleasure to the operation's guests.

More and more hospitality corporations realize the need for people to have relaxation and a personal life outside of the workplace. There are corporations that are committed to providing a five-day workweek for management, with two consecutive days off. The entrepreneur or single unit operator may not have as much flexibility, but can also profit from time away from work. A positive existence outside of the job contributes to the ability to maintain a mature perspective on the job.

It is essential that a manager, as a leader, serves as an effective role model for employees. As a mentor, his behavior should represent the behavior he expects of his employees. As a leader, if any misbehavior on the job or illegal use of drugs and alcohol occurs, it must not be ignored. Employee assistance programs (EAPs), drug screening of applicants, and random drug testing of workers are becoming more prevalent in the hospitality industry. Not only do these programs help employees (and managers) with existing problems, but these precautionary steps also help to discourage these issues from occurring at the workplace from the start. The hospitality industry is above average in comparison to other industries with regard to illicit drug use and heavy alcohol use. The stress of long hours, gratuity-dependent pay, and meeting the demands of guests can potentially lend themselves to higher incidences of drug and alcohol abuse and self-medication in the hospitality industry. **Exhibit 13.1** compares food preparers, waiters, waitresses, and bartenders to full-time workers in other industries reporting drug and alcohol usage. While the data is not up to the minute in terms of its currency, it still points to the need for the leader to be mindful when dealing with employees and management who choose to engage in pastimes that may have an impact on job performance.

Just as a good managers implement food cost controls to ensure a more profitable and successful operation, they must also address various issues affecting employees' job performance, such as misconduct, harassment, or drug and alcohol abuse head on to correct the behavior. The leader will confront these issues with resolutions in mind, while the manager who lacks leadership skills will ignore such problems, thereby perpetuating lackluster results for the operation as a whole. A manager who takes on the role of a leader recognizes the big picture at all times and acknowledges that all facets of the operation are intertwined and work together to achieve a common good—the profitability and success of the operation. It is essential for leaders to be proactive by setting a good example through their actions and by encouraging subordinates to perform to their greatest potential at all times.

**Exhibit 13.1** Percentage of Full-Time Workers, Age 18–49, Reporting Current Illicit Drug and Heavy Alcohol Use, by Occupation Categories, 1994 and 1997

| | Current Illicit Drug Use | | | Heavy Alcohol Use[1] | | |
|---|---|---|---|---|---|---|
| Occupation Category | 1994 | 1997 | Diff.[2] | 1994 | 1997 | Diff.[2] |
| **Total** | 7.6 | 7.7 | 0.1 | 8.4 | 7.5 | −0.9 |
| Executive, Administrative & Managerial | 5.5 | 8.9 | 3.4 | 6.5 | 7.1 | 0.6 |
| Professional Specialty | 5.1 | 5.1 | 0.1 | 4.3 | 4.4 | 0.1 |
| Technicians & Related Support | 5.5 | 7.0 | 1.5 | 6.2 | 5.1 | −1.1 |
| Sales | 11.4 | 9.1 | −2.3 | 8.3 | 4.1 | −4.2[b] |
| Administrative Support | 5.9 | 3.2 | −2.8[a] | 3.5 | 5.1 | 1.7 |
| Protective Service | 3.2 | 3.0 | −0.3 | 6.3 | 7.8 | 1.5 |
| **Food Preparation, Waiters, Waitresses & Bartenders** | **11.4** | **18.7** | **7.3** | **12.2** | **15.0** | **2.8** |
| Other Service | 5.6 | 12.5 | 6.9 | 5.1 | 11.4 | 6.4 |
| Precision Production & Repair | 7.9 | 4.4 | −3.5 | 13.1 | 11.6 | −1.5 |
| Construction | 15.6 | 14.1 | −1.4 | 17.6 | 12.4 | −5.2 |
| Extractive & Precision Production | 8.6 | 4.4 | −4.1 | 12.9 | 5.5 | −7.4[a] |
| Machine Operators & Inspectors | 10.5 | 8.9 | −1.6 | 13.5 | 9.0 | −4.6 |
| Transportation & Material Moving | 5.3 | 10.0 | 4.7 | 13.1 | 10.8 | −2.3 |
| Handlers, Helpers & Laborers | 10.6 | 6.5 | −4.1 | 15.7 | 13.5 | −2.2 |

[1] Heavy alcohol use is defined as drinking five or more drinks on the same occasion on each of at least five days in the previous 30 days.
[2] Diff. refers to the percentage difference between 1997 and 1994.
[a] Difference between 1994 and 1997 is statistically significant at the .05 level.
[b] Difference between 1994 and 1997 is statistically significant at the .01 level.

Source: *Office of Applied Studies, SAMHSA, National Household Survey on Drug Abuse, 1994 and 1997.*

## Communication and Leadership

Effective communication within the workplace helps to ensure that theoretical plans are put into practice. Clear communication with patrons is essential to help determine what their needs are to ensure they are satisfied. Effectively communicating in the workplace helps to make certain that we are able to practice what we plan. Listening to our guests and employees is crucial. Some barriers to effective listening include the following:

- *Language.* Are the sender and the listener using words that both understand to mean the same thing? No matter what their first language is or what slang terms are used, it is important that they are both clear about the meaning behind the message. Employees who speak more than one language assist us in communicating with guests who speak that language. Managers who speaks more than one language are in a better position to provide hospitality and to lead their diverse workforce.

- *Culture.* Do the sender and listener typically share the same cultural practices, or are they diverse in their use of spatial boundaries, eye contact, tone of voice, body language, or gender issues? Do they possess biases, prejudices, or assumptions about what the other might mean or how the message might be interpreted? Another barrier closely related to culture is communication style. If the style of communication is one that is disliked or interpreted negatively, the message will not be clearly understood.

- *Environment.* Do the sender and listener both have a lack of distractions to keep them focused on the message? If any members of the process are distracted by noise, other more pressing demands, lack of privacy, or just poor timing, the message may not be clearly understood.

- *Assumptions.* Do the sender and listener feel free to ask clarifying questions, seek feedback, and interact honestly yet tactfully? An old-style autocratic boss who yells mumbled orders to employees might not have his or her directions followed because crew members are too fearful to make certain that they understood the requests correctly. The supervisor who assumes that everyone knows what is meant by a "good job" may find the desired results lacking.

Communication in restaurants is generally verbal, but we correspond in written form with schedules, e-mails, posted notices, newsletters, employee manuals, standard recipes, and forms. Writing can be hindered by lack of purpose and planning. Writing must be clear and succinct in a style that is appropriate for the message to ensure the reader understands. It is always important to proofread and be aware of the tone of voice exhibited through what has been written. For instance, sarcastic humor often falls flat, and accusatory writing tends to put employees immediately on the defense, particularly in business communication. Also, providing too much detail in a written document or memo can be just as confusing as not providing enough. You should also never assume that every employee knows how to read. Just because someone speaks English, does not mean that the person reads it. That person may have had assistance in filling out the job application form. An adult who cannot read may not admit this readily. Assisting these employees with adult literacy courses offered in the community can prove to be a very valuable and motivating benefit.

Meetings at the workplace provide an opportunity for group communication. They provide a forum for both managers and employees to ask and answer questions, solicit feedback, and generate enthusiasm. It is important that these meetings remain focused at all times so the goals of each meeting are achieved and issues at hand are resolved as a result of the meeting discussions. Preshift meetings help to inform servers of specials and the kitchen of guest forecasts. Expectations of respectful, cooperative communications between the front and back of the house help to set a hospitable tone for the operation, and create a work climate that is conducive to employee satisfaction and higher morale. Managers who do not actively promote such interaction are not truly deserving of the title *leader*. Other types of meetings include those where announcements are made, ideas are generated, or training is conducted. It is always good to be clear about the nature of meetings when requesting that employees attend. Employees who arrive to a meeting anticipating that feedback will be elicited and considered will feel frustrated when the meeting is actually intended to announce a policy change. Informational meetings, as well as meetings where announcements are made, should be touted as such. Explaining the reasons for changes to employees helps to achieve compliance among them. Paying employees for the time they spend attending various meetings is the only ethical way to expect their receptive presence and participation. Being mindful of employees' schedules when planning a meeting that is not part of a scheduled work shift is another way of ensuring employee buy-in.

## Motivation and the Hospitality Leader

Psychological theory states that motivation comes from within. Leaders realize the various strengths of each employee and strive to capitalize on them in an effort to create an environment where the goals of the employees and the organization are harmonious, or in some instances, even identical. For example, employees who work to provide for a family may be motivated by a schedule that allows them to attend their children's school events. Taking time off in conjunction with some school vacations may also be important to employees with families. Other employees might be paying off debt and rebuilding credit. These employees might be motivated by a schedule that allows for as many shifts as is practical within the labor budget, since there would be more opportunity to earn bonuses and for economic advancement. Still others might be seeking leadership opportunities, such as taking on the role of shift supervisor, being coached to train newcomers, and contributing ideas for the betterment of the organization. In order for the leader to become aware of employees' aspirations on the job, the leader must take time to get to know each team member. Once the leader realizes what motivates each team member, these motivating factors must be nurtured and maintained. A leader must also keep in mind that what motivates a person may change over time. An effective leader continually checks in with employees to help determine what contributes to each person's motivation through time.

Treating each person with dignity and respect is perhaps one idea that is universal in its application. This simple principle provides the mission statement for relationships with customers, coworkers, purveyors, and neighbors. Upholding this mission statement will help to eliminate instances of sexual harassment and discrimination, and to address complaints and conflicts. Referring to each person by the name that they wish to be called, with or without title, knowing who enjoys good-natured ribbing and who prefers a more formal relationship, and listening to each person while making a genuine effort to understand them all contributes to creating a positive work climate. While the word *diversity* has been used a good deal during the past couple decades, the notion that the difference between people is the spice in the stew of life is really at the heart of the matter. A leader does not discriminate among employees based on their cultural beliefs, religious viewpoints, and so on. There is no "us" and "them" mentality. Leaders set an example by acknowledging and respecting the various beliefs and types of behavior people embody based on their cultural upbringing. In order to ensure a successful operation where customers are devoted guests who come back again and again, it is essential for the employees to work collaboratively to achieve a common goal—to provide the utmost in service to their guests. Employees who are treated with dignity and respect at the workplace will be more inclined to better serve their customers. Keeping the workforce motivated and satisfied with their jobs typically leads to better job performance.

## Setting Service Charges versus Gratuities

The long-standing practice of guests bestowing (or not) gratuities at their discretion has come into question and is the subject of study and discussion. It may be viewed as reducing the server to an indentured status of servitude. Gratuities also may involve employees having to put up with guest behavior that would not otherwise be acceptable to the dignity of the server for fear of retaliation in the form of guests withholding their payment of a gratuity to the server. It is also perceived as awkward for guests to be in the position of primary employer of the server, since servers receive far more remuneration in the form of tips than in the paycheck from the restaurant itself.

Leaders may instead opt to establish service charges in their operations. When a service charge is levied, the guest still may be in a position to decline payment if service is inadequate—just as they may decline to pay for food that is inedible. One advantage to the service charge system is that an equitable amount has been established, taxed upon, and paid. Such a practice might also serve to ease some of the tension that has historically existed between the front and back of the house, as servers are not subject to poor pay due to actions beyond their control in the kitchen, nor are they tempted to "give food cost away" to bribe customers. Traditionalists are accustomed to the gratuity system and will likely defend it, while the service charge system has advocates among some industry leaders and members of academia. This discussion will be likely to continue to evolve, with a variety of views on which system is the right and most equitable one.

## Creating Community in the Workplace

Once a leader becomes acquainted with each employee and becomes aware of the contribution that each unique individual makes, this knowledge empowers the leader to focus on team building. With a diverse workforce, the leader must bring people with varying skill sets, backgrounds, and social maturity together. Everyone from prisoners on work release, and high school students engaged in a *ProStart* curriculum through the National Restaurant Association to college students on their way to an alternate career, as well as college graduates in Culinary Arts and Hotel-Restaurant Management, will need to work collaboratively to achieve success for the operation.

Creating a sense of community at the workplace helps to encourage an atmosphere where people do not dread the start of a new work day. This is achieved through the previously stated principles of self-care, communication, motivation, and treating employees with dignity and respect. When these principles are put into action, they help to evoke an atmosphere where quality is produced and employees are proud to be a part of the organization. Pride in achievement helps to build team spirit, as does a culture built on cooperation. A leader also plays a significant role in setting the tone of transactions with guests, employees, purveyors, and neighbors. It is important to maintain equitable systems for scheduling, work assignments, promotions, raises, and discipline, as these are essential building blocks for creating a climate of quality. Creating a pay system whereby everyone has a stake in the success of the operation as a whole can help to strengthen the teams involved in the daily operations. For example, when servers receive a share of a bonus for reaching food cost goals, they do not overserve bread, butter, and other nonrecyclable condiments. Kitchen staff who gets a portion of the service charge tend to be more aware of guest satisfaction and less likely to be resentful toward servers, who are the messengers to the guest.

Cross-training can also be of assistance to employees walking a mile in the shoes or clogs of another coworker performing alternate job responsibilities. Cross-training is also a helpful tool for scheduling purposes.

On the external front, organizing extracurricular activities such as bowling, volleyball, softball, or water pistol wars to encourage team building outside the workplace can create camaraderie among co-workers, which typically tends to extend to the workplace. Activities chosen should not encourage behavior that is not healthy. For example, a North Carolina–based restaurant has a Caribbean menu that includes an award-winning children's menu created by local elementary school kids. After one summer season of success, the owners took the entire crew (yes, that included bus and dish staff) to the island of St. John's U.S. Virgin Islands, where they rented a house and shared in the good year and island-themed concept. Needless to say, this went over well and helped with recruiting! This is in stark contrast with the practice of making servers independent contractors who would pay the restaurant for the *privilege* of working for gratuities many years ago. If sporting teams or a trip sound like too much of an investment, a single outing such as a picnic, dinner out on an off night, a visit to an amusement park, ball game, and so on could all be beneficial. Internally, employee preshift meals can also be an opportunity to build camaraderie. Sometimes a postshift meal cooked by servers and served by cooks can create appreciation.

## Problem Solving

Problem solving processes assist in avoiding crises and the need for crisis management and help to discourage the burnout some managers may experience as a result of constantly being in a putting-out-fires mode of operation. One way to begin solving problems is to be aware of their existence in the first place. Although *management by walking around* encourages the manager on duty to ask guests, "How is everything?" and servers are required to do the *check back*, it may be safe to say that there is often a perception that neither the manager nor server truly wants to know what is wrong. If they were sincerely concerned, the manager might ask instead, "What could be done to make your meal more enjoyable?" and the server, "What can I provide to enhance your enjoyment this evening?" Additionally, some form of market feedback, be it the traditional tabletop comment cards, a Website, or toll-free number for a structured review of the experience, or mystery shoppers would be in place to help the establishment to determine what it could do to improve the guest experience as well as what is being done currently that is appreciated.

Although guests are typically a restaurant's source for brutally honest feedback about their experience within the establishment, employees often have ideas for improvement. Staff who are in contact with guests know what they like, dislike, and what they are asking for that is not on the menu due to existing policy or lack of particular inventory. Policies should never be written for the benefit of the house but should focus on the pleasure of the guest, which will eventually result in the success of the house. For example, a no-substitution policy makes the kitchen run more smoothly but may not satisfy what the guests want or need. If enough patrons desire seafood pan seared, that method should be added to the menu so that it is not a special request. If a few guests want the chilled seafood platter without mussels but with added shrimp, they ought to be able to order it with the understanding that it will require an adjustment in price—one that is communicated by the server and agreed upon up front by the guest before the order is placed. Other requests can be accommodated with the caveat that additional time will be required, or servers may suggest an alternative: "We are not able to prepare grilled shrimp (our shrimp size is too small), but we can broil the shrimp or offer you grilled fish or chicken." "The lasagna is made with meat sauce but we do have vegetarian baked ziti." Practical suggestions regarding product delivery should be sought from all who support servers. Each job should be designed so that someone who has successfully passed their specific job training period should be able to reasonably perform the job based on a forecast of the number of guests expected. Naturally, when things go out of whack (four tour buses pull up to an establishment at a traditionally slow time without warning), it is expected that it will be hectic and that everyone will need to pitch in. Each shift should not, however, feel like a fleet of tour buses have arrived. There are also other problems that are potentially far more serious. Leaders backed by a solid team can effectively tackle those, too. The team provides a source of ideas; the manager who is a leader is not constantly putting out fires and can think through weighty matters. Leaders with a record of successes are more likely to be backed by corporate headquarters or the bank when help is needed. They are also more likely to head off problems before they become too serious, since they detects warning signs sooner rather than later. The process model for problem solving consists of the following seven steps:

1. *Define the problem.* It may not be as it first appears. Ask questions and reach an understanding of exactly what is wrong. In conjunction with defining the problem, identify who or what is affected. This may help to determine what systems are not functioning properly.
2. *Determine the root cause.* The existing problem may be a result of a process or a person. The process might not be functioning as it should or might not have been designed well. The root cause is the place where the problem begins.
3. *Determine alternative solutions and consequences.* Brainstorm and list all possible reactions to the problem, including taking no action. Once a list has been generated, each item must be examined by answering the following questions:
   - What are the consequences?
   - Do the benefits outweigh the costs?
   - Is it sound?
   - Will it fix the system flaw(s)?
   - Will it work?
4. *Select the best solution.* Be deliberate in this process, taking a careful look at all the angles.
5. *Develop an action plan.* Create a step-by-step strategy to avoid recurrence of this problem or similar ones.
6. *Put the plan in place.* Communicate to all affected parties and follow up to determine how effective the solution is in solving the problem.
7. *Document.* History is doomed to repeat itself. By recording the process, others can learn from the problems encountered and avoid repeating actions with unwanted consequences.

When managers are not leaders, and fail to solve problems effectively through inaction or inadequate solutions, the consequences may be more far ranging than simply continuing to live with a problem. Problems that are not solved may worsen and manifest into crises. Employees who are at the receiving end of complaints or engaged in a struggle will suffer from decreased morale, which may lead to apathy and employee turnover. Profits may decrease through loss of sales or increased costs. In addition to fiascos that are the result of unsolved problems, some crises are not within complete control of the leader. Hurricanes, robberies, or other disasters can strike. Planning can help to alleviate some of the burden by establishing procedures to follow for each of a number of scenarios that could occur. Having guidelines provides a stabling influence and helps to keep something from spinning out of control. For example, it would be far more preferable to be robbed and have all people safe and sound than to lose anyone to needless acts of heroics or violence. By laying out policies and procedures and incorporating crisis management into training programs, it is possible to reduce the chances that an unpleasant event will escalate into a human tragedy or business catastrophe.

# Ethics

Personal ethics are the rules of conduct by which we choose to abide. Ethics represent some form of *the golden rule:* Treat others as you want to be treated. This precept is inherent in all of the major world religions. Business ethics are those behaviors governing the way supervisors or other representatives of the operation conduct business. The following ten *Ethical Principles for Hospitality Managers* were adapted from Josephson Institute of Ethics' "Core Ethical Principles:"

1. *Honesty.* We are honest and truthful and do not mislead or deceive others by misrepresentations.
2. *Integrity.* We demonstrate the courage of our convictions by doing what we know is right even when there is pressure to do otherwise.
3. *Trustworthiness.* We are trustworthy and candid in supplying information and correcting misrepresentation of fact. We do not create justifications for escaping promises and commitments.
4. *Loyalty.* We demonstrate loyalty to our operations through our devotion to duty, and to colleagues by friendship in adversity. We avoid conflicts of interest, do not use or disclose confidential information, and respect proprietary information should we change our place of employment.
5. *Fairness.* We are fair and equitable in all dealings, do not arbitrarily abuse power or take undue advantage of others' mistakes or difficulties. We treat all individuals with equality, tolerance, and respect for diversity and with an open mind.
6. *Concern and respect for others.* We are concerned, respectful, compassionate, and kind. We are sensitive to the personal concerns of colleagues and respect the rights and interests of all those who have a stake in our decisions.
7. *Commitment to excellence.* We pursue excellence when performing our duties and are willing to put more into our jobs than we can get out of them.

8. *Leadership.* We are conscious of the responsibility and opportunities of our position of leadership. We realize that the best way to instill ethical principles and awareness in our operations is by example.
9. *Reputation and morale.* We seek to protect and build the operation's reputation and morale of our employees by engaging in conduct that builds respect. We take whatever actions are necessary to correct or prevent inappropriate conduct of others.
10. *Accountability.* We are personally accountable for the ethical quality of our decisions as well as those of our subordinates.

We provide an intimate service for guests when we meet their needs for sustenance, dining pleasure, and social belonging. Their ability to place their trust in us is critical to our receiving their patronage. Today, the foodservice industry is faced with a number of ethical issues. By the time this text is published and in your hands, these topics will have evolved, and new ones may emerge. Regardless of what the issues are, the ten ethical principles serve as guidelines for making decisions about how to behave.

# Hospitality Ethical Issues

## Nutrition

Historically, the ethical responsibility of foodservice operations to provide nutritional value to patrons has been proportional to guest choices for dining. Patients, inductees, students, and inmates housed or stationed in institutional facilities were thought to have the least amount of alternatives available and, therefore, the duty to provide nutritional worth was deemed to be quite high. In the past, commercial operations were perceived as venues to visit for special occasions. The bulk of food preparation and consumption occurred at home and consisted of *home cooking* and the wholesomeness that such cooking implies.

Dining out used to be a special treat rather than the commonplace event of today. When the science of nutrition was still at the Four Food Groups stage of development, a burger topped with lettuce and tomato, served with a side of fries and a shake, was considered to be a nutritionally balanced selection. Years ago, Americans were more active on the job and in their daily lives, but circumstances have changed dramatically. Many people are now dining at restaurants as part of their daily existence. As a group, Americans are less active than in years past and tend to experience stressful situations that elicit the temptation to indulge in the unhealthy choices with little to no nutritional value, which physicians typically discourage. As a result, obesity has reached alarming proportions in the United States, as illustrated in **Exhibit 13.2.**

The Keystone Report, funded by the Food and Drug Administration, indicates that 64 percent of Americans are overweight, which includes the 30 percent who are obese. The annual medical cost associated with overweight Americans is estimated to be between $93 billion and $100 billion. This report also suggests a connection between the rise in obesity and an increase in the consumption of calories outside of the home (in restaurants). Studies such as this have enticed

**Exhibit 13.2**
Overweight and Obesity by Age: United States, 1960 to 2002

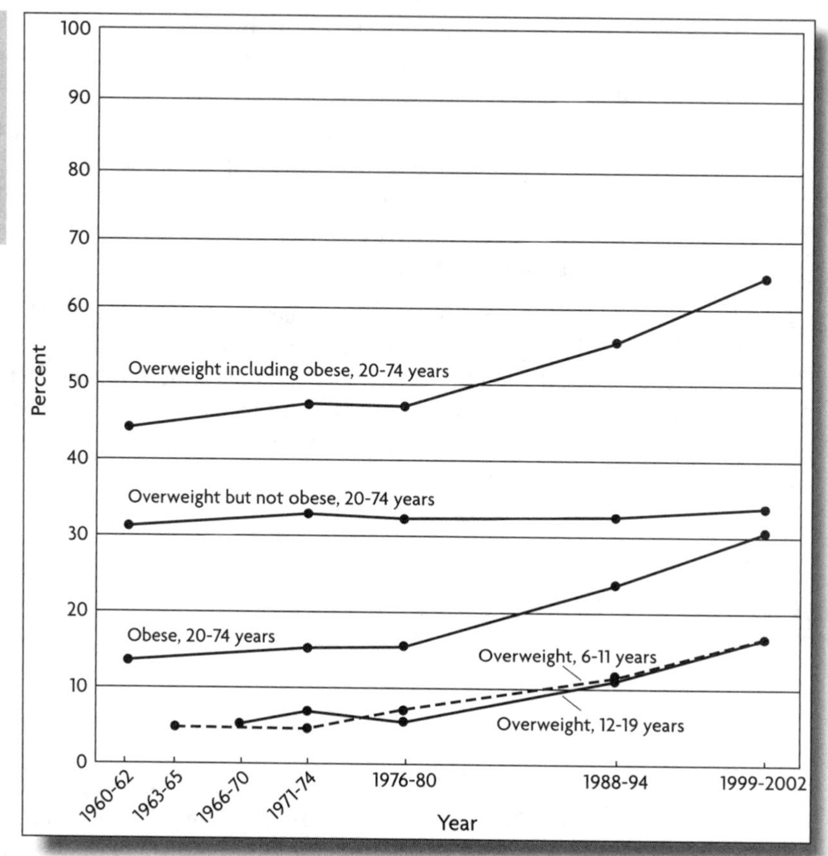


*Notes:* Percents for adults are age adjusted. For adults: overweight including obese is defined as a body mass index (BMI) greater than or equal to 25, overweight but not obese as a BMI greater than or equal to 25 but less than 30, and obese as a BMI greater than or equal to 30. For children: overweight is defined as a BMI at or above the sex- and age-specific 95th percentile BMI cut points from the 2000 CDC Growth Charts: United States. Obese is not defined for children. See Data Table for data points graphed, standard errors, and additional notes.

Sources: *Centers for Disease Control and Prevention, National Center for Health Statistics, National Health Examination Survey and National Health and Nutrition Examination Survey.*

consumer advocate groups to request a number of things of the foodservice industry, including mandatory labeling, reduced portion sizes, a retooling of marketing to emphasize lower calorie items, and an increase of menu choices that are healthier, which include bundling entrees with more fruits and vegetables.

The quick-service segment is perceived to be particularly limited in the nutritional options it provides. This perception has been thrust into the spotlight with the publication of the book (and film version of) *Fast Food Nation* and film *Supersize Me,* and our industry has been forced to examine itself, albeit defensively. As a result of this negative publicity, McDonald's employed Dr. Dean Ornish, a consultant in menu development, who recommended that McDonald's add the Fruit and Walnut Salad, as well as other salad entrees, to its menu. As a result, McDonald's reportedly sold 450 million salads in 2003, making this healthy addition not only the right thing to do but a good business decision as well. The idea is to introduce flavorful food items that happen to be healthy. Offering vibrant flavors, vegetarian options for those who are vegetarian all or some of the time, and emphasizing healthy fats and whole grains and a reduction in sodium go hand in hand with the renewed interest in world foods that has emerged. It is suggested that ceasing to look at health and nutrition as a negative will help us to craft the next generation of foodservice success stories. **Healthydiningfinder.com** is a Website linked through the National Restaurant Association (NRA) that helps restaurants to identify the healthy options that they offer on their menus. If necessary, **Healthy Dining Finder** will assist with product development.

Healthy Dining Finder will also analyze menu items and offers nutritional information. The site is available to consumers as a search mechanism to find all sorts of restaurants that offer healthy options. Partnering with manufacturers and research and development chefs can bridge the gap between deprivation and pleasure. Customers want and need to know that they can maintain or improve their health as they patronize our establishments. The ethical and profitable response is to provide guests with what they are asking for. Just as the industry has risen to meet the needs of a public that demands sanitary facilities and an assurance that the industry is helping to keep intoxicated drivers off the roads, many foodservice operations are striving to provide more nutritious options on menus to meet the needs and demands of their customers

## Labor

Managing a diverse and challenging workforce requires leadership and a strong sense of ethics. Our industry has long been sought after as the first rung on the ladder to the American dream. It is a strong employer of newly arrived immigrants. The National Restaurant Association provides a link to Daily Dose Language Systems Inc., which offers job-specific English programs for the workplace based on the concept of *language huddles*. These 10-minute daily huddles are highly interactive. The organization can be accessed at *http://www.dailydoselearning.com.* Immigrant labor is commonly used in foodservices and provides much needed labor to the industry. U.S. immigration policies are currently under much scrutiny, and the outcome of national policy changes will greatly affect the industry in the future.

## Piece of the Success Pie

As part of the self-analysis our industry has undergone in recent years, there has been an increasing examination of the costs and causes of turnover. Often, salaried and hourly employees feel no real stake in the success of the enterprise. Though management often receives bonuses based on the operation's performance results and their individual performance, the manager typically don't exhibit a great deal of loyalty to the operation. Some even complain that many of these bonuses are just out of their reach. As a result, many casual and fine-dining corporations are now offering employees an opportunity to become managing partners with a piece of ownership in the individual store at which they work. This is simply a logical extension of the franchising prospect that has long existed in the quick-service segment. Group health care plans, tuition reimbursement programs, and other benefit options are an investment in the workforce that can produce valuable outcomes in the form of decreased employee turnover and increased stability, loyalty, and motivation. Providing hourly employees with the means to achieve bonuses and having a genuine dovetailing of the needs of the individual and operation are also valuable recruiting and retention tools.

## Good Neighbor

Membership in the hospitality industry provides a network of like professionals who are working together toward the betterment of our industry and the communities in which we operate. The National Restaurant Association has state and local chapters that assist operators in developing their leadership skills through continuing education, fellowship, and political action. The Educational Foundation of the NRA also offers certification programs for high school and college-aged students to better prepare them for hospitality industry careers as well as to keep leaders current on sanitation and the responsible sale of alcoholic beverages. Similar organizations exist for chefs through the American Culinary Federation and Foodservice Educators Network International, Hospitality Education through the Council on Institution Hotel and Restaurant Education. Other civic groups include local chambers of commerce and service organizations. Those new to the industry will have much to learn while more seasoned leaders can serve as mentors and resources for others in the industry.

## Keeping it Green

Another ethical issue related to the hospitality industry is the environment. The restaurant industry (#1 electricity consumer in the retail sector) accounts for 33 percent of all U.S. retail electricity use. Additionally, this industry uses disposable products including polyurethane containers, water, and petroleum. The Green Restaurant Association (GRA), a 501(c)3 nonprofit, is a national environmental organization founded in 1990. The GRA helps restaurants and their customers become more environmentally aware and sustainable in ways that are convenient and cost-effective. There are five GRA's components:

1. *Research*
2. *Environmental Consulting*
3. *Education*
4. *Public Relations & Marketing*
5. *Community Organizing & Consumer Activism*

The GRA has a contract with restaurants who wish to become members and assists with marketing contracted restaurants to consumers who desire to express their environmental consciences with their purchasing power. The contract is illustrated in **Exhibit 13.3.**

## Charitable Contributions

The foodservice industry prides itself on its record of response to disasters by providing food to rescue workers and those in need of rescue. Its outpouring of assistance following the September 11th devastation and Hurricane Katrina helped to set a standard for giving. Perhaps it is the motivation of those who are attracted to the industry that provides hospitality that sets us apart. The

**Exhibit 13.3** Startup Membership Agreement and Green Restaurant Certification

# Green Restaurant Association℠

## Startup Membership Agreement

The Green Restaurant Association℠ (GRA) exists to help the restaurant industry become more environmentally sustainable in a convenient and cost-effective manner. By signing this contract, General Restaurant is accepting this invitation to become a member of the Green Restaurant Association℠ and a Certified Green Restaurant™.

### GRA's Environmental Guidelines

1. Pollution Prevention
2. Energy Efficiency & Conservation
3. Water Efficiency & Conservation
4. Recycling & Composting
5. Sustainable Food
6. Recycled, Tree-Free, Biodegradable & Organic Products
7. Chlorine-Free Paper Products
8. Non-Toxic Cleaning & Chemical Products
9. Green Power
10. Green Building & Construction
11. Employee Education

A start-up business qualifies to become a Certified Green Restaurant™ when it:
a. Signs this contract and commits to the GRA's Environmental Guidelines.
b. Uses no polystyrene foam (Styrofoam) products.
c. Recycles all products that are accepted by local waste collection companies.
d. Commits to completing four Environmental Steps per year of membership.

Upon signing the contract, each restaurant is entitled to Environmental Consulting of the Benefit Services. After becoming a Certified Green Restaurant™, each location will receive Marketing, Public Relations, and Education of the Benefit services.

### Location for this Contract

| Store Addresses | Store Addresses |
|---|---|
| | |
| | |
| | |
| | |
| | |
| | |
| | |
| | |
| | |
| | |

Printed on 100% Post Consumer Recycled, Process Chlorine Free Bleach Paper

*Courtesy of Green Restaurant Association℠*

**Exhibit 13.3** *(Continued)*

# Benefit Services

## Environmental Consulting

### • Environmental Assessment

A GRA specialist will determine what steps to implement into your restaurant & provide a report detailing the options.

### • 4-Step Environmental Consulting

The GRA will help implement the steps described in the environmental assessment report. The GRA Environmental Consultant will help you to open your establishment in the most environmentally responsible manner possible within the constraints of your budget and building needs. This includes food service and mechanical equipment, furniture, paint, building materials, solar energy, roofi ng products, fl ooring, waste diversion systems and more. The GRA will provide your contractors with information to implement building recommendations. The GRA Consultant will also provide solutions to help your locations make four additional environmental improvements each year, starting in the second year of the contract. It is our goal to help your restaurant be one of the most environmentally responsible establishments on this planet.

## Education

### • Employees

Restaurants will receive GRA's book, Dining Green, which explains the environmental impact of the restaurant industry and a path toward ecological sustainability. Restaurants receive kitchen signs for recycling, energy/water conservation, and the 4 Environmental Steps.

## Marketing and Public Relations

### • Signs

Certification Sign: Laminated color sign with GRA logo, displayed on storefront window.
Environmental Achievement™ Sign: Color sign that explains the environmental achievements of Certified Green Restaurants™ including saved trees, reduced electricity, garbage, and air pollution.
4-Step Sign™: Color sign showing the four Environmental Steps for the year.

### • Certification Logo License

Certified Green Restaurants™ have permission to use the Certification logo on items such as website, menus, brochures, advertisements, and apparel.

### • Listing in Certified Green Restaurant™ Guides

This restaurant guide is published and updated regularly on www.dinegreen.com.

**Exhibit 13.3** *(Continued)*

Membership Options

Please Check one of the options below.

| 1 Year Contract ☐ | 3 Year Contract ☐ | 5 Year Contract ☐ |
|---|---|---|
| | | |

**Mail one of the above amounts to:**
Green Restaurant Association[sm]
89 South St., suite LL02
Boston, MA 02111

**Full Payment due upon contract beginning**

This contract pertains to the location listed on page one of this contract. Each location commits to completing four new environmental steps per year, If a location doesn't make four new environmental steps for a contract year, it risks losing its status as a Certified Green Restaurant™. **Initial Here** ☐

Once General Restaurant signs this contract, it will not purchase polystyrene foam for the full term of this contract. **Initial Here** ☐

**The undersigned agree to the terms of this contract**

| X ______ | ______ | Michael J. Oshman (signature) | 12/19/06 |
|---|---|---|---|
| General Restaurant | Date | Green Restaurant Association[sm]<br>Michael Oshman, Executive Director | Date |

______
Your name and Title

______
Street Address

______
City, State, Zip

______
Phone

Contract Term: ___/___/___ to ___/___/___

**Exhibit 13.4** Restaurant Community Outreach: A Menu of Ideas

*Restaurant Community Outreach*

# A MENU OF IDEAS

**Philanthropy isn't just good for the community; it's good for your restaurant. Getting involved in community outreach helps boost employee morale, promotes good will among customers and can increase sales. In fact, nearly 80 percent of Americans believe companies should support a community's social needs. Are you doing your part?**

**Whether you're just starting out and need inspiration, or your restaurant wants to increase its philanthropic efforts, this Menu of Ideas can help motivate and guide your efforts.**

## *Starters*

*If you're new to community outreach work, think about these starters to "whet your palate" for bigger community outreach initiatives down the road.*

Donate money to support a children's hospital, United Way, local non-profit, museum, library, school, zoo, military support group or homeless shelter. Look for a creative way to involve your diners in the experience. For example, let them know you're donating 10 percent of sales or $1 off each menu item to a designated charity on a specific day.

Contribute to a food drive or a fundraising dinner. Contact your local shelter, and arrange to donate excess food to the homeless. Work with a local charity that delivers nutritious meals to people with AIDS or other life-threatening illnesses. More information: America's Second Harvest: www.secondharvest.org.

Ask employees to volunteer at a homeless shelter or soup kitchen. Recruit your head chef to be a guest chef.

Encourage your chef to teach low-income families to prepare nutritious meals on a budget.
More information: Share Our Strength: www.strength.org.

Build a float, distribute food samples or ask your employees to volunteer in community parades.

*www.restaurant.org/community*

**Exhibit 13.4** *(Continued)*

# *Chef's Specials*

*Once you've gained some experience with community outreach, you're ready to bite into something a little bigger, a little meatier. Consider this selection of ideas …*

Organize a themed dinner to benefit a local charity, hospital, museum or non-profit. Enlist your chef or a team of chefs from other restaurants to design a special-occasion menu, ask a DJ to provide music, and encourage local businesses to contribute items for a silent auction.

Organize a charity walk or galvanize the community to raise funds for cancer research, a new wing at your local hospital or another cause.

Create an annual coupon promotion campaign to benefit charity. Enlist your employees to sell $1 coupons at the restaurant or at local retailers. Customers exchange the coupons for free sandwiches or appetizers.

Organize a pledge drive. Create a competition among your units to see which store can raise the most money or can create the most unique fundraiser.

Sponsor a golf tournament to benefit one or more local charities. You'll need to plan well in advance and enlist the support of other businesses in your community to ensure your tournament runs smoothly.

Recruit employee volunteers to help rehabilitate a home through Habitat for Humanity, or refurbish a local day care, orphanage, school or rehabilitation center.

Organize a dinner to show your support for military servicemen and women. Ask local businesses to donate child care vouchers, prepaid phone cards and food for those families.

Set up a foundation to oversee your charitable giving. You'll be able to make tax-free gifts while supporting the community that supports you. More information: Council on Foundations, www.cof.org.

www.restaurant.org/community

**Exhibit 13.4** *(Continued)*

# *First Course*

***Pick a cause.*** Non-profits often approach restaurants first when they need support. To allocate your resources responsibly and not get bombarded with requests, try to establish a focus for your efforts. You might have a long-standing commitment to one non-profit organization or a specific cause, such as a sick child in the community or fire-gutted local landmark. Survey employees to assess the volunteer activities that interest them and the causes they support.

***Get buy-in from the top.*** Motivate your employees to get involved to ensure successful community outreach efforts. First, make sure the owner and/or senior-level executives at the restaurant endorse your efforts and demonstrate willingness to get their hands dirty and participate.

***Set a firm timetable.*** Restaurant operators rarely experience down time, which means you'll need to set a realistic timetable to plan and execute your initiative. Include enough time to properly promote the initiative and ensure a great turnout. Anticipate potential setbacks, such as an employee illness, event location problems or local business issues.

***Get employees involved.*** Your employees are your most vital resource—in the restaurant—and volunteering in the community. Solicit their ideas for potential community service projects. Use your company newsletter, email, staff meetings and/or Intranet to educate staff about the charity or cause you support, and provide details on the project the company is undertaking. Added incentives include paid time off to volunteer, appreciation luncheons and "volunteer" T-shirts, hats or other memorabilia. Provide stipends to cover parking, mileage, food and other expenses employees incur while volunteering. If you're hosting an event, ask your employees to arrange committee meetings, solicit sponsors, serve food, keep score, set up registration, distribute awards, sell raffle tickets, check coats and more. Or enlist their support to raise money for your target cause, and create a competition among store units to make it fun.

***Find partners to help.*** Identify strategic partners to help maximize the success of your efforts, educate them about the cause you support, and provide specific details on what you need from them. If you're hosting a benefit dinner, recruit your local distributor to donate the beer and wine. Ask a local DJ to contribute music. Team up with a local radio station to help get the word out about your event. Recruit local businesses to contribute items to raffle.

***Boost your turnout.*** Work with the charity you support to get their help promoting the event. Examples include newsletter and Web site articles, promotional letters to contacts in their database, joint press releases, etc. Ask local businesses to display fliers/cards, contact local radio stations to help publicize the event, and send out news releases.

***Get the word out about your efforts.*** Keep the community and the news media informed about your recent successes by describing your efforts in news releases, the charity you supported and the event's outcome. Examples: How much money you raised, the research it will support, etc. Work with the charity to maximize promotional efforts and use their connections to get media coverage.

***Use the National Restaurant Association as a resource.*** The National Restaurant Association can help restaurants get started or boost their community outreach efforts. You'll find Web links to organizations that support hunger prevention, youth development, improving health and improving communities. For more information, visit www.restaurant.org/community.

**www.restaurant.org/community**

**Exhibit 13.4** *(Continued)*

# Kids Menu

**Working with kids is one of the most rewarding ways to give back to your community. Whether you choose to sponsor local a Little League team or provide a restaurant internship to a high school student, the ideas below can help you make a lasting impact on your community's youth.**

Encourage kids to be active and support a Little League team. Donate uniforms, offer your restaurant for end-of-season banquets, or ask employees to volunteer for coaching duties.

Get involved with America's Promise, one of the leading youth development organizations. More information: americaspromise.org

Become involved with Big Brothers-Big Sisters of America, the oldest and largest youth mentoring organization in the United States. More information: www.bbbsa.org

Ask local YMCAs, Boys and Girls Clubs or other youth groups about how your restaurant can help them. You might be able to raise money for a good cause, donate food for an event or provide space for one of their activities or meetings.

Inspire students to learn about and consider a career in the restaurant industry. Participate in school career days, open houses and job-shadowing. Offer internships in your restaurant for local high-school students. Open your facility to elementary school field trips so students can learn how restaurants operate.

Offer space for tutoring sessions and after-school activities to local schools. Provide food and beverages.

Support school-lunch and school-breakfast programs for children with little to eat at home.

Motivate/reward local students who make honor roll or show perfect attendance with free pizza or dinner coupons. Acknowledge "Teacher Appreciation Day" by preparing a free lunch and delivering it to a school.

**www.restaurant.org/community**

foodservice industry is approached by numerous charitable and civic organizations seeking donations of food, gift certificates, catering, and sponsorships. Shelters for homeless and hungry people often request donations of leftovers and provide a pick-up service to simplify this process. Food donations can be a way to give back to the community and to build goodwill. Whether it be select groups that touch foodservice operators personally or relate to our markets (a fine-dining operation that provides meals for volunteers at a public radio pledge drive or a pizza parlor that sponsors a high school sports team), it can make a difference in the communities in which we operate. It is a good idea for restaurants to have policies and a plan in place for these arrangements. The National Restaurant Association offers guidelines as illustrated in **Exhibit 13.4.**

## SUMMARY

In a competitive industry such as the foodservice industry, solid leadership is a necessity. Leaders differentiate themselves from managers through their ability to get to know their team and to find out and act on what motivates the team members. Leaders are self-aware and take care of themselves in ways that are healthy and set an example for their team. They are conscious of the higher-than-average incidence of drug and alcohol abuse in the foodservice industry and are proactive in their approach to hiring, coaching, and providing discipline to their workforce.

Leaders are good communicators, due largely to their excellent listening skills. A leader values diversity and creates community in the workplace through team building. Each member of the team is valued for his or her contribution, and everyone is one of "us." A leader engages in planning rather than putting out fires. A leader pays careful attention to feedback from guests and team members so that problems can be spotted early and systems or training failures are addressed before a crisis develops. Unexpected events that may nevertheless be predicted as possible occurrences such as power outages, robberies, and a guest choking can be handled more effectively when procedures are laid out and readily available if and when they are ever needed.

*Ethical Principles for Hospitality Managers* provide guidelines for the delivery of quality and development of pride in what foodservice operations have accomplished. As members of the larger community, foodservice operations are aware of and concerned for the health of their guests and strive to provide menu items with nutritional value. Foodservice operators also realize the importance of acknowledging their employees' contributions to their success, and they like to give back to the communities in which their establishments prosper.

The foodservice industry provides exciting, rewarding professions that help to enrich the lives of others.

## Mini-Cases

**13.1** A party has been booked for a busy Saturday night in a private room for 100. The restaurant is running short staffed, as it is the busy season. One of four servers scheduled to work the party has called in to report that she has a stomach bug. You suspect that she is suffering from overindulgence in alcohol. The host of the party has called to state that one of the guests is allergic to peanuts, and nothing served that evening can contain even a trace of nuts. The set menu has a variety of items that are already purchased and set aside for the event. It has not been determined what the ingredients of each item are. Some appetizers and desserts are purchased already prepared, and the ingredients are unknown.

What will you do about the worker who called in and the shortage that it leaves for the party?

What will you do about accommodating the guest with the nut allergy?

Are there things that could have been done that may have helped to prevent these dilemmas? How might you incorporate these ideas into policies for your employees to abide by going forward?

**13.2** You are planning a festive holiday get together to celebrate a successful year of operation. Your staff consists of immigrants who perform prep tasks, dish room duties, and occupy your kitchen manager positions; teenage bussers; college student servers and delivery drivers; a twenty-something supervisory team; middle-aged day-prep people, including a few servers; your bartenders; and an executive chef. Most of these folks have significant others and the majority have children. You are aware, through your employee assistance plan, that you have about 10 percent of your crew who are doing good work on their sobriety; some of them have significant "clean" time.

What kind of party will you have that will be inclusive of all?

What sort of problems will you need to plan to head off?

Plan the menu, activities, hours, and entertainment, and present it in the form of an invitation.

**13.3** Using a restaurant concept that you may have created during the course of your class, have been in the employ of, or plan to create someday, describe the restaurant and the community in which this restaurant operates. Plan a way to give back to this community.

What sort of charity or service would you choose?

What kind of activity would you perform as a starter to get some experience?

How would you get employees involved in this effort?

What might be something bigger to do as you gain experience and momentum that fits with your operation and community?

Does your restaurant include children as guests? What sort of charity or donation associated with children might be suitable to your operation?

**13.4** Now that you have nearly completed the course, what are your thoughts on the profession of restaurant management in your future?

APPENDIX
A

# Accuracy in Menus

## Representation of Quantity

If standard recipes and portion control are strictly adhered to, no quantities of menu items should ever by misrepresented. For instance, it is perfectly acceptable to list precooked weight of a steak on a menu; double martinis must be twice the size of a regular drink; jumbo eggs must be labeled as such; petite and supercolossal are among the official size descriptions for olives.

Although there is no question about the meaning of a *three-egg omelet* or *all you can eat*, terms such as *extra large salad* or *extra tall drink* may invite problems if they are not qualified. Also remember the implied meaning *of* words: a bowl of soup should contain more than a cup of soup. There has been concern that restaurant serving sizes do not correlate with a *serving* as defined by nutritionists or physicians.

## Representation of Quality

Federal and state quality grades exist for many foods including meat, poultry, eggs, dairy products, fruits, and vegetables. Terminology used to describe grades include Prime, Grade A, Good, No. 1, Choice, Fancy, Grade AA, and Extra Standard.

Care must be exercised in preparing menu descriptions when these terms are used. In some uses, they imply a definite quality. An item appearing as "choice sirloin beef" should be USDA Choice Grade Sirloin Beef. One recognized exception is the term *prime rib*. Prime rib is a long-established, well-understood, and accepted description for a cut of beef (the "primal" ribs, the 6th to 12th) and does not represent the grade quality, unless USDA is used also.

Ground beef must contain no extra fat (no more than 30 percent), water, extenders, or binders. Seasonings may be added as long as they are identified. Federally approved meat must be ground and packaged in government-inspected plants.

## Representation of Price

If your pricing structure includes a cover charge, service charge, or gratuity, these must be brought to your customers' attention. If extra charges are made for special requests, guests should be told when they order.

Any restriction when using a coupon or premium promotion must be clearly defined. If a price promotion involves a multi-unit company, clearly indicate which units are or are not participating.

## Representation of Brand Names

Any product brand that is advertised must be the one served. A registered or copywritten trademark or brand name must not be used generically to refer to a product.

A house brand may be so labeled even when prepared by an outside source, if its manufacturing was to your specifications. Contents of brand-name containers must be the labeled product.

## Representation of Product Identification

Because of the similarity of many food products, substitutions are often made. These substitutions may be due to stockouts (also referred to in industry as *86),* the substitutions' sudden availability, merchandising considerations, or price. When substitutions are made, be certain these changes are reflected on your menu. Substitutions that *must* be spelled out as such include the following.

Maple-flavored syrup for maple syrup
Nondairy creamer for cream
Boiled ham for baked ham
Ground beef or chopped beef for ground sirloin
Capon for chicken
Veal pattie for veal cutlet
Ice milk for ice cream
Cod for haddock (or vice versa)
Powdered eggs for fresh eggs
Picnic-style pork for pork shoulder or ham
Light-meat tuna for white-meat tuna
Skim milk for milk
Pollack for haddock
Pectin jam for pure jam
Sole for flounder
Whipped topping for whipped cream
Processed cheese or cheese food for cheese
Chicken for turkey (or vice versa)
Nondairy cream sauce for cream sauce
Hereford beef for Black Angus beef
Peanut oil for corn oil (or vice versa)
Bonita for tuna fish
Blue cheese for Roquefort cheese
Beef liver for calf's liver (or vice versa)
Diced beef for tenderloin tips
Half & half for cream
Salad dressing for mayonnaise
Margarine for butter

## Representation of Points of Origin

A potential area of error is in describing the point of origin of a menu offering. Claims may be substantiated by the product, by packaging labels, invoices, or other documentation provided by your supplier. Mistakes are possible as sources of supply change and availability of product shifts. The following are common assertions of points of origin.

Lake Superior whitefish
Bay scallops
Gulf shrimp
Idaho potatoes

Florida orange juice
Maine lobster
Imported Swiss cheese
Smithfield ham
Wisconsin cheese
Puget Sound sockeye salmon
Danish blue cheese
Louisiana frog legs
Alaskan king crab
Colorado brook trout
Imported ham
Colorado beef
Florida stone crabs
Long Island duckling
Chesapeake Bay oysters

There is widespread use of geographic names used in a generic sense to describe a method of preparation of service. Such terminology is readily understood and accepted by the customer and their use should in no way be restricted. Examples of acceptable terms follow.

Russian dressing
French toast
New England clam chowder
Country fried steak
Denver sandwich
Irish stew
French dip
Country ham
Swiss steak
French fries
German potato salad
Danish pastries
Russian service
French service
English muffins
Manhattan clam chowder
Swiss cheese

## Representation of Menu Descriptive Terms

A difficult area to clearly define as right or wrong is the use of merchandising terms. "We serve the best gumbo in town" is understood by the dining-out public for what it is—boasting for advertising sake. However, to use the term "we use only the finest beef" implies that USDA prime beef is used, as a standard exists for this product.

Advertising exaggerations are tolerated if they do not mislead. When ordering a *milehigh pie* a customer would expect a pie heaped tall with meringue or similar fluffy topping, but to advertise a *footlong hot dog* and to serve something less would be in error.

Mistakes are possible in properly identifying steak cuts. Use industry standards such as provided in the National Association of Meat Purveyors *Meat Buyer's Guide*.

*Homestyle* or *our own* are suggested terminology rather than *homemade* in describing menu offerings prepared according to a home recipe. Most foodservice sanitation ordinances prohibit the preparation of foods in home facilities. The means of food preparation is often the determining factor in the customer's selection of a menu entree. Absolute accuracy is a must. (See **Exhibit A.1**.)

## Representation of Means of Preservation

The accepted means of preserving foods are numerous, including canned, chilled, bottled, frozen, and dehydrated. If you choose to describe your menu selections with these terms, they must be

**Exhibit A.1** Menu Descriptive Adjectives

Aged
Aromatic
Array of
Artfully
Assorted
Assortment
Authentic
Baked
Beautiful
Bite-size
Bitter
Bland
Blazed
Blended
Blunt
Burnt
Caked
Caramelized
Char-broiled
Cheesy
Chilled
Chocolate
Cinnamon
Classic
Clove
Coated
Cold
Cool
Crafted
Creamed
Creamy
Crisp
Crunchy
Dazzling
Deep-fried
Delectable
Delicious
Delight
Delightful
Distinctive
Doughy
Dressed
Drizzle, drizzled
Dry
Dull
Elastic
Encrusted
Epicurean, taste, treat
Ethnic
Extraordinary
Famous
Fantastic
Fizzy
Flaky
Flavorful
Fleshy
Fluffy
Fragile
Free-range*
Fresh
Fried
Frosty
Frozen
Fruity
Full, full-bodied
Furry
Garlic
Garlicky
Generous
Gingery
Glazed
Golden
Gourmet
Greasy
Grilled
Gritty
Heaping
Hearty
Hot
Ice-cold
Icy
Indulgent
Infused
Intriguing
Juicy
Kosher, Salt, meet etc.*
Large
Lavish
Layered
Leathery
Light/lite*
Lightly-breaded
Low*
Lukewarm
Luscious
Marinated
Messy
Mild
Minty
Mixture
Moist
Mouth-watering
Mushy
Natural
Nutmeg
Nutty
Oily
Organic*
Peppery
Perfection
Pickled
Plump
Poached
Popular
Pounded
Prepared
Prickly
Pulpy
Pungent
Puréed
Rancid
Reduced*
Refresh
Rich
Ripe
Roasted
Rotten
Rubbery
Salty
Sandy
Satin, satiny
Sautéed
Savory
Scrumptious
Sea, salt
Seared
Seasoned
Sharp
Silky
Simmered
Sizzling
Skillfully
Slippery
Small
Smoky
Smothered
Soothing
Sour
Special
Spicy
Spoiled
Spongy
Sprinkled
Stale
Steamed
Sticky
Strawberry, flavored etc.
Stuffed
Succulent
Sugary
Superb
Sweet
Syrupy
Tangy
Tantalizing
Tart
Tasteful
Tasteless
Tasty
Tender
Tepid
Terrific
Thick
Thin
Toasted
Tossed
Tough
Traditional
Vanilla, vanilla flavored
Velvety
Warm
Waxy
Whipped
Wonderful
Wooly
Yummy
Zesty

* Terms that have to meet Truth in Menu Labeling Laws.

*Courtesy of Nancy Miller-Randal, Pam Czaja, and Monroe Community College, August 25, 2006.*

accurate. Frozen orange juice is not fresh, canned peas are not frozen, and bottled applesauce is not canned.

## Representation of Verbal and Visual Presentation

When your menu, wall placards, or other advertising contains a pictorial representation of a meal or platter, it should portray the actual contents with accuracy. Following are several examples of *visual misrepresentation*.

- Using mushroom pieces in a sauce when the picture shows mushroom caps
- Using sliced strawberries on a shortcake when the picture shows whole strawberries
- Using numerous thin sliced meat pieces when the picture shows a single thick slice
- Using four or five shrimp when the picture shows six
- Omitting vegetables or other entree extras when the picture shows them
- Using a plain bun when the picture shows a sesame topped bun

Examples of verbal misrepresentation include the following.

- A server asking whether a guest would like sour cream or butter with a potato, but serving an imitation sour cream and margarine
- A server telling guests that menu items are prepared on the premises when in fact they are purchased preprepared

## Representation of Dietary or Nutritional Claims*

Potential public health concerns are real if misrepresentation is made of the dietary or nutritional content of food. The Center for Science in the Public Interest has put out a Diner's Guide to Restaurants Menu Health and Nutrition claims. This section is taken from that guide with italics added by *Management by Menu* authors to indicate public distrust of restaurant menu claims:

### What the Law Requires

- Claims must meet FDA standards. Restaurants that make health or nutrition claims, like "heart-healthy" or "low fat," must meet FDA standards. For example, if a restaurant claims its mashed potatoes are "low fat," a typical serving of the food must contain no more than 3 grams of fat.
- Nutrition information must be provided upon request. Nutrition information need not appear on the menu, but must be provided upon request. It may be provided in writing, such as on a flier, brochure, poster, or notebook, or may be provided orally by a statement from a waiter or other employee.

- Full nutrition information is not required. Restaurants only need to provide information that pertain to the particular claim. For example, a restaurant that makes a "low sodium" claim must tell consumers who ask that the food contains no more than 140 mg of sodium in a typical serving. However, the restaurant is not required to provide information about calories, fat, or other nutrients.
- Laboratory analysis is not required. Nutrient levels do not need to be determined by laboratory analyses or certified by third parties. Instead, the levels may be calculated from nutrient data bases or cookbooks. However, restaurants are required to explain how nutrient levels are determined upon request.
- Only menu items for which claims are made are affected. Restaurants are only required to provide nutrition information for foods for which health or nutrition claims are made.

## *Diner Beware—Claims to Look Out For*

- Low fat: Most foods may be described as low fat if there are no more than 3 grams of fat in a standard serving. *Standard servings have been established by the FDA to reflect the amount of food that is typically consumed. Since restaurants often serve foods much larger than the standard serving, a low-fat food may actually contain large amount of fat.*

For example, ice cream may be called low fat as long as there are no more than 3 grams of fat in a standard half cup serving. However, restaurants may offer portions several times larger than the standard serving size. So a two-cup serving of low-fat ice, cream may contain up to 12 grams of fat!

Also, keep in mind that even a small-sized serving of low fat main dishes like hamburgers, pizza, or sandwiches can have more than 3 grams of fat per serving.
And remember, low in fat does not always mean low in calories.

- Light: *Light* is commonly used to mean many different things. It may describe a food's taste, color, or texture, or it may indicate that the food's calorie, fat, or sodium content has been significantly reduced. Menus must clearly indicate what *light* is intended to convey. If the meaning is not clearly explained, diners should ask for clarification. If light is used to indicate a reduction in calories, fat, or sodium, information about those nutrients must be provided upon request.
- Cholesterol free: *Cholesterol free* claims are very popular on restaurant menus, but can also be very misleading. Keep in mind that cholesterol free does not mean fat free. Foods like meat, poultry, and seafood contain cholesterol-even if they are fried in cholesterol free oil.

Also, saturated fat and trans fat can raise the level of cholesterol in your blood. Cholesterol-free foods may contain saturated fat. The FDA only allows foods that are low in saturated fat to be described as cholesterol free, *but watch out-many restaurants may not comply with this requirement.*

The FDA allows foods with significant amounts of trans fats to be called cholesterol free. To avoid trans fats, limit foods prepared with vegetable shortening or partially hydrogenated oils.

- Sugar free: Some foods, especially desserts, may be described as *sugar free*. But keep in mind that "sugar free" does not mean "calorie free" or "fat free." If a food described as sugar free is now low-calorie or reduced-calorie, the menu must say so.
- Healthy: Food described as *healthy* must be low in fat and saturated fat and may not be high in cholesterol or sodium. *However, there are no limits on the amount of sugar or calories that a healthy food may contain.*
- Heart claims: Claims like heart-healthy," "heart smart," and "heart" symbols imply that a food may be useful in reducing the risk of heart disease. When such claims are made, the food must be low in fat, saturated fat, and cholesterol, and must not be high in sodium. *But keep in mind that restaurants may not always comply with these restrictions.*

## Making Healthful Choices

- Ask for nutrition information: When menus make claims like "healthy," "light," or "heart smart," ask for nutrition information. This information must be provided upon request. However, keep in mind that the law does not require that the food be tested in a laboratory, so the numbers may not be precise.
- Watch out for large serving sizes: Serving sizes on food labels are required to reflect the amount of food that is normally consumed. *But restaurants are free to determine their own serving sizes. A CSPI survey found that restaurants often serve from two to three times more than food labels list as a serving.*

For example, the official serving size of a tuna salad sandwich is 4 ounces and it contains 340 calories. The typical restaurant serving is 11 ounces, and it contains 720 calories.

Keep in mind that restaurant foods are probably fattier than you think: The nutrient content of restaurant meals is extremely difficult to assess. A survey conducted by CSPI and researchers at New York University found that trained dietitians underestimated the calorie content of five restaurant meals by an average of 37 percent. The underestimated fat content by 49 percent.

For example, the dietitians estimated, on average, that a tuna fish sandwich provided 374 calories and 18 grams of fat, while the sandwich actually contained about 720 calories and 43 grams of fat! *If well-educated nutrition professionals consistently and substantially underestimated the calorie and fat content of restaurant meals, it's clear that ordinary consumers also have trouble guessing what's in their meals.*

# Number of Portions Available from Standard Containers

## Hot and Cold Food Vessels

| Portion size | 2½ oz | 3 oz | 3½ oz | 4 oz | 5 oz | 6 oz | 7 oz | 8 oz | 10 oz | 12 oz |
|---|---|---|---|---|---|---|---|---|---|---|
| | | | | | NUMBER OF PORTIONS | | | | | |
| No. 2 can | 7 | 6 | 5 | 5 | 4 | 3 | 2 | 2 | 2 | 1 |
| No. 2½ can | 10 | 9 | 8 | 6 | 5 | 4 | 4 | 3 | 3 | 2 |
| 1 quart | 13 | 11 | 9 | 8 | 6 | 5 | 5 | 4 | 3 | 3 |
| 5 lb tin (80 ounces) | 32 | 27 | 23 | 20 | 16 | 13 | 11 | 10 | 8 | 7 |
| 7 lb tin (No. 10 can) | 45 | 37 | 32 | 28 | 22 | 19 | 16 | 14 | 11 | 9 |
| 1 gallon | 51 | 43 | 37 | 32 | 26 | 21 | 18 | 16 | 13 | 11 |
| 10 lb can | 64 | 53 | 46 | 40 | 32 | 27 | 23 | 20 | 16 | 13 |

# Souffle Cups, Creamers, Etc.

| Portion size | ¾ oz | 1 oz | 1½ oz | 2 oz | 3 oz | 3¾ oz | 5 oz | 5½ oz |
|---|---|---|---|---|---|---|---|---|
| | **NUMBER OF PORTIONS** | | | | | | | |
| No. 2 can | 24 | 18 | 12 | 9 | 6 | 5 | 4 | 3 |
| No. 2½ can | 34 | 26 | 17 | 13 | 9 | 7 | 5 | 5 |
| 1 quart | 43 | 32 | 21 | 16 | 11 | 9 | 6 | 5 |
| 5 lb tin (80 ounces) | 106 | 80 | 53 | 40 | 27 | 21 | 16 | 15 |
| 7 lb tin (No. 10 can) | 148 | 112 | 75 | 56 | 37 | 30 | 22 | 20 |
| 1 gallon | 171 | 128 | 85 | 64 | 43 | 34 | 26 | 22 |
| 10 lb can | 212 | 160 | 107 | 80 | 53 | 32 | 29 | |

**PORTION CUPS**

| Cup number | 050 | 075 | 100 | 125 | 200 | 250 | 325 | 400 | 550 |
|---|---|---|---|---|---|---|---|---|---|
| Cup size | ½ oz | ¾ oz | 1 oz | 1¼ oz | 2 oz | 2½ oz | 3¼ oz | 4 oz | 5½ oz |
| | **NUMBER OF PORTIONS** | | | | | | | | |
| No. 2 can | 36 | 24 | 18 | 14 | 9 | 7 | 5 | 5 | 3 |
| No. 2½ can | 52 | 34 | 26 | 20 | 13 | 10 | 8 | 6 | 5 |
| 1 quart | 64 | 43 | 32 | 26 | 16 | 13 | 10 | 8 | 6 |
| 5 lb tin (80 ounces) | 160 | 106 | 80 | 64 | 40 | 32 | 24 | 20 | 15 |
| 7 lb tin (No. 10 can) | 224 | 148 | 112 | 90 | 56 | 45 | 34 | 28 | 20 |
| 1 gallon | 256 | 171 | 128 | 102 | 64 | 51 | 39 | 32 | 22 |
| 10 lb can | 320 | 212 | 160 | 128 | 80 | 64 | 50 | 40 | 29 |

# Menu Evaluation

## Menu Profitability

| | Excellent | Good | Fair | Poor | Comments |
|---|---|---|---|---|---|
| 1. Does the menu have an adequate number of high gross profit items? | | | | | |
| 2. Is there a good selection of popular items? | | | | | |
| 3. Is there a good balance between high- and low-priced items and no concentration of either? | | | | | |
| 4. Does pricing meet competition? | | | | | |
| 5. Are menu prices changed frequently enough to reflect costs? | | | | | |
| 6. Are portion costs based on reliable cost information? | | | | | |
| 7. Is portion size in line with guest references? | | | | | |
| 8. Are menu items selected with a view toward reducing waste and other risks? | | | | | |
| 9. Are menu items selected to reflect labor requirements? | | | | | |
| 10. Are menu items selected to reflect energy needs? | | | | | |
| 11. Does the menu encourage a higher check average? | | | | | |
| 12. Can items be controlled in cost? | | | | | |

## Presentation of Wording

| | Excellent | Good | Fair | Poor | Comments |
|---|---|---|---|---|---|
| 1. Are menu items described accurately and truthfully? | | | | | |
| 2. Does the menu avoid indicating weight, size, using pictures and other factors that can cause problems in patron interpretation? | | | | | |
| 3. Are effective descriptive words used to indicate menu item qualities? | | | | | |
| 4. Is the choice of words adequate to describe the item? Is overkill avoided? Is the wording simple and easy to understand? | | | | | |
| 5. Are truth-in-menu requirements met? | | | | | |
| 6. If pictures are used, are they exact replicas of what is delivered? | | | | | |
| 7. Does the wording do a good job of merchandising? | | | | | |
| 8. Do menu items appear at consistent quality, size, etc. | | | | | |
| 9. Are nutritional claims honest representations backed up with required data? | | | | | |

## Menu Comprehension

| | | | | | |
|---|---|---|---|---|---|
| 1. Do menu items stand out? | | | | | |
| 2. Is print legible and easy to read quickly? | | | | | |
| 3. Are words and reading matter not crowded? | | | | | |
| 4. Is the menu free from clutter? | | | | | |
| 5. Are menu items presented in a logical order, usually the order of eating? | | | | | |
| 6. Are prices presented clearly? | | | | | |
| 7. Are items easy to find? | | | | | |
| 8. Will patrons know what to expect from items from the menu listing? | | | | | |
| 9. Are any unexplained words used? | | | | | |

| | Excellent | Good | Fair | Poor | Comments |
|---|---|---|---|---|---|
| 10. Is there adequate space between menu items so there is no confusion? | | | | | |
| 11. Does the print stand out from the background? | | | | | |
| 12. Are headings prominent and of sufficient size? | | | | | |

## Physical Support

| | Excellent | Good | Fair | Poor | Comments |
|---|---|---|---|---|---|
| 1. Is the menu appropriate for kitchen equipment production capability? | | | | | |
| 2. Is the menu appropriate for servers to do a proper job? | | | | | |
| 3. Is the layout of the operation adequate to meet menu needs? | | | | | |
| 4. Is there adequate storage to support the menu? | | | | | |
| 5. Are menu items matched to the ability of kitchen employees to produce them? | | | | | |
| 6. Is the distance between the kitchen and dining area reasonable? | | | | | |
| 7. Is lighting adequate to read the menu? | | | | | |
| 8. Is a comfortable environment provided for guest comfort? | | | | | |
| 9. Do patrons have adequate space? | | | | | |

## Menu Mechanics

| | Excellent | Good | Fair | Poor | Comments |
|---|---|---|---|---|---|
| 1. Is the cover durable, easily cleaned, and attractive? | | | | | |
| 2. Is the menu on strong, sturdy paper? | | | | | |
| 3. Is the menu easily read and understood? | | | | | |
| 4. Can one find things easily? | | | | | |
| 5. Does the menu appear neat and clean? | | | | | |
| 6. Is the menu free of cross-outs and handwritten changes? | | | | | |

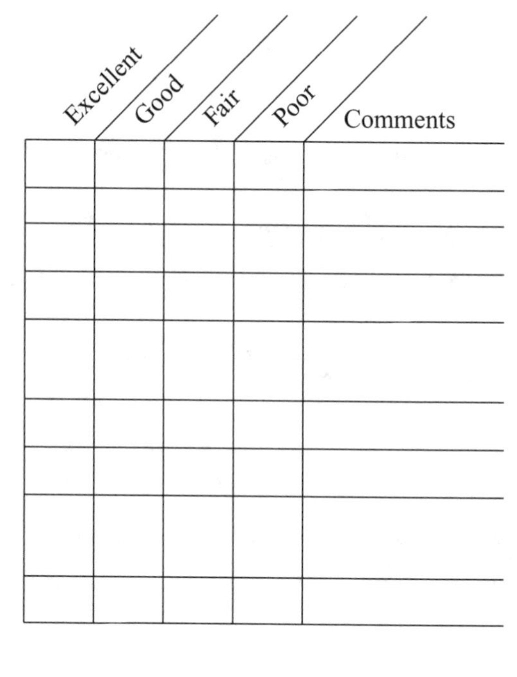


7. Are color combinations effective?
8. Is the shape appropriate?
9. Is it of proper size to be easily handled?
10. Are decorative features appropriate?
11. Are the style of the menu and wording in keeping with the decor, atmosphere, and logo of the operation?
12. Does the menu have good symmetry and form?
13. Does the menu look professional?
14. Is accurate information given on operation time, address and telephone number, special catering, and so on?
15. Is the logo prominent on the menu?

## Item Selection

1. Is the selection of items balanced and varied?
2. Do items offer variety of form, color, taste, and temperature?
3. Are seasonal foods offered?
4. Is preparation of items varied among broiling, frying, steaming, boiling, etc.?
5. Are items suited to the preparation skill of employees and servers?
6. Has simplicity been preserved and not too much that is lavish or ornate attempted?
7. Is there a wide enough selection to appeal to most patrons?
8. Is the sales mix effective?
9. Are specials prominent but not overemphasized?
10. Are items balanced in their popularity?
11. Are health concerns of patrons met with a variety of appealing, health-conscious, choices?

## Menu and Item Presentation

| | Excellent | Good | Fair | Poor | Comments |
|---|---|---|---|---|---|
| 1. Are items served in the proper dish or the right packaging? | | | | | |
| 2. Are they well garnished? | | | | | |
| 3. Are items attractive? | | | | | |
| 4. Are items offered in the order of eating or in an otherwise logical manner? | | | | | |
| 5. Are items not offered by order of price? | | | | | |
| 6. Are long columns broken up in some way? | | | | | |
| 7. Are items that management most wants to sell put in the most prominent places? | | | | | |
| 8. Are children's menus offered, and do they suit children's need? | | | | | |
| 9. Is the use of clip-ons or inserts well done (not repeating menu items, covering other material, etc.)? | | | | | |
| 10. Do menu items reflect a manageable inventory? | | | | | |
| 11. Are high-risk items limited? | | | | | |
| 12. Do most items have a high volume potential? | | | | | |
| 13. Is there good presentation of high gross-profit items? | | | | | |
| 14. Is attention given to nutritional concerns? | | | | | |
| 15. Is there a proper balance between à la carte and table d'hôte items? | | | | | |
| 16. Are ingredients available on the market without great price fluctuations? | | | | | |
| 17. If it is a cycle menu, is the cycle repeated at a reasonable length of time? | | | | | |
| 18. Is there good and adequate presentation of alcoholic beverages? | | | | | |

To score: Analyze the menu, evaluating each factor in the menu evaluation form and scoring 4 for excellent, 3 for good, 2 for fair, and 1 for poor. Do not score factors that are not applicable, but keep track of those omitted.

When scoring is completed, total the numbers. Next, count the factors not considered applicable and deduct the number from 85 (the total factors in this menu evaluation form). Multiply the result by 4, which will be the total of a perfect score. Divide the actual score given the menu by the perfect score. This will give a menu evaluation percentage.

Anything from 90 percent to 100 percent is excellent; anything from 60 percent to 69 percent is poor. Most likely, any menu scoring below 80 percent should be redone.

The comment column should be used to indicate how a factor might be improved if it is not satisfactory.

# Menu Factor Analysis

The following information is known:

| Salad Item | Number Sold | Selling Price | Total Sales | Food Cost | Total Food Cost | Item Gross Profit | Total Gross Profit |
|---|---|---|---|---|---|---|---|
| Crab Louis | 25 | $6.00 | $_______ | $2.10 | $_______ | $_______ | $_______ |
| Chef's | 40 | 5.00 | _______ | 1.25 | _______ | _______ | _______ |
| Fruit | 30 | 5.50 | _______ | 2.10 | _______ | _______ | _______ |
| Vegetable | 45 | 4.25 | _______ | 1.10 | _______ | _______ | _______ |
| Total | 140 | | $_______ | | $_______ | $_______ | $_______ |

Calculate the popularity index, sales index, food cost index, and gross profit index in column 1 below and then convert to the proper factors in column 2. The expected index for all four in popularity, sales, food cost, and gross profit is 0.25.

| Crab Louis salad | (1) | (2) |
|---|---|---|
| Popularity | _______ | _______ |
| Sales | _______ | _______ |
| Food cost | _______ | _______ |
| Gross profit | _______ | _______ |
| **Chef's salad** | | |
| Popularity | _______ | _______ |
| Sales | _______ | _______ |
| Food cost | _______ | _______ |
| Gross profit | _______ | _______ |

| Fruit salad | (1) | (2) |
|---|---|---|
| Popularity | _______ | _______ |
| Sales | _______ | _______ |
| Food cost | _______ | _______ |
| Gross profit | _______ | _______ |
| **Vegetable salad** | | |
| Popularity | _______ | _______ |
| Sales | _______ | _______ |
| Food cost | _______ | _______ |
| Gross profit | _______ | _______ |

Study the factors. What do they tell management about popularity, sales dollars, food costs, and contribution to gross profit?

| Salad Item | Total Sales | Total Food Costs | Item Gross Profit | Total Gross Profit |
|---|---|---|---|---|
| Crab Louis | $150.00 | $ 52.50 | $ 3.90 | $ 97.50 |
| Chef's | 200.00 | 50.00 | 3.75 | 150.00 |
| Fruit | 165.00 | 63.00 | 3.40 | 102.00 |
| Vegetable | 191.25 | 49.50 | 3.15 | 141.75 |
| Total | $706.00 | $215.00 | $14.20 | $491.00 |

| | Indexes | Factors |
|---|---|---|
| **Crab Louis salad** | | |
| Popularity | 0.179 | 0.716 |
| Sales | 0.212 | 0.848 |
| Food cost | 0.244 | 0.976 |
| Gross profit | 0.198 | 0.792 |
| **Fruit salad** | | |
| Popularity | 0.214 | 0.856 |
| Sales | 0.234 | 0.936 |
| Food cost | 0.293 | 1.172 |
| Gross profit | 0.208 | 0.832 |

| | Indexes | Factors |
|---|---|---|
| **Chef's salad** | | |
| Popularity | 0.286 | 1.144 |
| Sales | 0.283 | 1.132 |
| Food cost | 0.233 | 0.932 |
| Gross profit | 0.305 | 1.220 |
| **Vegetable salad** | | |
| Popularity | 0.321 | 1.284 |
| Sales | 0.271 | 1.084 |
| Food cost | 0.230 | 0.920 |
| Gross profit | 0.289 | 1.156 |

Comments:
Crab Louis salad has poor popularity, sales, and gross profit. Its food cost is about right but does not help generate profit. It needs to be given more sales emphasis. Perhaps it also needs better presentation on the menu.

Chef's salad does well. It has a good food cost and also good popularity, sales, and gross profit.

Fruit salad is another problem. It does not have good popularity and is slightly below expected sales. It has a high food cost, and this contributes to its low gross profit margin. The salad should be changed some to lower its food cost. Perhaps the price could be raised slightly to solve the problem.

Vegetable salad holds its own in every category. In fact, it is the best of the four. Do not change.

# A Brief History of Foodservice

| Year | Region | General History | Foodservice Events |
|---|---|---|---|
| Before 4500 B.C. | Denmark and Orkney Isles | Stone Age | First volume feeding seen by 10,000 B.C. |
| | Switzerland | New Stone Age | Swiss Lake Dwellers eating in groups by 5000 B.C. |
| 3000–500 B.C. | Egypt | Age of the Pharaohs | Art depicts food prepared for and served to large groups. |
| 2500–1000 B.C. | Mohenjo-Daro Pakistan | | First evidence of restaurant-type facilities, including ovens and stoves. |
| 2200–1000 B.C. | China | | Roadside inns appear for travelers, restaurants appear in larger cities. |
| 900–500 B.C. | Middle East | Assyrian Empire | Evidence of production of both beer and wine. King Solomon holds great feasts. |
| 500–300 B.C. | Europe | Classic Greek Civilization | Inns, restaurants, copyrights for recipes, and first known school for chefs all appear. |
| 300 B.C.–320 A.D. | Europe | Roman Empire | Romans hold lavish banquets. Tabernas (taverns) appear. |
| 200 B.C. | | | The first cookbook is written, which is resurrected after the Dark Ages. |
| 1300s | England | Middle Ages | Chaucer and others write about inns for travelers and pilgrims. Monasteries develop various liqueurs still known to us today. Rise of guilds, whose members produce specialty foods. |
| 1300s | France | Middle Ages | First reintroduction of a complex meal with separate courses. |
| 1491–1547 | England | Reign of Henry VIII | Henry VIII, known for his rotund figure and voracious appetite, encourages elaborate dining. |
| 1400s | Italy | European Renaissance | Great merchants like Marco Polo expand trade and bring new foods and spices to Europe. Rise of gourmet recipes and elaborate banquets. |

| Year | Region | General History | Foodservice Events |
| --- | --- | --- | --- |
| 1500s | France | Age of Discovery | Catherine de Medici of Italy marries Henry II, king of France and brings Italian cooks and recipes to the French court. Starts French tradition of fine dining. |
| | England | | Mary, Queen of England, brings Spanish dining traditions to England. |
| 1589–1610 | France | Colonial Expansion | Henry IV, king of France, encourages nobility to become gourmets. Beginnings of sauce making as an art. Good chefs are prized by nobility. |
| | | | First coffeehouses appear and quickly spread throughout Europe. |
| 1600s | France | Age of Reason | Bourbon kings bring French cuisine to its grandest heights. Louis XIV (the Sun King) builds his palace at Versailles. Foods are named after members of the nobility. |
| 1700s | France | Age of Enlightenment | Louis XV marries Polish princess who, like de Medici, supervises kitchens and sets new standards of excellence. Very elaborate meals, with 100 or more dishes served, become common. |
| | Russia | | Catherine the Great of Russia introduces French language and customs to the court. Introduces concepts of appetizers, dishes like caviar, etc., to European tables. |
| 1760 | France | | Boulanger opens *restaurants* serving "restorative" soups. Legal fight with guilds results in legalization of the restaurant concept. |
| | | | Chef Carême simplifies and codifies the royal cuisine, and trains many famous chefs. |
| 1792 | France | French Revolution | First books appear detailing the ideal life of a gourmet. First gourmet magazine appears. |
| 1800–1815 | France | Napoleonic Period | While not much of a gourmet himself, Napoleon's wife and counselors are. They inspire a rebirth of imperial style of food presentation and service. |
| | | | About 500 restaurants are now operating in Paris. |
| | | | Appert invents the canning process, used by Napoleon to feed his army. |
| 1760–1800s | England and Europe | Industrial Revolution | The Industrial Revolution changes the way people live, work, and eat. With mass production, items once reserved only for the rich become available to a large new economic class, the "middle class." |
| | | | With the Industrial Revolution comes a faster-paced lifestyle. Eating out becomes both affordable and necessary. |
| 1700s–early 1800s | United States | Colonial Period | President Thomas Jefferson, remembering the foods he ate in France while he was an ambassador, appoints a French chef to the White House kitchens. |
| 1818 | New York City | | There are eight hotels in New York City. There will be more than 100 hotels by 1846. |

| Year | Region | General History | Foodservice Events |
|---|---|---|---|
| 1800–1850 | England | | Great chefs trained by Carême serve the English elite. They introduce French eating styles to England and create first English-language cookbooks of French recipes. |
| | | | Private clubs, like the Reform Club, become popular with English gentlemen. |
| | | | Alexis Soyer, chef at the Reform Club, invents the mobile army field kitchen, feeding over 10,000 Irish citizens a day during the potato famine. |
| 1850–1938 | Europe | Age of Grand Hotels | César Ritz and Auguste Escoffier serve the wealthiest Europeans in their grand hotels and restaurants. Escoffier revolutionizes the Kitchen by applying principles used in industry and introducing personnel changes. He codifies and further simplifies haute cuisine. His idea is copied for restaurants. More than 1,000 chefs trained by Escoffier revolutionize the entire foodservice industry. |
| 1850 | Chicago | | There are more than 150 hotels in Chicago. |
| 1849–1892 | United States | California Gold Rush | Gold is discovered in California in 1849. By the end of the Civil War in 1865, extremely luxurious hotels become more common. |
| 1890–1915 | United States and Europe | Victorian Era | Delmonico's and Rector's are famous New York City restaurants. Fine dining reaches levels not seen again until after World War II. |
| | | | Restaurant chains, such as Harvey House and Schrafft's, bring consistent, good food to the masses. |
| 1920s | United States | | With Prohibition, the *speakeasies*, usually bar or nightclubs, become popular. Some famous speakeasies, like The 21 Club, later become famous fine-dining establishments. |
| 1946 | United States | Post-World War II | The National School Lunch Act introduces the start of large-scale public feeding programs. |
| 1950s | United States | | Quick-service chains, such as McDonald's and Kentucky Fried Chicken, appear. |
| 1960s | France | | French chefs develop nouvelle cuisine. Chef Bocuse, Gaston Lenetre, and the Trosgros brothers define the new cuisine with healthier foods that are lower in fats, starches, and sugars. |
| 1970s | United States | | American chefs discover regional specialties. New awareness of locally grown produce creates an explosion of innovation in the finedining field. |
| 2000s | United States | | An exciting time in the culinary world, Culinology emerges as a field of study. American wines grow in quality and popularity. More Americans are eating out than any time in history. Health concerns emerge in the form of obesity lawsuits with films like *Super Size Me* and *Fast Food Nation* taking our industry to task. Restaurateurs show heart in response to Hurricane Katrina. Quick-service and casual restaurants show explosive growth. |

# NOTES

## CHAPTER 1

1. The earliest record of ice cream is that it was made for the Persian kings by freezing cream, honey, and other flavorings in snow in the high mountains. This was then packed in snow and taken by runner down the mountain to the king's court. The Carthaginians learned how to make ice cream from the Persians and carried the knowledge to the Sicilians who, in turn, brought it to Florence. Benjamin Franklin, when he was ambassador to France, liked it so much that he brought the recipe to the United States, where both Martha Washington and Dolly Madison made history by serving it at the White House.
2. Alexander of Russia died of eating a dish of poisonous mushrooms several years after Carême left him to go to the Rothschild household.
3. Root W, de Rouchemont R. *Eating in America, A History,* NY: Ecco Press 1981, 2nd edition, pp. 321–351.
4. This did not occur until the 1920s and 1930s. As late as the 1940s, foodservice operations used electricity only for lights.
5. *www.leye.com*

## CHAPTER 2

1. Much of the material in recreational feedings was originally supplied by Professor Mickey Warner from his 1982 Masters degree thesis, "Recreational Foodservice Management," School of Hospitality Management, Florida International University.
2. Locally operated programs existed before this. In 1884, in Boston, Mrs. Ellen H. Richards, a home economist, started what was probably the first program. It was not until after World War II, however, that a broader need for such programs was shown when figures for malnutrition and physical defects, and some data obtained on the nutritional status of schoolchildren, were analyzed.

## CHAPTER 4

1. John Rosson, "Menus Still Have a Long Way to Go." *National Restaurant Association News*, Volume 3, Number 5, May 1983, pp. 15–17.

## CHAPTER 5

1. The frozen, preblanched strips usually contain around 11 percent fat when purchased. Thus, the finished potatoes will contain around 17 percent fat when sold.
2. *Standards. Principles, and Techniques in Quantity Food Production, Third Edition*, by Lendal H. Kotschevar, published by Van Nostrand Reinhold Company, New York, has a large number of tables indicating portions and yields.

## CHAPTER 8


1. Michael Hurst owned the very successful 15th Street Fisheries in Florida, and taught at Florida International University. He was a past president of the National Restaurant Association-and driving force behind the Salute to Excellence student event at the NRA show in Chicago.
2. David K. Hayes and Lynn Huffman, "Menu Analysis: A Better Way." *Cornell Quarterly*, Vol. 25, No. 4, February 1985, pp. 64–70.


## CHAPTER 9

1. Commissions range from around 10 to 20 percent, or a cork price may be given, such as $0.59 for each bottle of wine (cork) sold.
2. Wine inventories usually do not turn over more than four times per year.

## CHAPTER 13


1. Adapted from NRAEF *Manage First Hospitality and Restaurant Management Competency Guide*
2. Christine Jaszay and Paul Dunk, *Ethical Decision-Making in the Hospitality Industry*, 2006, pp. 2–3 (Reprinted by permission from Pearson Education, Inc., Upper Saddle River, NJ)
3. Ferrell, *Business Ethics: Ethical Decision Making and Cases,* 2000, 4th edition, Houghton Miflin
4. Jaszay, *Ethical Decision-Making in the Hospitality Industry*, 2006, Pearson Prentice Hall
5. Schlosser, *Fast Food Nation: The Dark Side of the All-American Meal,* 2002, Houghton Miflin
6. National Restaurant Association Educational Foundation, *Hospitality and Restaurant Management Competency Guide,* 2007, Pearson Prentice Hall
7. Kotschevar, *Presenting Service,* 2007, 2nd edition, Wiley
8. Marvin, *Restaurant Basics*, 1992, Wiley
9. Shriver, *Managing Quality Services,* 1988, Educational Institute of the American Hotel & Motel Association
10. Evans, *The Management and Control of Quality*, 1993, West Publishing
11. Parsa, *Why Restaurants Fail,* 2005 Cornell Quarterly, Volume 46, Number 3
12. asbdc.ualr.edu/drugfree/dfwp_fa.ppt Arkansas Restaurant Association power point presentation on establishing a Drug (and Alcohol) free workplace
13. *http://asbdc.ualr.edu/drugfree/dfwpfood.pdf* ARA booklet on establishing a Drug (and Alcohol) Free Workplace
14. *http://restaurantedge.com* Is there a social imperative to provide healthy menu selections in the restaurant industry? December 2003
15. Nation's Restaurant News, November 21, 2005 *Feeding the Needs of Health-Savvy Customers,* by Pamela Parseghian, Ron Ruggless and Bret Thorn
16. QSR Magazine, Issue 89, May 2006 by Jamie Hartford; *Dr Dean Ornish's Prescription for the Quick Service Restaurant Industry*
17. National Center for Health Statistics, *Chart Book on Trends in the Health of Americans,* 2005
18. Philadelphia Inquirer, May 25, 2006, Carpentar, Dave, Associated Press *McDonald's CEO Decries Fast Food "Fiction"*
19. New York Times, May 10, 2006, Severson, *An Interview with Eric Schlosser: A Food Crusader's Alarm Is Supersized*
20. *www.healthydiningfind.com*
21. *www.dinegreen.com*
22. *http://pubs.niaaa.nih.gov/publications/arh26-1/49-57.htm*
23. Office of Applied Studies, SAMHSA, *National Household Survey on Drug Abuse,* 1994 and 1997

# GLOSSARY

**Accuracy-in-menu guidelines** Guidelines that specify general terminology for menus to ensure that what is listed on a menu is what is actually served.

**Acid-base ratio** Balance in the body between alkali and acid; an individual must maintain an almost neutral acid-base ratio to stay alive; the body does this automatically, except in the course of some diseases.

**Actual (cost) pricing** Pricing method based on actual costs, including purchasing, labor, and operating costs.

**À la carte** Literally "to the card." The term used to indicate that a menu item is ordered with few or no accompaniments and has its own price.

**Apéritif** Alcoholic beverage drunk before a meal used to whet the appetites.

**Aquaculture** Raising fish or other marine life in an artificial environment as close to the natural environment as possible.

**Aquavit** Clear Scandinavian liquor that is very high in proof, often flavored with caraway seeds; it can be seasoned with other items, such as orange peel or cardamom.

**Assets** Property, capital, and other resources available to meet cost or debts as needed.

**As purchased (AP) cost** As the product is purchased, in the raw state before production.

**Baby boomer** Person born in the period from after World War II through the early 1960s.

**Back of the house** Area where food production occurs, as distinct from the service area, which is called the front of the house.

**Bain marie** Small steam table.

**Base pricing** Pricing based on what a certain market will pay and what is needed to cover costs.

**Batch cooking** Cooking foods in small batches so foods will be at their peak freshness when served.

**Bid** Price quote made by a supplier, using a written form.

**Bin card** Card attached to a storage area that indicates what items are stored and their quantities.

**Blank check buying** Asking a purveyor to supply an item without knowledge of the price.

**Bouillon** From the French verb meaning "to boil"; a soup usually made from a rich beef stock.

**Break even** Point at which an operation neither makes a profit nor loses money.

**Buffet** Type of service in which guests are served foods from a long table.

**Café** Literally "coffee"; a place where beverages and light meals are obtained at a moderate cost.

**Cafeteria** Type of foodservice where patrons select their own food from a counter.

**California menu** Menu that offers breakfast, lunch, dinner, and snack at any hour.

**Call brand** Popular brand of liquor often requested by name by patrons.

**Call sheet** Market list on which suppliers are named; used for calling for prices and market information (also called quotation sheet).

**Calorie** One unit of heat energy.

**Canapé** Small, open-faced sandwich served as an appetizer or hors-d'oeuvre.

**Capital costs** Funds spent to supply land, building equipment, and other investments.

**Captive market** Market that must, because of circumstances, eat at a particular foodservice.

**Carafe** Container used for serving small amounts of wine.

**Carbohydrate** Family of organic compounds that the body needs to produce energy.

**Cast type** Type that is set by pouring molten lead into molds that form letters.

**Catering** Dispensing food away from the production facilities for parties or special occasions.

**Central commissary** Large kitchen that prepares food for satellite service areas.

**Chain** Group of foodservice units related to each other through some corporate or other business group; often they have similar themes or menus.

**Check average** Total dollar sales divided by the number of patrons served.

**Checker** Back-of-the-house employee who checks food orders as they leave the kitchen.

***Chef du parti*** Chef in charge of a particular production section, such as the broiler section, roasting section, or sauce section.

**Cholesterol** Substance necessary for some vital functions in the body; some forms can cause arterial problems that lead to strokes, heart attacks, and other health problems.

**Clip-on** Temporary attachment to a menu to announce special items.

**Club** Organization that often has a private or special membership and that serves food and drinks.

**Commercial feeding operation** Operation serving food and beverages for profit.

**Consommé** Soup made of rich, concentrated stocks or broths, seasoned with herbs and spices, and often garnished with vegetables, dumplings, or other items.

**Contract feeding operation** Organization that has a contract to produce and serve food for an organization involved in some other business activity.

**Convection oven** Oven with a fan that circulates heated air, allowing for efficiency and heat distribution.

**Convenience foods** Foods prepared to such a state that very little further preparation is needed for service.

**Cordial** Sweet, aromatic liquor.

**Cost allowance** Cost of food groups allocated to feed individuals based on a budgeted figure.

**Cost-plus buying** Making an arrangement with a supplier to purchase items at cost plus a set markup.

**Cost-plus-profit pricing** Adding the desired profit to the total cost of an item to arrive at a selling price.

**Counter service** Service style in which guests are served at a counter rather than at a table.

**Cover charge** Basic charge added to a guest's bill, usually for entertainment or some special feature that the enterprise offers in addition to food.

**Current ratio** Current assets divided by current liabilities.

**Cycle menu** Menu for a certain number of days, weeks, or meals that is repeated after a set amount of time.

**Daily receiving report** Summary of items received each day by the receiving department.

**Dietetics** Application of the science of nutrition through diet.

**Dietitian** Individual trained to apply the science of nutrition through diet.

**Differentiation** Act of making an enterprise and/or menu unique in the marketplace so patrons are drawn there, rather than to the competition.

**Direct labor cost** That part of the labor budget that is spent directly preparing or serving a menu item.

**Directs** Deliveries that go directly into use and not into inventory and are charged on the day received.

**Draft (draught) beer** Beer drawn from a keg and propelled by a carbonating unit.

**Drive-in** Foodservice that serves food to people in vehicles in an area provided for their parking or at a drive-up window.

**Du jour** French term meaning "of the day." It is often used on menus with daily specials. A du jour menu is a daily menu.

**Edible portion (EP)** Portion of food that is edible and that must be separated from any inedible portion, such as fat or bone.

**Entree** Main dish on a menu, as opposed to accompaniments.

**Executive chef** Head of the chefs in a kitchen.

**Feasibility study** Compilation of data made before opening a new operation to see how successful a foodservice business is expected to be.

**Fixed costs** Costs that do not fluctuate with changes in business volume.

**Floater** Check held by a server and used more than once, usually as a means of stealing money.

**Food cost** Cost of the portion of food used in a menu item.

**Food cost pricing** Basing a selling price on the cost of the food by using a percentage markup.

**Forecasting** Estimating future sales, costs, and other factors.

**Front of the house** Service and sales area of a foodservice operation.

**Grade** Standards of quality established for some foods.

**Gross profit** Sales revenue less cost of materials sold.

**Gross profit pricing** Pricing method based on the cost of profit.

**Haute cuisine** Literally *high cooking;* refers to fine and elegant food and service.

**Health care feeding** Foodservices typical of hospitals, rest homes, and convalescent centers where diet and health considerations are stressed.

**Hospitality industry** Overall industry that provides lodging, food, and entertainment through hotels, restaurants, airlines, resorts, etc.

**House wine** Bulk-stocked wine usually taken from a jug, cask, or other large container. It can also be a specially bottled wine that the house services as its own.

**Hydrogenation** Adding hydrogen to make an unsaturated fat into a saturated one. Hydrogenation will change a liquid oil into a solid fat.

**Institutional feeding** Foodservice in institutions such as schools, prisons, and industries.

**Invoice** Form given by a delivery person to the purchaser that indicates what is delivered and other information about the delivery.

**Issuing** Function of giving food, materials, and supplies to employees from inventory.

**Jigger** Small measure used for alcoholic beverages.

**Lamination** Cover, such as a stiff plastic, to give paper a sturdy feel and texture.

**Light beer** Beer that contains fewer calories and less alcohol than normal.

**Limited menu** Menu offering only a few items.

**Liqueur** Liquor often made from a brandy or grain alcohol base that is usually sweetened and flavored.

**Liquidity** Financial solvency of a business.

**Logo** Distinctive mark or emblem that differentiates an operation from others and identifies it to patrons.

**Low-sodium diet** Diet in which the salt is restricted.

**Maitre d'hôtel** Literally "master of the hostel or place." This is the person in charge of the service in a dining room.

**Malnutrition** Poor nutrition; a diet that lacks adequate nutrients.

**Marginal analysis pricing** Pricing method based on maximum net profit.

**Markup** Amount added to a basic cost to arrive at a selling price.

**Meal pattern** (1) Kind of foods served along with their progression through a meal; (2) Federal guidelines for elementary and secondary school foodservices that establish what should be included in school lunches.

**Meal plan** Sequence of different foods in a meal without mention of specific food items, only of their kind, such as meat or fruit.

**Menu** List of food offerings and major operational document.

**Menu factor analysis** Method of analysis in which items are tested for their popularity, sales revenue, food cost, and gross profit.

**Mineral** Group of elements used by the body to produce bones, teeth, and other substances.

***Mise en place*** Literally "to put in place." It involves getting everything ready for a job or task to be done.

**Negotiated buying** Purchasing after negotiation with a number of suppliers.

**Noncommercial operation** Foodservice that functions without the objective of making a profit; often the unit is part of some other type of enterprise.

**Non-cost pricing** Pricing method based on factors other than cost, such as on what the market will bear.

**Occasion meals** Meals that have special significance, such as the celebration of a wedding or a holiday.

**Par stock** Established minimum stock level that must be on hand at all times.

**Perpetual inventory** Inventory that is maintained solely through records.

**Physical inventory** Inventory that is conducted by hand.

**Polyunsaturated fat** Fat that has two or more unsaturated double bonds of carbon.

**Precosting** Calculating costs ahead of actual operation by forecasting sales or the cost of menu items for those sales.

**Prime cost** Food cost plus direct labor cost.

**Printout** Hard copy of information that has been fed into a computer.

**Productivity** Amount of work or output produced by human resources.

**Promotion** Selling efforts aimed at getting new business or moving special items.

**Proof** Alcoholic content of liquor. American proof is based on two times the percentage of alcohol.

**Proprietorship** Ownership of a business by an individual.

**Protein** Nutrient that provides amino acids for building body tissues.

**Purchase order** Formal presentation from the buyer of a purchase requirement in written form.

**Quotation sheet** Market list on which suppliers are named; used for calling for prices and market information (also called call sheet).

**Ration allowance** Rationing of specific amounts of certain foods based on nutritional requirements and budgetary constraints.

**Receiving** Function of receiving and checking deliveries from suppliers.

**Réchaud** Small heating unit used to keep foods warm in French service.

**Recipe costing** Calculating total and portion costs of a recipe.

**Recipe forecasting** Estimating portions from a standardized recipe.

**Recreational foodservice** Foodservice established to serve individuals at a recreational park, sports stadium, athletic club, theme park, theater, etc.

**Restaurant** Foodservice that operates commercially to sell food and beverages to patrons.

**Retail host** Foodservice operation in a retail establishment.

**Return on investment (ROI)** Profit made on what owners have invested in the business.

**Rosé** Pink or reddish wine, often slightly sweet, with a very delicate flavor.

**Roughage** Fiber or bulk in food that provides for efficient digestion and elimination of wastes.

**Sales mix** Selection pattern of patrons in choosing menu items.

**Satellite feeding** Foodservices that heat up and serve foods that are prepared elsewhere, often at a central commissary.

**Saturated fat** Fat that has all its carbon bonds filled with hydrogen.

**Seated service** Foodservice that serves patrons who are seated at tables.

**Service bar** Bar that dispenses alcoholic beverages to servers who serve guests.

**Shot glass** Measure for dispensing liquor.

**Silk-screening** Method of printing that uses a stencil to transfer print onto an object.

**Smorgasbord** Scandinavian buffet.

**Snack bar** Counter selling snack items, to be eaten there or taken away.

**Snifter** Large, bowl-shaped glass into which only a small amount of brandy or other aromatic liqueur is poured.

**Social caterer** One who caters meals to social affairs such as receptions, private parties, and other special occasions.

**Solvency ratio** Ratio calculated from assets and liabilities to indicate how liquid (able to meet financial obligations) an operation is.

***Sommelier*** Wine steward.

**Sous chef** Assistant to the executive chef.

**Specification** Written statement of all characteristics needed in an item to be purchased.

**Spirit** Distilled alcoholic beverage.

**Standard(ized) recipe** Recipe that gives a known quality and quantity at a known cost.

**Supply and demand** Two factors in economics that are influential in establishing the price of commodities. A large supply usually lowers price, while a large demand raises it.

**Table d'hôte menu** Meal or a group of foods sold together at one price.

***Tastevin*** Small silver cup used for tasting wine.

**Tavern** Operation that sells mainly alcoholic beverages.

**Truth-in-menu guidelines** Group of laws and guidelines that require that what a menu offers must be truthful in describing and delivering what it offers for sale.

**Turnover (customer)** Number of times a seat in a foodservice operation is filled during a meal period, day, etc.

**Unsaturated fat** Fat that has double carbon bonds that can pick up other substances, such as nitrogen or sulfur.

**Value analysis** Study used to assess how well a function was performed.

**Value perception** Perceived value of an item in comparison to its cost.

**Variable cost pricing** Pricing method based on the variable costs of the menu item, usually based on a set food cost markup.

**Variable costs** Costs that change with changes in business volume.

**Vending** Service of foods using vending machines.

**Vitamins** Nutrients that in small amounts can help regulate the body and promote essential body functions.

**Walk-out** Individual or group of customers that has walked out of the operation without paying the bill.

**Wine steward** Individual who discusses and recommends wine to patrons (*sommelier*).

**Working capital** Assets minus liabilities.

# INDEX

**C**

**F**

## G

## M

**N**

**Q**



**R**

**T**

**U**



**V**



**W**

**Y**



**Z**